Polymer Membranes for Gas Separation

Polymer Membranes for Gas Separation

Editors

Elsa Lasseuguette
Bibiana Comesaña-Gándara

MDPI • Basel • Beijing • Wuhan • Barcelona • Belgrade • Manchester • Tokyo • Cluj • Tianjin

Editors

Elsa Lasseuguette
School of Engineering
University of Edinburgh
Edinburgh
United Kingdom

Bibiana Comesaña-Gándara
Institute of Sustainable
Processes (ISP)
University of Valladolid
Valladolid
Spain

Editorial Office
MDPI
St. Alban-Anlage 66
4052 Basel, Switzerland

This is a reprint of articles from the Special Issue published online in the open access journal *Membranes* (ISSN 2077-0375) (available at: www.mdpi.com/journal/membranes/special_issues/ polymer_membr_gas_separation).

For citation purposes, cite each article independently as indicated on the article page online and as indicated below:

LastName, A.A.; LastName, B.B.; LastName, C.C. Article Title. *Journal Name* **Year**, *Volume Number*, Page Range.

ISBN 978-3-0365-3396-4 (Hbk)
ISBN 978-3-0365-3395-7 (PDF)

© 2022 by the authors. Articles in this book are Open Access and distributed under the Creative Commons Attribution (CC BY) license, which allows users to download, copy and build upon published articles, as long as the author and publisher are properly credited, which ensures maximum dissemination and a wider impact of our publications.

The book as a whole is distributed by MDPI under the terms and conditions of the Creative Commons license CC BY-NC-ND.

Contents

Preface to "Polymer Membranes for Gas Separation" vii

Elsa Lasseuguette and Bibiana Comesaña-Gándara
Polymer Membranes for Gas Separation
Reprinted from: *Membranes* **2022**, *12*, 207, doi:10.3390/membranes12020207 1

Yu Zang, Yinghui Lun, Masahiro Teraguchi, Takashi Kaneko, Hongge Jia and Fengjuan Miao et al.
Synthesis of Cis-Cisoid or Cis-Transoid Poly(Phenyl-Acetylene)s Having One or Two Carbamate Groups as Oxygen Permeation Membrane Materials
Reprinted from: *Membranes* **2020**, *10*, 199, doi:10.3390/membranes10090199 5

Fidel E. Rodríguez-González, Germán Pérez, Vladimir Niebla, Ignacio Jessop, Rudy Martin-Trasanco and Deysma Coll et al.
New Poly(imide)s Bearing Alkyl Side-Chains: A Study on the Impact of Size and Shape of Lateral Groups on Thermal, Mechanical, and Gas Transport Properties
Reprinted from: *Membranes* **2020**, *10*, 141, doi:10.3390/membranes10070141 19

Quan Liu, Long Cheng and Gongping Liu
Enhanced Selective Hydrogen Permeation through Graphdiyne Membrane: A Theoretical Study
Reprinted from: *Membranes* **2020**, *10*, 286, doi:10.3390/membranes10100286 37

Federico Begni, Elsa Lasseuguette, Geo Paul, Chiara Bisio, Leonardo Marchese and Giorgio Gatti et al.
Hyper Cross-Linked Polymers as Additives for Preventing Aging of PIM-1 Membranes
Reprinted from: *Membranes* **2021**, *11*, 463, doi:10.3390/membranes11070463 49

Clara Casado-Coterillo, Aurora Garea and Ángel Irabien
Effect of Water and Organic Pollutant in CO_2/CH_4 Separation Using Hydrophilic and Hydrophobic Composite Membranes
Reprinted from: *Membranes* **2020**, *10*, 405, doi:10.3390/membranes10120405 65

Nándor Nemestóthy, Péter Bakonyi, Piroska Lajtai-Szabó and Katalin Bélafi-Bakó
The Impact of Various Natural Gas Contaminant Exposures on CO_2/CH_4 Separation by a Polyimide Membrane
Reprinted from: *Membranes* **2020**, *10*, 324, doi:10.3390/membranes10110324 77

Bouchra Belaissaoui, Elsa Lasseuguette, Saravanan Janakiram, Liyuan Deng and Maria-Chiara Ferrari
Analysis of CO_2 Facilitation Transport Effect through a Hybrid Poly(Allyl Amine) Membrane: Pathways for Further Improvement
Reprinted from: *Membranes* **2020**, *10*, 367, doi:10.3390/membranes10120367 87

Francesco M. Benedetti, Maria Grazia De Angelis, Micaela Degli Esposti, Paola Fabbri, Alice Masili and Alessandro Orsini et al.
Enhancing the Separation Performance of Glassy PPO with the Addition of a Molecular Sieve (ZIF-8): Gas Transport at Various Temperatures
Reprinted from: *Membranes* **2020**, *10*, 56, doi:10.3390/membranes10040056 111

Leandri Vermaak, Hein W. J. P. Neomagus and Dmitri G. Bessarabov
Recent Advances in Membrane-Based Electrochemical Hydrogen Separation: A Review
Reprinted from: *Membranes* **2021**, *11*, 127, doi:10.3390/membranes11020127 . . . . **145**

Preface to "Polymer Membranes for Gas Separation"

Over the past decade, polymeric membranes have been widely investigated for a variety of industrial gas separation applications. In today's competitive and ever-changing environment, membrane gas separation is now widely accepted as an economic process to produce moderate purity stream gases.

This Special Issue on "Polymer Membranes for Gas Separation" of the journal *Membranes* aims to assess the state-of-the-art and future developments in the field of polymeric membranes. Various topics have been discussed, including the synthesis and characterization of novel membrane materials, membrane aging, and the impact of process conditions on transport phenomena, with the desire to improve the gas separation process in all the articles. There are nine contributions, namely, eight articles and one review, in this Special Issue.

The editors would like to thank the authors for their high-quality outputs and Lydia Li for her patient and motivating help and assistance.

Elsa Lasseuguette, Bibiana Comesaña-Gándara
Editors

Editorial

Polymer Membranes for Gas Separation

Elsa Lasseuguette [1,*] **and Bibiana Comesaña-Gándara** [2,*]

1. School of Engineering, University of Edinburgh, Robert Stevenson Road, Edinburgh EH9 3FB, UK
2. Institute of Sustainable Processes (ISP), University of Valladolid, 47011 Valladolid, Spain
* Correspondence: E.Lasseuguette@ed.ac.uk (E.L.); bibiana.comesana@uva.es (B.C.-G.)

Citation: Lasseuguette, E.; Comesaña-Gándara, B. Polymer Membranes for Gas Separation. *Membranes* **2022**, *12*, 207. https://doi.org/10.3390/membranes12020207

Received: 21 January 2022
Accepted: 5 February 2022
Published: 10 February 2022

Publisher's Note: MDPI stays neutral with regard to jurisdictional claims in published maps and institutional affiliations.

Copyright: © 2022 by the authors. Licensee MDPI, Basel, Switzerland. This article is an open access article distributed under the terms and conditions of the Creative Commons Attribution (CC BY) license (https://creativecommons.org/licenses/by/4.0/).

Over the past decade, polymeric membranes have been widely investigated for a variety of industrial gas separation applications. In today's competitive and ever-changing environment, membrane gas separation is now widely accepted as an economic process to produce moderate purity stream gases.

This Special Issue on "Polymer Membranes for Gas Separation" of the journal *Membranes* aims to assess the state-of-the-art and future developments in the field of polymeric membranes. Various topics have been discussed, including the synthesis and characterization of novel membrane materials, membrane aging, and the impact of process conditions on transport phenomena, with the desire to improve the gas separation process in all the articles. There are nine contributions, namely, eight articles and one review, in this Special Issue.

Zang et al. [1] synthetized new polymers based on phenylacetylene monomers having one or two carbamate moieties in order to enhance the O_2/N_2 separation performance. The presence of two carbamate groups induced a better membrane-forming ability and a higher oxygen solubility and diffusivity thanks to the cistransoid conformation, which is flexible. Thus, these materials presented a greater oxygen permeability compared to the one without carbamate group, 420 Barrer and 3 Barrer, respectively.

In the same way, Rodriguez-Gonzalez et al. [2] synthetized a new set of polyimides in order to study the impact of different chemical moieties (acyclic alkyl-*N*-carbamoyl group with different alkyl chains) on gas properties. For all gases, the authors noticed a decrease in the alkyl chains as they were shortened. These results were correlated with the calculated FFV and the obtained d-spacing values.

Liu et al. [3] described how graphdiyne could make an excellent candidate for hydrogen purification. In their paper, they investigated the gas permeation of four pure gases (H_2, N_2, CO_2, and CH_4) and binary mixtures through a theoretical study based on MD and DFT calculations. Thanks to its uniform pore size (2.1 Å) and atomic thickness, the graphdiyne presented approximatively infinite selectivities of H_2 over N_2, CO_2, and CH_4. Moreover, with a slight presence of surface charges, the authors showed that the H_2 permeance of the binary H_2/CO_2 could be increased up to $2 \times 8 \times 10^5$ GPU, which is several orders of magnitude greater than the existing experiments.

Begni et al. [4] developed new materials to enhance gas separation over time. They synthesized new hypercrosslinked polymers based on the Friedel–Crafts reaction between a tetraphenyl methane monomer and a bromomethyl benzene monomer in order to use them as fillers in PIM-1 mixed matrix membranes. According the reaction process, two fillers have been obtained, with different particle sizes and surface areas, 498 nm and 823 m^2/g and 120 nm and 990 m^2/g, respectively. The authors showed that the presence of the fillers induced a slowdown of the membrane's physical aging, which is greater for the smaller filler. After almost three years of aging, mixed matrix membranes retained approximatively 40% of the initial CO_2 permeability while pure PIM-1 showed a reduction of 85%. ^{13}C spin-lattice relaxation time studies showed that this slow-down in aging is due to the interactions between the PIM-1 chain and the HCP fillers.

The enhancement of gas separation performance could also be achieved by investigating the effect of the process conditions, such as process temperature or the presence of water or other pollutants in the feed. Casado-Coterillo et al. [5] studied the effect of the presence of water and organic pollutants in CO_2/CH_4 separation within a hydrophilic and hydrophobic composite membrane. The authors showed that the permeance of the hydrophobic membrane was affected by the presence of damp impurities with an increase in CO_2 permeance, while the permeance and selectivity of the hydrophilic membrane were almost invariable in the presence of humid streams. By addition of toluene, used as model organic pollutants, the differences were enhanced. Moreover, the CO_2 permeances decreased with increasing CO_2 concentration in the feed in a more remarkable way for the hydrophobic PDMS composite membrane than for the hydrophilic IL–CS-based composite membrane, which may be attributed to the water-facilitated transport through the hydrophilic membrane. By consequence, the tuning up of the hydrophilic/hydrophobic character of the membrane surface can be an effective way of improving facilitated transport properties and improving membrane performance in CO_2 capture applications.

Nemestothy et al. [6] also studied the impact of the presence of pollutants on CO_2/CH_4 separation for a polyimide membrane. Commercial hollow fibers based on polyimide have been tested in the presence of naturally occurring contaminants of natural gases, namely, hydrogen sulfide, dodecane, and the mixture of aromatic hydrocarbons (benzene, toluene, xylene). The authors showed that all of the investigated pollutants had an impact on the membrane's performance but in different ways and to different extents. Hydrogen sulfide increased the permeability of both CO_2 and CH_4, and the CO_2/CH_4 selectivity had a decreasing tendency as a function of increasing H_2S exposures. In the case of dodecane, the permeability of CO_2 and CH_4 decreased moderately by increasing the degree of exposure, while the CO_2/CH_4 selectivity, according to tendencies, was left unaffected. By contrast, the larger exposures of aromatic hydrocarbons caused the increase in gas permeabilities; however, the corresponding trends indicated only marginal changes in the CO_2/CH_4 selectivity.

Belaissaoui et al. [7] showed as well that by playing on the selective layer thickness and the carrier concentration of a facilitated transport membrane, it is possible to enhance the separation performance. Their analyses are based on experimental measurements of CO_2 and N_2 fluxes through a hybrid fixed-site carrier membrane, based on poly(allyl amine) matrix and on analytical solutions of the facilitation factor mathematically described by means of differential equations expressing a steady-state nonlinear diffusion reaction problem. The dedicated parametric analysis demonstrated that decreasing the selective layer thickness to 0.1 µm together with doubling of the total carrier concentration would theoretically shift the membrane performance far above the Robeson upper bound for the CO_2/N_2 pair. However, this potential path for membrane performance improvement has to be weighted by the possible depletion in the reaction complex effective diffusivity.

Benedetti et al. [8] prepared new mixed matrix membranes based on poly(2,6-dimethyl-1,4-phenylene oxide) (PPO) and particles of the size-selective Zeolitic Imidazolate Framework 8 (ZIF-8). The aim was to increase the permselectivity properties of pure PPO. The authors showed that the addition of 45%wt ZIF8 improved the separation performance of PPO by an increase of 800% for CO_2 and He permeability coefficients. The temperature increase also yielded a simultaneous increase of permeability and selectivity, indicating that such membranes can have potential for applications at high temperatures.

Finally, Vermaak et al. [9] described in their review an overview of membrane-based electrochemical hydrogen separation technologies. Electrochemical membranes are seen as a promising alternative to pressure-driven membranes. Electrochemical membranes are known to generate electricity (fuel cells) or to apply it (water electrolysis), and they are also used to purify/enrich and compress hydrogen streams. They detailed the working principle of electrochemical hydrogen separation and discussed the impact of condition processes, such as temperature, gas mixture, and catalysts, on the separation performance.

In conclusion, the findings and critical discussions from these contributions highlight the importance of membrane materials and the processes for gas separation.

Funding: This research received no external funding.

Acknowledgments: The editors would like to thank the authors for their high-quality outputs and Lydia Li for her patient and motivating help and assistance.

Conflicts of Interest: The authors declare no conflict of interest.

References

1. Zang, Y.; Lun, Y.; Teraguchi, M.; Kaneko, T.; Jia, H.; Miao, F.; Zhang, X.; Aoki, T. Synthesis of Cis-Cisoid or Cis-Transoid Poly(Phenyl-Acetylene)s Having One or Two Carbamate Groups as Oxygen Permeation Membrane Materials. *Membranes* **2020**, *10*, 199. [CrossRef] [PubMed]
2. Rodríguez-González, F.E.; Pérez, G.; Niebla, V.; Jessop, I.; Martin-Trasanco, R.; Coll, D.; Ortiz, P.; Aguilar-Vega, M.; Tagle, L.H.; Terraza, C.A.; et al. New Poly(imide)s Bearing Alkyl Side-Chains: A Study on the Impact of Size and Shape of Lateral Groups on Thermal, Mechanical, and Gas Transport Properties. *Membranes* **2020**, *10*, 141. [CrossRef] [PubMed]
3. Liu, Q.; Cheng, L.; Liu, G. Enhanced Selective Hydrogen Permeation through Graphdiyne Membrane: A Theoretical Study. *Membranes* **2020**, *10*, 286. [CrossRef] [PubMed]
4. Begni, F.; Lasseuguette, E.; Paul, G.; Bisio, C.; Marchese, L.; Gatti, G.; Ferrari, M.-C. Hyper Cross-Linked Polymers as Additives for Preventing Aging of PIM-1 Membranes. *Membranes* **2021**, *11*, 463. [CrossRef] [PubMed]
5. Casado-Coterillo, C.; Garea, A.; Irabien, Á. Effect of Water and Organic Pollutant in CO_2/CH_4 Separation Using Hydrophilic and Hydrophobic Composite Membranes. *Membranes* **2020**, *10*, 405. [CrossRef] [PubMed]
6. Nemestóthy, N.; Bakonyi, P.; Lajtai-Szabó, P.; Bélafi-Bakó, K. The Impact of Various Natural Gas Contaminant Exposures on CO_2/CH_4 Separation by a Polyimide Membrane. *Membranes* **2020**, *10*, 324. [CrossRef] [PubMed]
7. Belaissaoui, B.; Lasseuguette, E.; Janakiram, S.; Deng, L.; Ferrari, M.-C. Analysis of CO_2 Facilitation Transport Effect through a Hybrid Poly(Allyl Amine) Membrane: Pathways for Further Improvement. *Membranes* **2020**, *10*, 367. [CrossRef] [PubMed]
8. Benedetti, F.M.; De Angelis, M.G.; Degli Esposti, M.; Fabbri, P.; Masili, A.; Orsini, A.; Pettinau, A. Enhancing the Separation Performance of Glassy PPO with the Addition of a Molecular Sieve (ZIF-8): Gas Transport at Various Temperatures. *Membranes* **2020**, *10*, 56. [CrossRef] [PubMed]
9. Vermaak, L.; Neomagus, H.W.J.P.; Bessarabov, D.G. Recent Advances in Membrane-Based Electrochemical Hydrogen Separation: A Review. *Membranes* **2021**, *11*, 127. [CrossRef] [PubMed]

Article

Synthesis of Cis-Cisoid or Cis-Transoid Poly(Phenyl-Acetylene)s Having One or Two Carbamate Groups as Oxygen Permeation Membrane Materials

Yu Zang [1,*], Yinghui Lun [2], Masahiro Teraguchi [3], Takashi Kaneko [3], Hongge Jia [1], Fengjuan Miao [4], Xunhai Zhang [1] and Toshiki Aoki [1,3]

1. Key laboratory of polymer matrix composites, Heilongjiang Province, College of Materials Science and Engineering, Qiqihar University, Wenhua Street 42, Qiqihar, Heilongjiang 161006, China; jiahongge11@hotmail.com (H.J.); zhangxunhai@163.com (X.Z.); toshaoki@eng.niigata-u.ac.jp (T.A.)
2. Department of Materials and Chemical Engineering, Hunan Institute of Technology, Hengyang, Hunan 421002, China; lunyinghui_888@163.com
3. Chemistry and Chemical Engineering, Graduate School of Science and Technology, Niigata University, Ikarashi 2-8050, Nishi-ku, Niigata 950-2181, Japan; teraguti@eng.niigata-u.ac.jp (M.T.); kanetaka@gs.niigata-u.ac.jp (T.K.)
4. College of Communications and Electronics Engineering, Qiqihar University, Wenhua Street 42, Qiqihar, Heilongjiang 161006, China; miaofengjuan@qqhru.edu.cn
* Correspondence: zangyu@qqhru.edu.cn

Received: 24 July 2020; Accepted: 21 August 2020; Published: 25 August 2020

Abstract: Three new phenylacetylene monomers having one or two carbamate groups were synthesized and polymerized by using (Rh(norbornadiene)Cl)$_2$ as an initiator. The resulting polymers had very high average molecular weights (Mw) of 1.4–4.8 × 10^6, with different solubility and membrane-forming abilities. The polymer having two carbamate groups and no hydroxy groups in the monomer unit showed the best solubility and membrane-forming ability among the three polymers. In addition, the oxygen permeability coefficient of the membrane was more than 135 times higher than that of a polymer having no carbamate groups and two hydroxy groups in the monomer unit with maintaining similar oxygen permselectivity. A better performance in membrane-forming ability and oxygen permeability may be caused by a more extended and flexible cis-transoid conformation and lower polarity. On the other hand, the other two new polymers having one carbamate group and two hydroxy groups in the monomer unit showed lower performances in membrane-forming abilities and oxygen permeabilities. It may be caused by a very tight cis-cisoid conformation, which was maintained by intramolecular hydrogen bonds.

Keywords: polyphenylacetylene; cis-cisoid conformation; cis-transoid conformation; carbamate group; membrane-forming ability; solubility; oxygen permeation membrane

1. Introduction

π-Conjugated polymers like polyacetylenes [1–3] have aroused interest because of their noteworthy physical properties, such as conductivity, organomagnetism, and optical nonlinear susceptibility. Among them, poly(substituted acetylene)s such as poly(substituted phenylacetylene)s are useful because of their stability in the air, possibility of a variety of derived structures, and good performances as separation membrane materials [4–12]. Poly(substituted phenylacetylene)s are generally rigid polymers and show good oxygen permselectivities. In addition, they are soluble and suitable as oxygen permeation membrane materials. Most of poly(substituted acetylene)s reported take cis-transoid conformation. On the other hand, the poly(substituted acetylene with two hydroxyl groups)s we reported take cis-cisoid conformation [13–15]. As a result, they are expected to have a more rigid backbone than cis-transoid poly(substituted acetylene)s. Therefore, cis-cisoid poly(substituted acetylene)s are very promising as better oxygen permeation membrane materials if their processability is good. In this paper, to discuss this factor—that is, the rigidity of the backbone—we compared the two acetylene polymers having these different conformations.

In general, polymers obtained by the polymerization of monosubstituted acetylenes using a (Rh(norbornadiene)Cl)$_2$ (norbornadiene = nbd) catalytic system take a cis-transoidal loosely helical conformation [16–19]. We have been reporting poly(phenylacetylene)s taking a cis-cisoidal tightly helical conformation from monomers having two hydroxy groups, such as **4** (in Chart 1), using a similar catalytic system [13,14]. The cis-cisoidal tightly helical conformation of poly(**4**) was kept by an intramolecular hydrogen bond between the OH groups [14]. Although the monomer is very valuable, because it is the only monomer to give such polymers, the structures of suitable monomers giving such polymers are very limited. We also reported a phenylacetylene with two amido groups that can make hydrogen bonds instead of hydroxy groups as the second suitable monomer [15]. Therefore, it is important to find other suitable monomers having another functional groups that can make hydrogen bonds instead of hydroxy groups.

Since amino groups can make hydrogen bonds similarly to hydroxyl groups, monomers having amino groups are very promising. In addition, since amino groups are basic and important functional groups, therefore, polymers with amino groups are also important as reagents, catalysts, biocompatible or biodegradable materials, carbon dioxide permselective membranes, and so on [20–27]. However, the direct synthesis of amino group-containing poly(substituted acetylene)s by polymerizing the corresponding amino group-containing monomers using a rhodium complex as a catalyst has some problems. For example, the amino groups in monomers interact with the Rh catalyst so strongly [28,29] that the polymerization can be disrupted to yield only low M_W polymers. Even if the polymerization proceeded, the resulting polymer should be insoluble due to the strong hydrogen bonds. Therefore, amino groups should be protected before rhodium complex-catalyzed polymerization. A carbamate group is a typical protecting group for amines, and it can weaken hydrogen bonds and makes the monomer more hydrophobic. In addition, since it has C=O and NH groups, it can still make hydrogen bonds. Therefore, carbamate group-containing poly(phenylacetylene)s can show similar characteristics and better solubility and membrane-forming properties than amino-containing poly(phenylacetylene)s. However, the solubility and the membrane-forming ability of polymers with cis-cisoidal tightly helical conformations such as poly(**4**) were not the best, although they were applied to oxygen permselective membranes [30]. It may be due to the tight cis-cisoidal helical main chain and rigid rod structures inducing some crystalline domains [30].

In this study, in order to obtain carbamate group-containing poly(phenylacetylene)s having cis-cisoid or cis-transoid main chains that show high solubility and good membrane-forming abilities, we carried out the synthesis and polymerization of two kinds of novel phenylacetylene monomers—that is, two new monomers with one carbamate group and two hydroxy groups (Figure 1, **1,2**) and a new monomer containing two carbamate groups and no hydroxy groups (Figure 1, **3**). Then, we discuss some properties as oxygen permeation materials, such as the solubility, membrane-forming ability, and oxygen permeability of the resulting new polymers. The effects of the conformation of the polymer main chains on the properties are discussed.

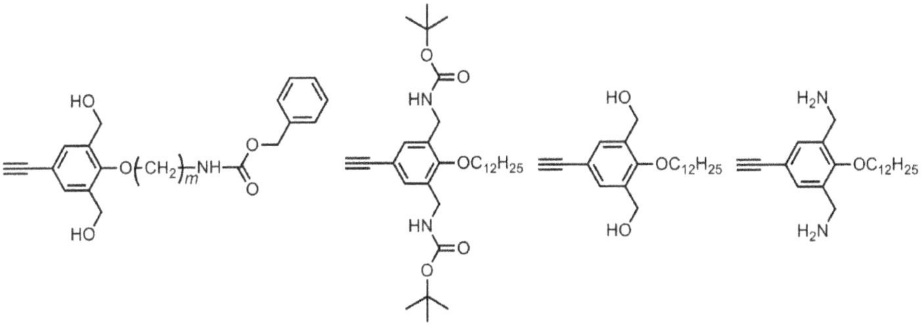

Figure 1. Chemical structures of new one or two carbamate-containing phenylacetylenes (**1–3**) and (**3** and **5**).

2. Materials and Methods

2.1. Materials

All the solvents used for synthesis and polymerization of the monomers were distilled as usual. The polymerization initiator, (Rh(nbd)Cl)$_2$ (nbd = 2,5 norbornadiene), purchased from Aldrich Chemical (Tokyo, Japan), was used as received.

2.2. Measurements

^{1}H NMR (400 MHz) spectra were recorded on a JEOL LEOLEX-400 spectrometer (JEOL, Akishima, Japan). The average molecular weights (M_n and M_w) were evaluated by gel permeation chromatography (GPC) by using JASCO liquid chromatography instruments (JASCO, Tokyo, Japan) with PU-2080, DG- 2080-53, CO-2060, UV-2070, and two polystyrene gel columns (Shodex KF-807 L, tetrahydrofuran (THF) eluent, polystyrene calibration, TCI, Tokyo, Japan). The infrared spectra (IR) were recorded on FT-IR-4200 (JASCO) (JASCO, Tokyo, Japan). UV-vis spectra were measured with a JASCO V-550 spectropolarimeter (JASCO, Tokyo, Japan).

2.3. Synthesis of Monomer **1**

2.3.1. N-Benzyloxycarbonyl-2-aminoethanol (**10**, m = 2)

Benzyl chloroformate (6.40 mL, 45.4 mmoL) in diethyl ether (14.0 mL) was added dropwise to a solution of 2-aminoethanol (2.70 mL, 45.4 mmoL) in 10% aqueous Na$_2$CO$_3$ (54.0 mL) at 0 °C and stirred for 1.5 h. The reaction mixture was acidified with 10% HCl at 0 °C to give precipitates that were filtered and washed with H$_2$O to give **10** as a white crystal. [31] Yield: 35.2% (3.06 g). ^{1}H NMR (400 MHz, CDCl$_3$, TMS, δ): 7.32 (m, 5H, Ph*H*), 5.15 (br, 1H, CH$_2$N*H*CO), 5.09 (s, 2H, PhC*H*$_2$OCO), 3.70 (q, 2H, OHC*H*$_2$CH$_2$), 3.35 (q, 2H, CH$_2$C*H*$_2$NH), 2.17 (br, 1H, CH$_2$O*H*).

2.3.2. N-Benzyloxycarbonyl-2-bromoethylamine (**11**, m = 2)

A flask was charged with **10** (9.76 g, 50.0 mmoL), methanesulfonyl chloride (4.65 mL, 60.0 mmoL,), and CH$_2$Cl$_2$ (150 mL). To this stirring solution, Et$_3$N (9.01 mL, 65.0 mmoL) was added. Stirring was continued for 45 min, and then, LiBr (43.5 g, 500 mmoL) and acetone (150 mL) were added. The reaction mixture was stirred for an additional 21.5 h, and then, the solvents were removed by rotary evaporation. The contents were partitioned between Et$_2$O (100 mL) and H$_2$O (65.0 mL), and the Et$_2$O layer was washed with brine, dried over anhydrous Na$_2$SO$_4$, filtered, evaporated to dryness, and got the brown liquid product. Yield: 90.9% (2.98 g). [32] ^{1}H NMR (400 MHz, CDCl$_3$, TMS, δ): 7.35 (m, 5H, Ph*H*), 5.18 (br, 1H, CH$_2$N*H*CO), 5.10 (s, 2H, PhC*H*$_2$OCO), 3.60 (q, 2H, CH$_2$C*H*$_2$NH), 3.45 (t, 2H, BrC*H*$_2$CH$_2$).

2.3.3. 4-(N-Benzyloxycarbonyl-2-ethylamino)benzyloxy-3,5-bis(hydroxymethyl)phenylacetylene (1, m = 2)

The solution of **11** (0.736 g, 2.87 mmoL), **9** (0.500 g, 2.81 mmoL), and potassium carbonate (0.620 g, 4.49 mmoL) in N,N-dimethylformamide (DMF) (15.0 mL) was refluxed for 48 h and cooled to room temperature. Then, the mixture was filtered, and the solvent in the filtrate was removed by evaporation. The crude product was purified by silica-gel column chromatography to give 1 as a white solid. Yield: 23.1% (0.231 g). Retention volumes (R_f) = 0.20 (ethyl acetate/hexane = 1/1). ^{1}H NMR (400 MHz, CDCl$_3$, TMS, δ): 7.45 (s, 2H, PhH), 7.32 (m, 5H, PhH), 5.60 (br, 1H, CH$_2$NHCO), 5.10 (s, 2H, PhCH_2OCO), 4.62 (d, 4H, Ph(CH_2OH)$_2$), 4.01 (t, 3H, PhOCH_2CH$_2$), 3.58 (q, 2H, CH$_2$CH_2NH), 3.02 (s, 1H, HC≡C), 2.03 (t, 2H, Ph(CH$_2$OH)$_2$). IR (cm^{-1}, KBr): 3380 (OH), 3311 (NH), 3298 (H–C≡), 1692 (C=O), 1267 (C–O), 1051 (C–N). (For the synthesis of **9**, see S1.1–S1.4 in the Supporting Information.)

2.4. Synthesis of Monomer 2

2.4.1. N-(Benzyloxycarbonyl)-6-amino-1-hexanol (10, m = 6)

The synthesis procedure for N-(benzyloxycarbonyl)-6-amino-1-hexanol (**10**, $m = 6$) was similar to N-benzyloxycarbonyl-2-aminoethanol (**10**, $m = 2$) to give a white crystal. Yield: 44.8% (4.82 g). ^{1}H NMR (400 MHz, CDCl$_3$, TMS, δ): 7.35 (m, 5H, PhH), 5.08 (s, 2H, PhCH_2OCO), 4.72 (br, 1H, CH$_2$NHCO), 3.61 (q, 2H, HOCH_2CH$_2$), 3.19 (q, 2H, CH$_2$CH_2NH), 1.55–1.31 (m, 8H, HOCH$_2$(CH_2)$_4$CH$_2$).

2.4.2. N-(Benzyloxycarbonyl)-6-bromohexylamine (11, m = 6)

The synthesis procedure for N-(benzyloxycarbonyl)-6-bromohexylamine (**11**, $m = 6$) was similar to N-benzyloxycarbonyl-2-bromoethylamine (**11**, $m = 2$) to give a brown crystal. Yield: 73.7% (2.30 g). ^{1}H NMR (400 MHz, CDCl$_3$, TMS, δ): 7.35 (m, 5H, PhH), 5.08 (s, 2H, PhCH_2OCO), 4.72 (br, 1H, CH$_2$NHCO), 3.40 (t, 2H, BrCH_2CH$_2$), 3.17 (q, 2H, CH$_2$CH_2NH), 1.83 (m, 2H, BrCH$_2$CH_2CH$_2$), 1.55–1.31 (m, 6H, BrCH$_2$CH$_2$(CH_2)$_3$CH$_2$).

2.4.3. 4-(N-Benzyloxycarbonyl-6-hexylamino)benzyloxy-3,5-bis(hydroxymethyl)phenylacetylene (2, m = 6)

The synthesis procedure for monomer **2** was similar to monomer **1** to give a white solid. Yield: 32.2% (0.370 g). R_f = 0.30 (ethyl acetate/hexane = 1/1). ^{1}H NMR (400 MHz, CDCl$_3$, TMS, δ): 7.45 (s, 2H, PhH), 7.32 (m, 5H, PhH), 5.07 (s, 2H, PhCH_2OCO), 4.81 (br, 1H, CH$_2$NHCO), 4.67 (d, 4H, Ph(CH_2OH)$_2$), 3.85 (t, 2H, PhOCH_2CH$_2$), 3.20 (q, 2H, CH$_2$CH_2NH), 3.02 (s, 1H, HC≡C), 2.16 (t, 2H, Ph(CH$_2$OH)$_2$), 1.77 (m, 2H, PhOCH$_2$CH_2CH$_2$), 1.52-1.36 (m, 6H, CH$_2$(CH_2)$_3$CH$_2$NH). IR (cm^{-1}, KBr): 3380 (OH), 3356 (NH), 3304 (H–C≡), 1681 (C=O), 1268 (C–O), 1060 (C–N).

2.5. Synthesis of Monomer 3

2.5.1. 4-Dodecyloxy-3,5-bis(hydroxymethyl)phenylacetylene (4)

According to the literature [23], **4** was synthesized to give a white solid. Yield: 69.6% (1.25g). R_f = 0.24 (ethyl acetate/hexane = 1/4). ^{1}H-NMR(400MHZ, CDCl$_3$, TMS, δ): 7.46 (s, 2H, PhH), 4.68 (d, 4H, Ph(CH_2OH)$_2$), 3.88 (t, 2H, OCH_2CH$_2$(CH$_2$)$_9$CH$_3$), 3.00 (s, 1H, HC≡C), 1.96 (t, 2H, (CH$_2$OH)$_2$), 1.79 (dm, 2H, OCH$_2$CH_2CH$_2$), 1.50-1.20 (m, 18H, OCH$_2$CH$_2$(CH_2)$_9$CH$_3$), 0.883 (t, 3H, OCH$_2$CH$_2$(CH$_2$)$_9$CH_3). IR (KBr): 3300 (OH), 3250 (H–C≡), 2840 (CH), 2095 (C≡C).

2.5.2. 4-Dodecyloxy-3,5-bis(bromomethyl)phenylacetylene (12)

The solution of **4** (1.00 g, 2.89 mmoL), carbon tetrabromide (3.26 g, 9.82 mmoL), and triphenylphosphine (2.27 g, 8.67 mmoL) in dichloromethane (40.0 mL) was stirred at 0 °C for 4 h. Saturated solution of NaHCO$_3$ was added to the mixture, and then, liquid–liquid separation was carried out by tap funnel. The organic layer was dried over anhydrous MgSO$_4$, evaporated, and the residue was purified by column chromatography to give a white solid. Yield: 77.0% (1.05 g). R_f = 0.17 (hexane). ^{1}H NMR (400 MHz, CDCl$_3$, TMS, δ): 7.50 (s, 2H, PhH), 4.49 (s, 4H, Ph(CH_2Br)$_2$), 4.09 (t, 2H, OCH_2CH$_2$(CH$_2$)$_9$CH$_3$), 3.06 (s, 1H, HC≡C), 1.89 (m, 2H, OCH$_2$CH_2(CH$_2$)$_9$CH$_3$), 1.27–1.55 (m, 18H, OCH$_2$CH$_2$(CH_2)$_9$CH$_3$), 0.88 (t, 3H, OCH$_2$CH$_2$(CH$_2$)$_9$CH_3).

2.5.3. 4-Dodecyloxy-3,5-bis(azidomethyl)phenylacetylene (13)

A solution of **12** (0.500 g, 1.06 mmoL) and NaN$_3$ (275 mg, 4.24 mmoL) in DMF (4.50 mL) was stirred at room temperature. After stirring of the mixture for 48 h at room temperature, the mixture was poured into water and extracted with dichloromethane. The organic layer was washed with water, brine, and dried over anhydrous MgSO$_4$. The organic solvent was removed by evaporation, and the crude product was purified by silica-gel column chromatography to give a pale-yellow oil. Yield: 95.0% (403 mg) [33]. R_f = 0.4 (ethyl/hexane = 1/20). ^{1}H NMR (400 MHz, CDCl$_3$, TMS, δ): 7.50 (s, 2H, PhH), 4.49 (s, 4H, Ph(CH_2N$_3$)$_2$), 4.09 (t, 2H, OCH_2CH$_2$(CH$_2$)$_9$CH$_3$), 3.06 (s, 1H, HC≡C), 1.89 (m, 2H, OCH$_2$CH_2(CH$_2$)$_9$CH$_3$), 1.27–1.55 (m, 18H, OCH$_2$CH$_2$(CH_2)$_9$CH$_3$), 0.88 (t, 3H, OCH$_2$CH$_2$(CH$_2$)$_9$CH_3).

2.5.4. 4-Dodecyloxy-3,5-bis(aminomethyl)phenylacetylene (5)

To a solution of **13** (400 mg, 1.01 mmoL) in THF (15.0 mL), a mixture of LiAlH$_4$ (95.8 mg, 2.53 mmoL) and THF (10.0 mL) was added dropwise at 0° C. The mixture was stirred for 24 h at room temperature. When the reaction finished, H$_2$O was added dropwise at 0° C. A saturated aqueous solution of NaOH (10.0 mL) was added dropwise to the mixture and stirred for 30 min. After the mixture was filtered, the organic solvent was removed by evaporation. The crude product was washed with a saturated solution of NaCl. The organic layer was dried over anhydrous MgSO$_4$ and then evaporated to give a yellow liquid. Yield: 91.5% (0.541 g) [33]. ^{1}H NMR (400 MHz, CDCl$_3$, TMS, δ): 7.36 (s, 2H, PhH), 3.81 (s, 4H, Ph(CH_2NH$_2$)$_2$), 3.80(t, 2H, PhOCH_2CH$_2$), 3.00 (s, 1H, HC≡C), 1.81 (m, 2H, PhOCH$_2$CH_2CH$_2$), 1.57 (br, 4H, Ph(CH$_2$NH_2)$_2$), 1.27–1.55 (m, 18H, OCH$_2$CH$_2$(CH_2)$_9$CH$_3$), 0.88 (t, 3H, OCH$_2$CH$_2$(CH$_2$)$_9$CH_3).

2.5.5. 4-Dodecyloxy-3,5-bis(tert-butoxycarbonylamino)) Phenyl Acetylene (3)

A THF (12 mL) solution of **5** (20 mg, 0.058 mmoL), di-tert-butyl dicarbonate ((Boc)$_2$O) (27.9 mg, 0.128 mmoL), and 4-dimethylaminopyridine (DMAP) (0.7 mg) was refluxed for 6 h and then cooled to room temperature. THF was removed by evaporation, and the crude product was purified by silica-gel column chromatography to give monomer **3** as a white solid. Yield: 25.3% (7.10 mg). R_f = 0.45 (hexane/ethyl acetylene = 4/1). ^{1}H NMR (400 MHz, CDCl$_3$, TMS, δ): 7.34 (s, 2H, PhH), 4.84 (br, 2H, Ph(CH$_2$NHCO)$_2$), 4.32 (d, 4H, Ph(CH_2NH)$_2$), 3.77 (t, 2H, PhOCH_2CH$_2$), 3.01 (s, 1H, HC≡C), 1.80 (m, 2H, OCH$_2$CH_2CH$_2$), 1.44 (s, 18H, ((OCH_3)$_3$)$_2$), 1.27-1.55 (m, 18H, CH$_2$(CH_2)$_9$CH$_3$), 0.88 (t, 3H, CH$_2$CH_3). IR (cm^{-1}, KBr): 3370 (NH), 3298 (≡C–H), 1691 (C=O), 1480 (CH), 1168 (C–O), 1096 (C–N).

2.6. Polymerization of Monomers **1**–**3**

A typical procedure for **1** was as follows (Schemes 1 and 2): A dry THF (0.350 mL) solution of (Rh(nbd)Cl)$_2$ (0.322 mg, 0.700 μmol) and 1-phenethylamine (PEA) (17.9 μL, 0.141 mmoL) was added to a dry THF (0.350 mL) solution of **1** (25.0 mg, 0.070 mmoL). The reaction solution was stirred at room temperature for 8 h. The crude polymer was purified by reprecipitation of the THF solution into a large amount of methanol, and the formed solid was dried in vacuo to give poly(**1**).

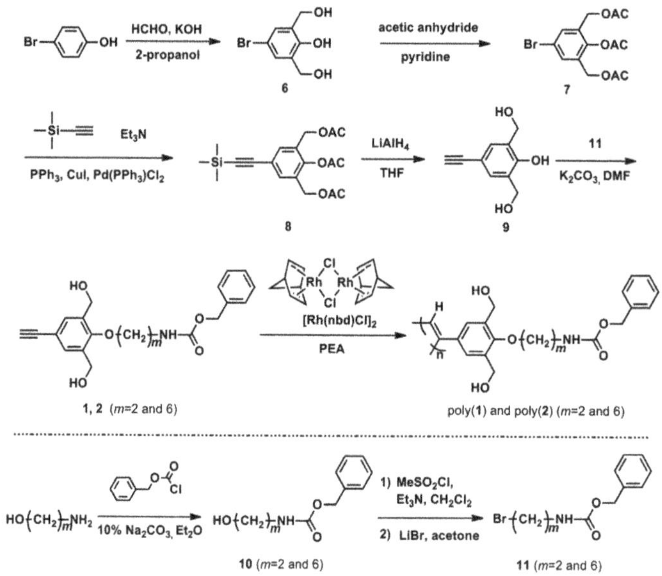

Scheme 1. Synthetic route to poly(**1**) and poly (**2**).

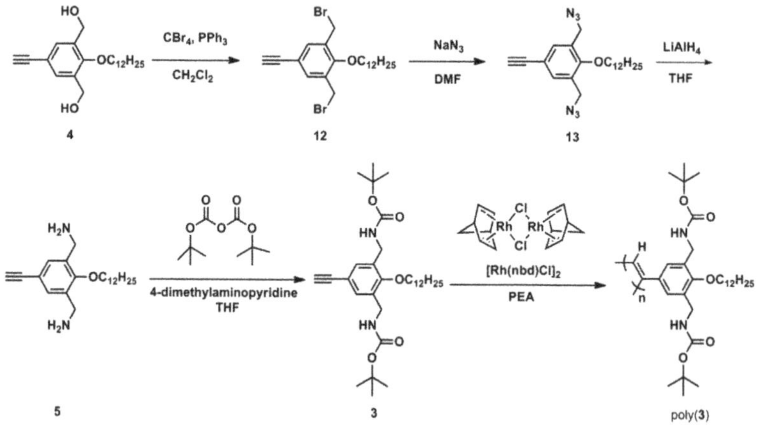

Scheme 2. Synthetic route to poly(**3**).

Other polymerizations of monomers **2** and **3** were carried out similarly. (Figures S1–S20)

Poly(**1**) ^{1}H NMR(400 MHz, dimethylsulfoxide-d_6 (DMSO-d_6), δ): 7.37 (br, PhH), 6.68 (br, *cis* proton in the main chain), 5.70 (br, CH$_2$NHCO), 5.07–4.63 (br, PhCH_2OCO, Ph(CH_2OH)$_2$), 4.28 (br, Ph(CH$_2$OH)$_2$). IR (cm^{-1}, KBr): 3334 (NH, OH), 1690 (C=O), 1480 (CH), 1261 (C–O), 1096 (C–N).

Poly(**2**) ^{1}H NMR(400 MHz, DMSO-d_6, δ): 7.38–7.27 (br, PhH), 6.73 (br, *cis* proton in the main chain), 5.76 (br, CH$_2$NHCO), 5.29–4.68 (br, PhCH_2OCO, Ph(CH_2OH)$_2$), 4.33 (br, Ph(CH$_2$OH)$_2$), 3.04 (br, CH$_2$CH_2NH, PhOCH_2CH$_2$), 1.46–1.30 (t, 6H, CH$_2$(CH_2)$_3$NH). IR (cm^{-1}, KBr): 3334 (NH, OH), 1690 (C=O), 1480 (CH), 1261 (C–O), 1096 (C–N).

Poly(**3**) ^{1}H NMR(400 MHz, CDCl$_3$, TMS, δ): 7.50 (br, PhH), 5.10 (br, Ph(CH_2NHCO)$_2$), 3.48 (br, Ph(CH$_2$NH)$_2$, PhOCH_2CH$_2$), 2.02 (br, PhOCH$_2$CH_2CH$_2$), 1.66–1.54 (br, ((OCH_3)$_3$)$_2$), 1.23–1.09

(br, CH$_2$(CH$_2$)$_9$CH$_3$), 0.86 (br, CH$_2$CH$_3$). IR (cm^{-1}, KBr): 3370 (N–H), 1691 (C=O), 1480 (CH), 1168 (C–O), 1096 (C–N).

2.7. Membrane Preparation

A typical membrane fabrication method for poly(**1**) was as follows: A solution of the poly(**1**) (0.060-10.0 wt%) in DMF (40.0 mg/mL) was cast on a poly(tetrafluoroethylene) sheet (4 cm^2). After evaporating of the solvent for 12 h at 25 °C, the membranes were detached from the sheet and dried in a vacuum oven for 24 h at 60 °C. Thickness (L) of the membranes was 50.0–80.0 µm ± 0.5 µm. Other polymer membranes were prepared similarly.

2.8. Estimation of Polymers as Oxygen Permeation Membranes

2.8.1. Membrane Strength

Maximum flexural stresses (σ/kPa) of membranes were calculated according to the following equation:

$$\sigma = \frac{3FL}{2bd^2} \quad (1)$$

where F, L, b, and d are the load, length of the support span, width of membrane, and thickness of membrane, respectively (Figure S21).

2.8.2. Oxygen Permeation

Oxygen and nitrogen permeability coefficients (P_{O2} and P_{N2}: cm^3(STP)·cm·cm^{-2}·s^{-1}·cmHg^{-1}) (STP = standard temperature and pressure) and the oxygen separation factor ($\alpha = P_{O2}/P_{N2}$) were measured by a gas chromatographic method by using YANACO GTR-10, according to reference [34]. The P_{O2} and P_{N2} were calculated by the following equation:

$$P = \frac{Q \times L}{A \times \Delta P \times t} \quad (2)$$

where Q, L, A, ΔP, and t are the amount of the permeated gas, the thickness of the membrane, the permeation area of the membrane, the pressure difference across the membrane, and the permeation time, respectively. Disc-type membranes were used. The A and L of the membranes were 3.14 cm^2 and around 60-80 µm, respectively. The ΔP was 1 atm, and the measurement temperature was 25 °C.

The diffusion coefficient (D: cm^2·σ^{-1}) was calculated by the time-lag method represented by $D = L^2/6\theta$, where L (cm) is the thickness of the membrane, and θ(s) is the time-lag.

3. Results and Discussion

3.1. Synthesis of Monomers 1–3

To obtain polymers having one or two carbarmate groups, two different types of new phenylacetylene monomers were synthesized according to Schemes 1 and 2. The first type of the monomers contains two hydroxy groups and one carbarmate group (Figure 1, **1,2**). By five-step reactions from 4-bromophenol, monomers **1** and **2** were synthesized successfully (Scheme 1). The two novel monomers were purified by silica-gel column chromatography, and the total yields of monomers **1** and **2** were 5.5% and 7.7%, respectively. The second type of the novel monomers contains two carbarmate groups (Figure 1, **3**). It was synthesized by a four-step reaction from **4** according to our previous report [13] (Scheme 2). Monomer **3** was purified by silica-gel column chromatography, and the total yield was 7.6%. In addition, monomer **5** was synthesized from **4** by a three-step reaction and was purified by vacuum-drying as a yellow liquid in a total yield of 69.2% (Scheme 2). The polarity of the monomers showed a significant impact on the polymerization results, such as yields, solubility, membrane-forming ability, and so on (Tables 1 and 2). The decreasing order of polarity of the five monomers was **5 > 1 > 2 > 4 > 3**, judging from the retention volumes (R_f) of the monomers on the silica-gel thin-layer chromatography (TLC) using ethyl acetate/hexane (1/1) as an eluent (Table 1). When the two hydroxy (in monomer **4**) or amino groups (in monomer **5**) were replaced with two carbamate groups, the polarity of the monomer (monomer **3**) decreased largely.

Table 1. The polymerization results of monomers 1–5 [a].

No.	Monomer [b]	R_f [c]	Solvent	Yield (%) [d]	M_w (×10^6) [e]	M_w/M_n [e]
1	1	0.20	THF	52.4	4.80	2.32
2	2	0.30	THF	38.0	1.40	6.50
3	3	0.93	toluene	72.4	2.90	4.31
4	4	0.80	toluene	43.2	3.10	5.40
5	5	0.00	toluene	4.60	_[f]	_[f]

[a] At room temperature for 8 h, (Monomer) = 0.100 mol/L, (Monomer)/((Rh[nbd]Cl)$_2$) = 100, and ((Rh[nbd]Cl)$_2$)/(1-phenethylamine) = 1/200. [b] For the codes, see Chart 1. [c] Retention volumes of the monomers on silica-gel thin-layer chromatography (TLC) using eltylacetate/hexane = 1/1 as an eluent. [d] Methanol insoluble part. [e] Determined by the gel permeation chromatography (GPC) correlating polystyrene standard with a THF eluent. [f] No data due to insolubility of the polymer (Table 2).

Table 2. Solubility of poly(**1**)–poly(**5**) and the characterization of their membranes.

No.	Polymer	Solubility [a]			Membrane-Forming Ability [b]	Maximum Flexural Stress (×10^3) (σ/KPa) [c]	Color [d]
		Toluene	THF	DMF			
1	poly(**1**)	−	+	+	+ [e]	0.968	deep red
2	poly(**2**)	−	+	+	++ [e]	2.40	deep red
3	poly(**3**)	++	++	−	+++ [f]	53.6	orange
4	poly(**4**)	+	+	−	+ [f]	4.29	deep red
5	poly(**5**)	−	−	−	− [g]	− [g]	yellow [h]

[a] ++: Highly soluble, +: soluble, and -: insoluble. [b] +++: Tough, ++: flexible, and +: brittle. [c] See Figure S21. [d] In a membrane state (see Figure 2). [e] The membranes were fabricated by solvent cast using a DMF solution. [f] The membranes were fabricated by solvent cast using a toluene solution. [g] No data due to the insolubility. [h] In a powder state.

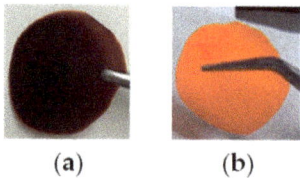

(a)   (b)

Figure 2. Photographs of membranes from (**a**) poly(**2**) and (**b**) poly(**3**).

3.2. Synthesis of poly(**1**)–poly(**3**)

Monomers **1–3** were polymerized by using (Rh(nbd)Cl)$_2$/1-phenethylamine(PEA) (nbd = norbornadiene) as a catalytic system to give poly(**1**)–poly(**3**) (Schemes 1 and 2). The polymerization results are shown in Table 1. The yields of the resulting polymers were 38.0–72.4%, and they had very high molecular weights of 1.4–4.8×10^6. Polymerization yields of **1** and **2**, which contain two hydroxy groups and one carbamate group, were lower (Table 1, 1,2, 52.4% and 38.0%) than that of **3**, which has two carbamate groups instead of two hydroxy groups (Table 1, 3, 72.4%). This result may be because the high polarity of **1** and **2** decreased the efficiency of the rhodium catalytic system during the polymerization. In addition, the solubility of the monomers and polymers of **1** and **2** were lower than that of **3**. It may be another reason for the low yields of poly(**1**) and poly(**2**). The yield of poly(**4**) was lower similar than those for poly(**1**) and poly(**2**) due to the two hydroxy groups (Table 1, 4, 43.2%). For poly(**5**), the yield was quite low; only 4.6% of the monomer was converted to the polymer (Table 1, 5). Since monomer **5** has two amino groups, it may be interacting with the rhodium catalyst to prevent the polymerization. Due to the high polarity, the resulting poly(**5**) was insoluble in the common solvent.

3.3. Effects of the Main Chain Conformation on the Solubility and Membrane Strengths

The three new synthetic polymers (poly(**1**)–poly(**3**)) showed different solubilities, as shown in Table 2. Poly(**1**) and poly(**2**) have two hydroxy groups and one carbamate group in the monomer unit and different lengths of methylene spacers (m) between the carbamate and phenoxy group. They showed low solubility in THF and DMF and insolubility in toluene. On the other hand, poly(**3**), which has two carbamate groups instead of the two hydroxy groups in poly(**1**) and poly(**2**), showed good solubility not only in THF but, also, in toluene (Table 2, 3). Judging from the solubilities for the polymers and R_f values for the monomers in the TLC analysis (Table 1), **3** and poly(**3**) were more hydrophobic than **1** and poly(**1**) and **2** and poly(**2**).

Poly(**1**)–poly(**3**) showed different membrane-forming abilities, as shown in Table 2. Although self-standing membranes could be fabricated from the DMF solution of poly(**1**) and poly(**2**), they were brittle and weak. The maximum flexural stresses for poly(**1**) and poly(**2**) were 0.968 and 2.40×10^3 KPa (Table 2, 1,2), respectively. Since poly(**2**) showed a little better membrane-forming ability than poly(**1**), the longer spacer (m = 6) in poly(**2**) enhanced the flexibility. On the other hand, the membrane strength of poly(**3**) was much higher (the maximum flexural stress was 53.6×10^3 KPa) and 55 times higher than that of poly(**1**) (Table 2, 3).

Since poly(**1**) and poly(**2**) membranes having carbamate groups showed the same red color as poly(**4**) having no carbamate groups (Table 2 and Figure 2), the main chains of the two new polymers having two hydroxy groups in the monomer unit had a very tight cis-cisoid conformation similar to poly(**4**) [16–19]. We previously reported that this conformation tended to decrease the solubility of the polymers and flexibility of the polymer membranes. On the other hand, poly(**3**) having no hydroxy groups and carbamate groups was orange (Table 2). It was suggested that the polymer had a more extended and flexible cis-transoid conformation, which tended to increase the solubility of the polymers and flexibility of the polymer membranes. In order to confirm the main chain conformations of the three new polymers having one or two carbamate groups (poly(**1**), poly(**2**), and poly(**3**)), UV-vis spectra were measured for them, together with poly(**4**) having no carbamate groups (Figure 3). Since poly(**1**) and poly(**2**) showed a similar UV-vis pattern to poly(**4**), whose main chain had a very tight cis-cisoid conformation maintained by intramolecular hydrogen bonds reported by our group [13,14], we concluded that poly(**1**) and poly(**2**) took very tight cis-cisoid conformations. On the other hand, the UV-vis spectrum of poly(**3**) showed different absorption bands from those of poly(**1**) and poly(**2**). The peak around 480nm indicates that the main chain of poly(**3**) consists of a more extended and flexible cis-transoid conformation (Appendix A, 1st item).

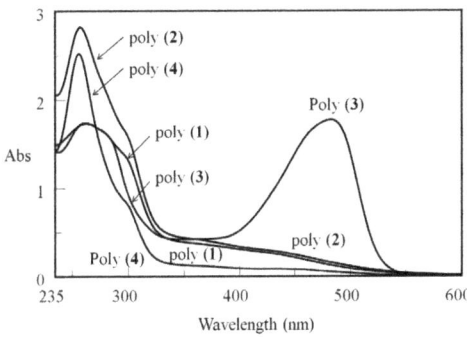

Figure 3. UV spectra of poly(**1**)–poly(**4**) in $CHCl_3$ (c = 2.00mmoL/L).

In order to discuss hydrogen bonds that can support the very tight cis-cisoid conformations in the new polymers, the IR spectra of poly(**1**) and poly(**2**), together with poly(**4**), which has two hydroxy groups in its monomer unit, were measured in CHCl$_3$ (2.00 mmoL/L) (Figure 4). The stretching vibration bands of O–H were observed around 3336 and 3337 cm^{-1} for poly(**1**) and poly(**2**), respectively. Since they were similar to the stretching vibration band of O–H for poly(**4**) having no carbamate groups, it is suggested that poly(**1**) and poly(**2**) also have similar hydrogen bonds to poly(**4**). Therefore, they could have cis-cis conformation. In order to discuss hydrogen bonds between the carbamate groups in poly(**3**), the IR spectra in CHCl$_3$ were measured in different concentrations (0.50–8.0 mmoL/L) (Figure 5). No intramolecular hydrogen bonds were found, because the stretching vibration band of N–H around 3361 cm^{-1} has almost no shift by changing the concentration. Therefore, poly(**3**) could not take a very tight cis-cisoid conformation supported by hydrogen bonds between the carbamate groups, and instead, took a more extended and flexible cis-transoid conformation.

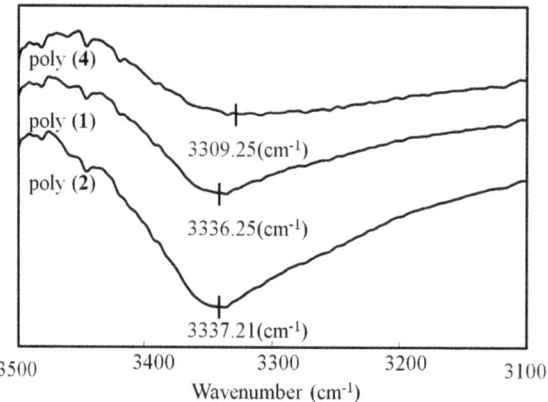

Figure 4. Solution IR spectra of poly(**1**), poly(**2**), and poly(**4**) in CHCl$_3$ (2.00 mmoL/L) at room temperature.

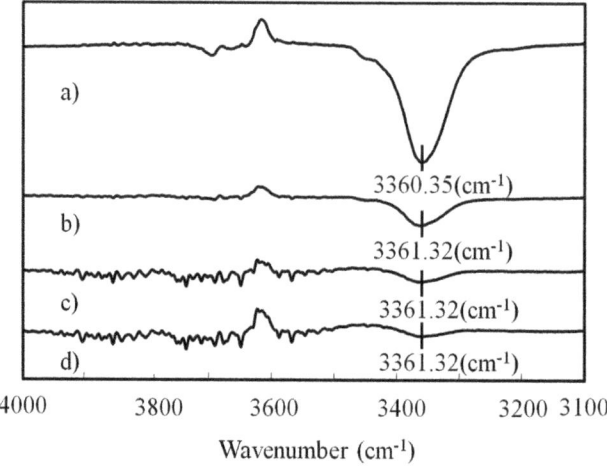

Figure 5. Solution IR spectra of poly(**3**) in CHCl$_3$ at room temperature: (**a**) 8.0 mmoL/L, (**b**) 2.0 mmoL/L, (**c**) 1.0 mmoL/L, and (**d**) 0.50 mmoL/L.

In conclusion, poly(3) had the best solubility and the best membrane-forming ability among the three new polymers, because it had a flexible cis-transoid conformation without hydrogen bonds, while poly(1) and poly(2) had less solubility and less membrane-forming abilities than poly(3), because they had rigid cis-cisoid conformations maintained by hydrogen bonds (Appendix A, 2nd item). In addition, the polarity of the polymers also affected the solubility. Poly(1) and poly(2) having two hydroxy groups showed higher polarity than poly(3). Therefore, the solubility and the membrane-forming ability of poly(3) was the best.

3.4. Oxygen Permeability of the Membranes from the New Polymers

The oxygen and nitrogen permeability coefficients (P_{O2} and P_{N2}) through the membranes of polyphenylacetylenes (poly(2) and poly(3)) having one or two carbamate groups were newly determined, as shown in Table 3 and Figure S22. (Poly(1) was too weak to resist one atom pressure difference during the oxygen permeation measurement.) The polymer membranes of poly(2) and poly(3) showed much higher permeability coefficients (P_{O2} = 188 and 420, respectively) than that of poly(4) (P_{O2} = 3.09) having no carbamate group. Poly(3) showed about 135 times higher P_{O2} than poly(4) and about 2.2 times higher P_{O2} than poly(2). In other words, by introducing one relatively hydrophobic carbamate group to poly(4), poly(2) had a much higher P_{O2} than poly(4), and by replacing two hydroxy groups in poly(4) with two relatively bulky carbamate groups, poly(3) had a much higher P_{O2} than poly(4). In addition, poly(2) and poly(3) having higher P_{O2} than poly(4) showed only a small drop in oxygen permselectivity (P_{O2}/P_{N2}) compared with poly(4) (See Table 3 and Figure S22).

Table 3. Oxygen permeation behavior of the membranes of poly(2)–poly(4) [a].

No.	Membrane [a]	P_{O2} (Barrer) [b]	P_{O2}/P_{N2}	D_{O2} [c]	D_{O2}/D_{N2}	S_{O2} [d]	S_{O2}/S_{N2}
1	poly(2)	188	2.56	11.6	1.05	16.1	2.44
2	poly(3)	420	2.70	184	1.06	2.28	2.54
3	poly(4)	3.09	3.04	3.41	1.25	0.909	2.44

[a] For the codes, see Chart 1. [b] 1 barrer = 10^{-10} cm^3(STP) cm cm^{-2} s^{-1} cmHg^{-1}. [c] In × 10^{-8} cm^2/s. [d] In × 10^{-2} cm^3(STP) cm^3 cmHg^{-1}.

To discuss the reason for this change in P_{O2} and P_{O2}/P_{N2}, the oxygen diffusion coefficient (D_{O2}) and D_{O2}/D_{N2} values were determined from the time lags (Table 3). The higher P_{O2} values of the poly(2) and poly(3) membranes were caused by the higher D_{O2} values. Since the P_{O2}/P_{N2} values mainly depend not on D_{O2}/D_{N2} values, which were almost a unit (=1.0), but S_{O2}/S_{N2} values, which were almost the same (=2.5), the P_{O2}/P_{N2} values (=D_{O2}/D_{N2} × S_{O2}/S_{N2}) did not change largely. In other words, the permeability depends on the diffusivity, and the selectivity depends on the solution selectivity (Figure S23).

In conclusion, poly(3) taking a cis-transoid conformation gave a higher P_{O2} because of a higher D_{O2} than poly(2) and poly(4) taking cis-cisoid conformations. The three polymers showed similar P_{O2}/P_{N2} values, because P_{O2}/P_{N2} mainly depend on S_{O2}/S_{N2}, which were almost similar among the three polymers.

4. Conclusions

Three new phenylacetylene monomers (1–3) having one or two carbamate groups were successfully synthesized, and they gave polymers in 38.0–72.4% yields by polymerization using (Rh(norbornadiene)Cl)$_2$ as a catalyst (Appendix A, 3rd item). The polymers had very high average molecular weights (Mw) of 1.4–4.8 × 10^6, with different solubility and membrane-forming abilities. Poly(3) having two carbamate groups and no hydroxy groups in the monomer unit showed the best solubility and membrane-forming ability among the three new polymers. In addition, the oxygen permeability coefficient (P_{O2}) of the membrane of poly(3) was 420 barrer, which was more than 135 times higher than that of poly(4) having no carbamate group and hydroxy groups with maintaining a similar oxygen permselectivity (P_{O2}/P_{N2}). The better performance in membrane-forming ability and oxygen permeability for poly(**3**) may be caused by the more extended and flexible cis-transoid conformations and lower polarity (Figures S24 and S25). In other words, a higher solubility gave good dense membranes without defects. On the other hand, the other two polymers having one carbamate group and two hydroxy groups in the monomer unit (poly(**1**) and poly(**2**)) showed less performance in membrane-forming abilities and oxygen permeabilities. It may be caused by a very tight cis-cisoid conformation that was maintained by intramolecular hydrogen bonds and a higher polarity.

Supplementary Materials: The following are available online at http://www.mdpi.com/2077-0375/10/9/199/s1: Synthesis of compounds 6–9. Scheme S1: Polymerization of poly(**4**) and poly(**5**). Figures S1–S13: ^{1}H NMR spectra of monomers and polymers. Figures S14–S19: IR spectra of monomers and polymers. Figure S20: UV spectra of monomers. Figure S21: Measurement of a maximum flexural stress (σ/pa). Figure S22: Relationship between α and P_{O2} through the membranes of poly(**2**)–poly(**4**). Figure S23: The possible separation mechanism of O_2/N_2 through the membranes. Figure S24: XRD of poly(**3**) and poly(**4**) in membrane state. Figure S25: The SEM images of (a) poly(**4**) and (b) poly(**3**) membranes.

Author Contributions: Investigation: Y.Z. and T.A.; experiments: Y.L.; writing—original draft: Y.Z. and T.A.; writing—review and editing: M.T. and T.K.; and measurements and analyses: H.J., F.M., and X.Z. All authors have read and agreed to the published version of the manuscript.

Funding: This research was funded by the Fundamental Research Funds in Heilongjiang Provincial Universities, grant number 135409201, the General Project of the Hunan Education Department, grant number 19C0514, the Doctoral Research Startup Project of the Hunan Institute of Technology, grant number HQ19013, and the College Student Innovation Training Program of the Hunan Institute of Technology, grant number CX2019059.

Conflicts of Interest: The authors declare no conflict of interest.

Appendix A

1. We have previously reported these cis-cisoid and cis-transoid conformation can be determined by UV-vis spectroscopy because the conjugation length of the former is shorter than that of the latter [13,14,18,19]. For the extensive consideration, see Ref. [19].
2. If these polymers made hydrogen bonds *inter*molecularly, it also affected their solubility largely. However, we have already reported that these polymers from the monomers having two hydroxy groups (such as **1**, **2** and **4**) had *intra*molecular hydrogen bonds because the long alkyl groups could prevent from forming *inter*molecular hydrogen bonds. Therefore, the effects of the *inter*molecular hydrogen bonds on the solubility is thought to be not large.
3. Since the synthesis of these monomers needed multi-step synthesis, the total yields were not high (5.5–7.7%).

References

1. Ito, T.; Shirakawa, H.; Ikeda, S. Simultaneous polymerization and formation of polyacetylene film on the surface of concentrated soluble Ziegler-type catalyst solution. *J. Polym. Sci. Part A Polym. Chem.* **1974**, *12*, 11–20. [CrossRef]
2. Shirakawa, H.; Louis, E.J.; MacDiarmid, A.G.; Chiang, C.K.; Heeger, A.J. Synthesis of electrically conducting organic polymers: Halogen derivatives of polyacetylene, (CH)x. *J. Chem. Soc. Chem. Commun.* **1977**, 578–580. [CrossRef]
3. Chiang, C.K.; Fincher, C.R.; Park, Y.W.; Heeger, A.J.; Shirakawa, H.; Louis, E.J.; Gau, S.C.; MacDiarmid, A.G. Electrical conductivity in doped polyacetylene. *Phys. Rev. Lett.* **1977**, *39*, 1098–1101. [CrossRef]
4. MacDiarmid, A.G. "Synthetic metals": A novel role for organic polymers (Nobel Lecture). *Angew. Chem. Int. Ed.* **2001**, *40*, 2581–2590. [CrossRef]
5. Shirakawa, H. The discovery of polyacetylene film: The dawning of an era of conducting polymers (Nobel Lecture). *Angew. Chem. Int. Ed.* **2001**, *40*, 2574–2580. [CrossRef]
6. Heeger, A.J. Semiconducting and metallic polymers: The fourth generation of polymeric materials (Nobel Lecture). *Angew. Chem. Int. Ed.* **2001**, *40*, 2591–2611. [CrossRef]
7. Alagi, K. Helical polyacetylene: Asymmetric polymerization in a chiral liquid-crystal field. *Chem. Rev.* **2009**, *109*, 5354–5401.

8. Liu, J.; Lam, J.W.Y.; Tang, B.Z. Acetylenic polymers: Syntheses, structures, and functions. *Chem. Rev.* **2009**, *109*, 5799–5867. [CrossRef]
9. Rudick, J.G.; Percec, V. Induced helical backbone conformations of self-organizable dendronized polymers. *Acc. Chem. Res.* **2008**, *41*, 1641–1652. [CrossRef]
10. Percec, V.; Aqad, E.; Peterca, M.; Rudick, J.G.; Lemon, L.; Ronda, J.C.; De, B.B.; Heiney, P.A.; Meijer, E.W. Steric communication of chiral information observed in dendronized polyacetylenes. *J. Am. Chem. Soc.* **2006**, *128*, 16365–16372. [CrossRef]
11. Yashima, E.; Maeda, K.; Iida, H.; Furusho, Y.; Nagai, K. Helical polymers: Synthesis, structures, and functions. *Chem. Rev.* **2009**, *109*, 6102–6211. [CrossRef] [PubMed]
12. Shiotsuki, M.; Sanda, F.; Masuda, T. Polymerization of substituted acetylenes and features of the formed polymers. *Polym. Chem.* **2011**, *2*, 1044–1058. [CrossRef]
13. Motoshige, R.; Mawatari, Y.; Motoshige, A.; Yoshida, Y.; Sasaki, T.; Yoshimizu, H.; Suzuki, T.; Tsujita, Y.; Tabata, M. Mutual conversion between stretched and contracted helices accompanied by a drastic change in color and spatial structure of poly(phenylacetylene) prepared with a [Rh(nbd)Cl]$_2$-amine catalyst. *J. Polym. Sci. Part A Polym. Chem.* **2014**, *52*, 752–759. [CrossRef]
14. Wang, J.; Li, J.; Aoki, T.; Kaneko, T.; Teraguchi, M.; Shi, Z.; Jia, H. Subnanoporous highly oxygen pemselective membranes from poly(conjugated hyperbranched macromonomer)s synthesized by one-pot simultaneous two-mode homopolymerization of 1,3-bis(silyl)phenylacetylene using a single Rh catalytic system: Control of their structures and pemselectivities. *Macromolecules* **2017**, *50*, 7121–7136.
15. Liu, L.; Zhang, G.; Aoki, T.; Wang, Y.; Kaneko, T.; Teraguchi, M.; Zhang, C.; Dong, H. Synthesis of one-handed helical block copoly(substituted acetylene)s consisting of dynamic cis-transoidal and static cis-cisoidal block: Chiral teleinduction in helix-sense-selective polymerization using a chiral living polymer as an initiator. *ACS Macro Lett.* **2016**, *5*, 1381–1385. [CrossRef]
16. Jin, Y.; Aoki, T.; Kwak, G. Control of intamolecular hydrogen bonding in a conformation-switchable helical spring polymer by solvent and temperature. *Angew. Chem. Int. Ed.* **2020**, *59*, 1837–1844. [CrossRef] [PubMed]
17. Aoki, T.; Kaneko, T.; Maruyama, N.; Sumi, A.; Takahashi, M.; Sato, T.; Teraguchi, M. Helix-sense-selective polymerization of phenylacetylene having two hydroxy groups using a chiral catalytic system. *J. Am. Chem. Soc.* **2003**, *125*, 6346–6347. [CrossRef] [PubMed]
18. Liu, L.; Namikoshi, T.; Zang, Y.; Aoki, T.; Hadano, S.; Abe, Y.; Wasuzu, I.; Tsutsuba, T.; Teraguchi, M.; Kaneko, T. Top-down preparation method of self-supporting supramolecular polymeric membranes using highly selective photocyclicaromatization of cis-cisoid helical poly(phenylacetylene)s in membrane state. *J. Am. Chem. Soc.* **2013**, *135*, 602–605. [CrossRef]
19. Teraguchi, M.; Aoki, T.; Kaneko, T.; Tanioka, D. Helix-sense-selective polymerization of achiral phenylacetylenes with two N-alkylamide groups to generate the one-handed helical polymers stabilized by intramolecular hydrogen bonds. *ACS Macro Lett.* **2012**, *1*, 1258–1261. [CrossRef]
20. Sanda, F.; Endo, T. Syntheses and functions of polymers based on amino acids. *Macromol. Chem. Phys.* **1999**, *200*, 2651–2661. [CrossRef]
21. Bauri, K.; Roy, S.G.; De, P. Side-chain amino-acid-derived cationic chiral polymers by controlled radical polymerization. *Macromol. Chem. Phys.* **2015**, *217*, 365–379. [CrossRef]
22. Suyama, K.; Shirai, M. Photobase generators: Recent progress and application trend in polymer systems. *Prog. Polym. Sci.* **2009**, *34*, 194–209. [CrossRef]
23. Guo, X.; Facchetti, A.; Marks, T.J. Imide-and amide-functionalized polymer semiconductors. *Chem. Rev.* **2014**, *114*, 8943–9021. [CrossRef] [PubMed]
24. Gao, G.; Sanda, F.; Masuda, T. Synthesis and properties of amino acid-based polyacetylenes. *Macromolecules* **2003**, *36*, 3932–3937. [CrossRef]
25. Sogawa, H.; Shiotsuki, M.; Sanda, F. α-Propargyl amino acid-derived optically active novel substituted polyacetylenes: Synthesis, secondary structures, and responsiveness to ions. *J. Polym. Sci. Part A Polym. Chem.* **2012**, *50*, 2008–2018. [CrossRef]
26. Shirakawa, Y.; Suzuki, Y.; Terada, K.; Shiotsuki, M.; Masuda, T.; Sanda, F. Synthesis and secondary structure of poly(1-methylpropargyl-N-alkylcarbamate)s. *Macromolecules* **2010**, *43*, 5575–5581. [CrossRef]
27. Liu, R.; Shiotsuki, M.; Masuda, T.; Sanda, F. Synthesis and chiroptical properties of hydroxyphenylglycine-based poly(m-phenyleneethynylene-p-phenyleneethynylene)s. *Macromolecules* **2009**, *42*, 6115–6122. [CrossRef]

28. Saeed, I.; Khan, F.Z.; Shiotsuki, M.; Masuda, T. Synthesis and properties of carbamate- and amine-containing poly(phenylacetylenes). *J. Polym. Sci. Part A Polym. Chem.* **2009**, *47*, 1853–1863. [CrossRef]
29. Sanda, F.; Yukawa, Y.; Masuda, T. Synthesis and properties of optically active substituted polyacetylenes having carboxyl and/or amino groups. *Polymer* **2004**, *45*, 849–854. [CrossRef]
30. Zang, Y.; Aoki, T.; Teraguchi, M.; Kaneko, T.; Ma, L.; Jia, H. Synthesis and oxygen permeation of novel polymers of phenylacetylenes having two hydroxyl groups via different lengths of spacers. *Polymer* **2015**, *56*, 199–206. [CrossRef]
31. Yamada, Y.K.; Okada, C.; Yoshida, K.; Umeda, Y.; Arima, S.; Sato, N.; Kai, T.; Takayanagi, H.; Harigaya, Y. Convenient synthesis of 7′ and 6′-bromo-D-tryptophan and their derivatives by enzymatic optical resolution using D-aminoacylase. *Tetrahedron* **2002**, *58*, 7851–7861. [CrossRef]
32. Carrasco, M.R.; Alvarado, C.I.; Dashner, S.T.; Wong, A.J.; Wong, M.A. Synthesis of aminooxy and N-alkylaminooxy amines for use in bioconjugation. *J. Org. Chem.* **2010**, *75*, 5757–5759. [CrossRef] [PubMed]
33. Nativi, C.; Francesconi, O.; Gabrielli, G.; Simone, I.D.; Turchetti, B.; Mello, T.; Mannelli, L.D.C.; Ghelardini, C.; Buzzini, P.; Roelens, S. Aminopyrrolic synthetic receptors for monosaccharides: A class of carbohydrate-binding agents endowed with antibiotic activity versus pathogenic yeasts. *Chem. Eur. J.* **2012**, *18*, 5064–5072. [CrossRef] [PubMed]
34. Zang, Y.; Aoki, T.; Shoji, K.; Teraguchi, M.; Kaneko, T.; Ma, L.; Jia, H.; Miao, F. Synthesis and oxygen permeation of novel well-defined homopoly(phenylacetylene)s with different sizes and shapes of oligosiloxanyl side groups. *J. Membrane Sci.* **2018**, *56*, 26–38. [CrossRef]

© 2020 by the authors. Licensee MDPI, Basel, Switzerland. This article is an open access article distributed under the terms and conditions of the Creative Commons Attribution (CC BY) license (http://creativecommons.org/licenses/by/4.0/).

Article

New Poly(imide)s Bearing Alkyl Side-Chains: A Study on the Impact of Size and Shape of Lateral Groups on Thermal, Mechanical, and Gas Transport Properties

Fidel E. Rodríguez-González [1,†], Germán Pérez [2,†], Vladimir Niebla [1], Ignacio Jessop [3], Rudy Martin-Trasanco [4], Deysma Coll [5], Pablo Ortiz [6], Manuel Aguilar-Vega [7], Luis H. Tagle [1], Claudio A. Terraza [1,8,*] and Alain Tundidor-Camba [1,8,*]

1. Research Laboratory for Organic Polymers (RLOP), Department of Organic Chemistry, Pontificia Universidad Católica de Chile, Santiago 7810000, Chile; ferg@uc.cl (F.E.R.-G.); vniebla@uc.cl (V.N.); ltagle@uc.cl (L.H.T.)
2. Bureau Veritas Laboratories, 7150 Rue Frederick Banting, Saint-Laurent, Montreal, QC H4S 2A1, Canada; gmperez80@gmail.com
3. Organic and Polymeric Materials Research Laboratory, Departamento de Química, Universidad de Tarapacá, Av. General Velásquez 1775, P.O. Box 7-D, Arica 1000000, Chile; iajessop@uta.cl
4. Departamento de Química, Universidad Tecnológica Metropolitana, J. P. Alessandri 1242, Santiago 7810000, Chile; rudy.martint@utem.cl
5. Núcleo de Química y Bioquímica, Facultad de Estudios Interdisciplinarios, Universidad Mayor, Santiago 3830000, Chile; deysma.coll@mayor.cl
6. Centro de Nanotecnología Aplicada, Facultad de Ciencias, Universidad Mayor, Santiago 3830000, Chile; pablo.ortiza@mayor.cl
7. Laboratorio de Membranas, Unidad de Materiales, Centro de Investigación Científica de Yucatán A.C. (CICY), Chuburna de Hidalgo, Merida, Yucatán 97205, Mexico; mjav@cicy.mx
8. UC Energy Research Center, Pontificia Universidad Católica de Chile, Santiago 7810000, Chile
* Correspondence: cterraza@uc.cl (C.A.T.); atundido@uc.cl (A.T.-C.)
† Fidel E. Rodríguez-González and Germán Pérez contributed equally to this work.

Received: 5 June 2020; Accepted: 1 July 2020; Published: 4 July 2020

Abstract: A set of five new aromatic poly(imide)s (PIs) incorporating pendant acyclic alkyl moieties were synthesized. The difference among them was the length and bulkiness of the pendant group, which comprises of linear alkyl chains from three to six carbon atoms, and a *tert*-butyl moiety. The effect of the side group length on the physical, thermal, mechanical, and gas transport properties was analyzed. All PIs exhibited low to moderate molecular weights (Mn ranged between 27.930–58.970 Da, and Mw ranged between 41.760–81.310 Da), good solubility in aprotic polar solvents, except for PI-*t*-4, which had a *tert*-butyl moiety and was soluble even in chloroform. This behaviour was probably due to the most significant bulkiness of the side group that increased the interchain distance, which was corroborated by the X-ray technique (**PI-*t*-4** showed two *d*-spacing values: 5.1 and 14.3 Å). Pure gas permeabilities for several gases were reported (**PI-3** (Barrer): He(52); H_2(46); O_2(5.4); N_2(1.2); CH_4(1.1); CO_2(23); **PI-*t*-4** (Barrer): He(139); H_2(136); O_2(16.7); N_2(3.3); CH_4(2.3); CO_2(75); **PI-5** (Barrer): He(44); H_2(42); O_2(5.9); N_2(1.4); CH_4(1.2); CO_2(27); **PI-6** (Barrer): He(45); H_2(43); O_2(6.7); N_2(1.7); CH_4(1.7); CO_2(32)). Consistent higher volume in the side group was shown to yield the highest gas permeability. All poly(imide)s exhibited high thermal stability with 10% weight loss degradation temperature between 448–468 °C and glass transition temperature between 240–270 °C. The values associated to the tensile strength (45–87 MPa), elongation at break (3.2–11.98%), and tensile modulus (1.43–2.19 GPa) were those expected for aromatic poly(imide)s.

Keywords: aromatic poly(imide)s; bulky pendant groups; gas permeability; structure-property relationship

1. Introduction

One of the most important technological advances at the end of the 20th century was the industrial development of processes that include polymeric membranes for water treatment and gas separation [1]. Membrane-based gas separation technologies have not yet succeeded in displacing traditional processes, such as separation by absorbents or cryogenic distillation, even when different polymeric materials have been tested as filters [2]. Therefore, numerous research teams continue to make progress to obtaining new polymers or mixtures of them, with very high permeability to gases [3–6].

Aromatic poly(imide)s have excellent thermal and mechanical properties, high resistance to chemical agents, and the ability to form thin layers films [7–9], qualities that make them suitable materials for gas separation. The poly(imide) Matrimid is one of the widely commercial polymers used in the preparation of membranes [10].

Many scientific papers have examined the relationship between the chemical structure of polymers and their macroscopic properties, including their behaviour as gas separating membranes [11–14]. Those studies have focused on changing the size and shape of the side groups to tune the stiffness of poly(imide)s main chain or to introduce functional groups that could be chemically modified post-polymerization. The use of monomers with contorting units that prevented efficient packaging of the chains incorporated bulky substituents close to imide groups to avoid the free rotation of the chain has also been tested. With those structural modifications, it is possible to modulate not only the permeability of the material but also properties such as solubility (processability), thermal stability, and mechanical resistance. For example, Huang et al. synthesized two novel diphenyl ether diamines with one or two *tert*-butyl groups as building blocks for the synthesis of new aromatic poly(imide)s. The *tert*-butyl groups increased the solubility of poly(imide)s due to the increase in the interchain distances [15]. Yao et al. studied the effect of attaching different side groups on the properties of poly(amide-imide)s, showing good solubility and excellent comprehensive properties [16]. Liou et al. prepared processable aromatic poly(imide) membranes containing trimethyl substituted triphenylamine units. The poly(imide)s were readily soluble in polar solvents and exhibited an improvement of gas permeability [17].

In previous works, we synthesized a new set of fluorinated aromatic poly(imide)s based on aromatic diamines with various cycloalkyl pendant groups and 4,4′-(hexafluoroisopropylidene)diphthalic anhydride (6FDA) as a common monomer for all poly(imide)s. Thermal, mechanical, and gas transport properties were studied to evaluate the impact of the volume of C3, C5, C6, C8, and C10 (adamantyl) cyclic lateral groups [18,19]. All poly(imide)s showed similar thermal resistance, reflecting the independency between their overall thermal stability and the bulkiness of the monocyclic pendant groups. Regarding mechanical resistance, the results suggested that the bigger the pendant group, the easier the materials deformed, and more tension was required to break them. Additionally, the polymer with the bulkiest pendant group (adamantyl) exhibited the largest interchain space, which was corroborated by X-rays, leading to the most permeable membrane.

Continuing with this previous work, here we present a set of new fluorinated aromatic poly(imide)s based on the same dianhydride (6FDA) and aromatic diamines with acyclic alkyl fragments as side groups formed by R-NHCO-chains of three, four, five, and six carbon atoms. A *tert*-butyl substituent was also used as R moiety to study the effect of the ramification against linearity. Their physical, thermal, mechanical, and pure gas transport properties were measured, and the results were compared among themselves and with those previously reported.

2. Materials and Methods

2.1. Materials

Butylamine (99 +%), pentylamine (99%), and hexylamine (99%) were purchased from Acros Organics (Morris Plains, NJ, USA). 3,5-dinitrobenzoyl chloride (DNBC) (96.5%), propylamine (98%), *tert*-butylamine (99.5%), 4,4'-(hexafluoroisopropylidene) diphthalic anhydride (6FDA) (99%), anhydrous *N,N*-dimethylacetamide (DMAc) (99.8%), triethylamine (TEA) (99.5%), hydrazine monohydrate (80%), anhydrous pyridine (99.8%), and Pd/C (10%) were purchased from Sigma-Aldrich (Milwaukee, WI, USA). All other reagents and solvents were purchased commercially as analytical grade and used without further purification.

2.2. Monomer Synthesis and Characterization

Monomers were synthesized following the same procedure described in our previous paper [19]. Briefly, DNBC reacted with an excess of the corresponding amine (propylamine, butylamine, pentylamine, hexylamine, and *tert*-butylamine), using THF as solvent and TEA as base. The dinitro derivatives were obtained in yields between 86 and 89% (yield was calculated following the equation: Yield = ($mol_{exp}/mol_{theorical}$) × 100; Dinitro **M-3**: 86%; Dinitro **M-4**: 86%; Dinitro **M-*t*-4**: 89%; Dinitro **M-5**: 87% and Dinitro **M-6**: 86%). The 3,5-dinitro *N*-alkylbenzamide derivatives obtained were reduced to 3,5-diamine-*N*-alkylbenzamide derivatives by using monohydrate hydrazine (80%) in the presence of Pd/C and ethanol as solvent (Scheme 1). The amide-diamine monomers were purified by sublimation technique.

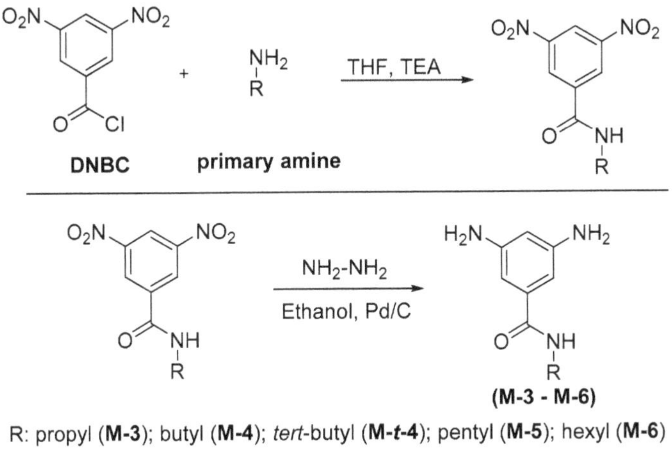

Scheme 1. Synthesis of amide-diamine monomers.

R: propyl (**M-3**); butyl (**M-4**); *tert*-butyl (**M-*t*-4**); pentyl (**M-5**); hexyl (**M-6**)

M-3: Yield: 93% (calculated in the same way for Dinitro derivatives). FT-IR (KBr, ν, cm^{-1}): 3420, 3388 (N–H, amino); 3326 (N–H, amide); 3049 (C–H, arom.), 2955, 2926, 2855 (C–H, aliph.); 1645 (C=O); 1602, 1550, 1472 (C=C); 746 (out-of-plane ring bending). ^{1}H NMR (400 MHz, DMSO-d_6, δ, ppm) 7.76 (t, *J* = 5.2 Hz, 1H, H-7); 6.22 (s, 2H, H-4); 5.92 (s, 1H, H-1); 4.91 (broad peak, 4H, H-2); 3.22 (m, 2H, H-8); 1.51 (m, 2H, H-9); 0.87 (t, *J* = 7.2 Hz, 3H, H-10). ^{13}C NMR (100 MHz, DMSO-d_6, δ, ppm) 167.04 (C-6); 148.60 (C-3); 136.55 (C-5); 102.25 (C-4); 101.80 (C-1); 41.51 (C-8); 22.11 (C-9); 11.40 (C-10). Elem. Anal. Calcd. for $C_{10}H_{15}N_3O$ (193.25), C, 62.15%; H, 7.82%; N, 21.74%. Found: C, 62.09%; H, 7.93%; N, 21.65%.

M-4: Yield: 92% (calculated in the same way for Dinitro derivatives). FT-IR (KBr, ν, cm^{-1}): 3418, 3389 (N–H, amino); 3324 (N–H, amide), 3049 (C–H, arom.), 2959, 2920, 2855 (C–H, aliph.); 1646 (C=O); 1602, 1551, 1473 (C=C); 745 (out-of-plane ring bending). ^{1}H NMR (400 MHz, DMSO-d_6, δ, ppm) 7.75 (d, J = 5.0 Hz, 1H, H-7); 6.22 (s, 2H, H-4); 5.92 (s, 1H, H-1); 4.91 (broad peak, 4H, H-2); 3.25 (m, 2H, H-8); 1.49 (m, 2H, H-9); 1.33 (m, 2H, H-10); 0.86 (t, J = 7.0 Hz, 3H, H-11). ^{13}C NMR (100 MHz, DMSO-d_6, δ, ppm) 166.02 (C-6); 148.61 (C-3); 136.56 (C-5); 102.30 (C-4); 101.85 (C-1); 38.97 (C-8); 29.30 (C-9); 22.22 (C-10); 13.71 (C-11). Elem. Anal. Calcd. for $C_{11}H_{17}N_3O$ (207.28), C, 63.74%; H, 8.27%; N, 20.27%. Found: C, 62.99%; H, 8.43%; N, 19.95%.

M-t-4: Yield: 92% (calculated in the same way for Dinitro derivatives). FT-IR (KBr, ν, cm^{-1}): 3419, 3388 (N–H, amino); 3328 (N–H, amide), 3046 (C–H, arom.), 2959, 2865 (C–H, aliph.); 1646 (C=O); 1601, 1552, 1472 (C=C); 746 (out-of-plane ring bending). ^{1}H NMR (400 MHz, DMSO-d_6, δ, ppm) 6.92 (s, 1H, H-7); 6.20 (s, 2H, H-4); 5.88 (s, 1H, H-1); 4.90 (broad peak, 4H, H-2); 1.35 (s, 9H, H-9). ^{13}C NMR (100 MHz, DMSO-d_6, δ, ppm) 166.02 (C-6); 148.63 (C-3); 136.53 (C-5); 102.23 (C-4); 101.79 (C-1); 50.62 (C-8); 28.33 (C-9). Elem. Anal. Calcd. for $C_{11}H_{17}N_3O$ (207.28), C, 63.74%; H, 8.27%; N, 20.27%. Found: C, 63.17%; H, 7.95%; N, 20.03%.

M-5: Yield: 87% (calculated in the same way for Dinitro derivatives). FT-IR (KBr, ν, cm^{-1}): 3418, 3389 (N–H, amino); 3326 (N–H, amide), 3050 (C–H, arom.), 2959, 2920, 2855 (C–H, aliph.); 1647 (C=O); 1602, 1551, 1473 (C=C); 745 (out-of-plane ring bending). ^{1}H NMR (400 MHz, DMSO-d_6, δ, ppm) 7.74 (t, J = 5.1 Hz, 1H, H-7); 6.21 (s, 2H, H-4); 5.92 (s, 1H, H-1); 4.92 (broad peak, 4H, H-2); 3.23 (m, 2H, H-8); 1.51 (m, 2H, H-9); 1.29 (m, 4H; H-10, H-11); 0.85 (t, J = 6.6 Hz, 3H, H-12). ^{13}C NMR (100 MHz, DMSO-d_6, δ, ppm) 166.10 (C-6); 148.57 (C-3); 136.52 (C-5); 102.21 (C-4); 101.84 (C-1); 38.90 (C-8); 29.96 (C-10); 28.23 (C-9); 22.40 (C-11); 13.80 (C-12). Elem. Anal. Calcd. for $C_{12}H_{19}N_3O$ (221.30), C, 65.13%; H, 8.65%; N, 18.99%. Found: C, 64.89%; H, 8.43%; N, 18.23%.

M-6: Yield: 91% (calculated in the same way for Dinitro derivatives). FT-IR (KBr, ν, cm^{-1}): 3419, 3388 (N–H, amino); 3327 (N–H, amide), 3049 (C–H, arom.), 2959, 2921, 2857 (C–H, aliph.); 1647 (C=O); 1602, 1550, 1474 (C=C); 745 (out-of-plane ring bending). ^{1}H NMR (400 MHz, DMSO-d_6, δ, ppm) 7.74 (d, J = 5.2 Hz, 1H, H-7); 6.21 (s, 2H, H-4); 5.92 (s, 1H, H-1); 4.92 (broad peak, 4H, H-2); 3.23 (m, 2H, H-8); 1.50 (m, 2H, H-9); 1.26 (m, 6H; H-10, H-11, H-12); 0.84 (t, J = 6.7 Hz, 3H, H-13). ^{13}C NMR (100 MHz, DMSO-d_6, δ, ppm) 166.09 (C-6); 148.57 (C-3); 136.53 (C-5); 102.22 (C-4); 101.83 (C-1); 38.85 (C-8); 31.25 (C-11); 28.86 (C-9); 26.54 (C-10); 22.41 (C-12); 13.84 (C-13). Elem. Anal. Calcd. for $C_{13}H_{21}N_3O$ (235.33), C, 66.35%; H, 8.99%; N, 17.86%. Found: C, 66.00%; H, 8.27%; N, 17.15%.

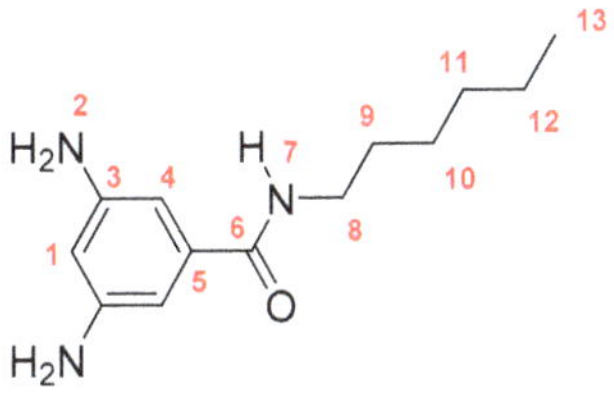

2.3. Polymer Synthesis and Characterization

A typical polymerization procedure for the synthesis of the poly(imide)s was followed (Scheme 2). To a three-necked round-bottomed flask equipped with mechanical stirrer and under the nitrogen atmosphere, a mixture of 2.0 mmol of 3,5-diamino-N-alkylbenzamide, 2.0 mmol 6FDA, and 4 mL of DMAc were added and stirred at room temperature for 6 h. After that, 1.0 mL of acetic anhydride and 0.8 mL of pyridine were added. The mixture was stirred for another two hours at room temperature and then raised and maintained at 60 °C for one more hour. Then, the mixture was cooled and poured in 300 mL of water under stirring. The white solid was filtered, thoroughly washed with methanol, and dried at 100 °C for 12 h.

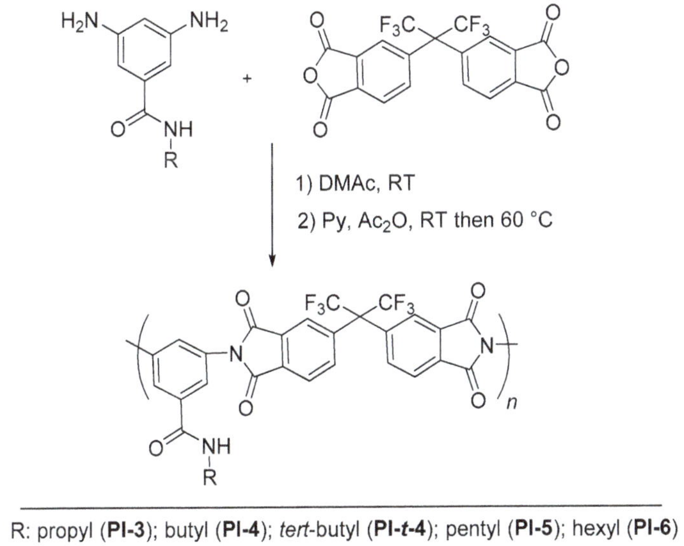

R: propyl (**PI-3**); butyl (**PI-4**); *tert*-butyl (**PI-*t*-4**); pentyl (**PI-5**); hexyl (**PI-6**)

Scheme 2. Synthesis of poly(imide)s.

PI-3: FT-IR (KBr, ν, cm^{-1}): 3417 (N–H, amide); 3093 (C–H, arom.); 2965, 2928, 2875, 2855 (C–H, aliph.); 1785 (C=O asym., imide); 1729 (C=O sym., imide); 1659 (C=O, amide); 1597, 1536, 1457 (C=C, arom. ring); 1356 (C-N, imide); 721 (imide ring deformation). ^{1}H NMR (600 MHz, DMSO-d_6, δ, ppm)

8.59 (broad peak, 1H, H-16); 8.20 (d, J = 7.8 Hz, 2H, H-9); 8.04 (s, 2H, H-4); 7.98 (broad peak, 2H, H-10); 7.78 (s, 2H, H-13); 7.74 (s, 1H, H-11); 3.23 (broad peak, 2H, H-17); 1.52 (m, 2H, H-18); 0.87 (t, J = 7.2 Hz, 3H, H-19). ^{13}C NMR (150 MHz, DMSO-d_6, δ, ppm) 167.80 (C-15); 166.20 (C-7); 165.99 (C-6); 138.01 (C-3); 136.48 (C-5); 136.40 (C-14); 133.78 (C-10); 133.25 (C-12); 133.08 (C-8); 129.05 (C-11); 126.81 (C-13); 124.97 (C-9); 124.31 (C-4); 123.95 (q, J = 286 Hz, C-1); 65.05 (hept, J = 25 Hz, C-2); 40.70 (C-17); 21.89 (C-18); 11.72 (C-19). Elem. Anal. Calcd. for $[C_{29}H_{17}F_6N_3O_5]_n$ $(601.46)_n$, C, 57.91%; H, 2.85%; N, 6.99%. Found: C, 57.13%; H, 2.53%; N, 6.65%.

PI-4: FT-IR (KBr, ν, cm^{-1}): 3403 (N–H, amide); 3084 (C–H, arom.); 2967, 2931, 2866 (C–H, aliph.); 1785 (C=O asym., imide); 1729 (C=O sym., imide); 1667 (C=O, amide); 1598, 1519, 1455 (C=C, arom. ring); 1356 (C–N, imide); 721 (imide ring deformation). ^{1}H NMR (600 MHz, DMSO-d_6, δ, ppm) 8.57 (broad peak, 1H, H-16); 8.20 (d, J = 7.9 Hz, 2H, H-9); 8.03 (s, 2H, H-4); 7.98 (broad peak, 2H, H-10); 7.78 (s, 2H, H-13); 7.74 (s, 1H, H-11); 3.27 (broad peak, 2H, H-17); 1.49 (m, 2H, H-18); 1.32 (m, 2H, H-19); 0.87 (t, J = 6.9 Hz, 3H, H-20). ^{13}C NMR (150 MHz, DMSO-d_6, δ, ppm) 166.75 (C-15); 166.22 (C-7); 165.97 (C-6); 138.02 (C-3); 136.50 (C-5); 136.40 (C-14); 133.79 (C-10); 133.25 (C-12); 133.12 (C-8); 129.07 (C-11); 126.81 (C-13); 124.91 (C-9); 124.33 (C-4); 124.01 (q, J = 286 Hz, C-1); 64.98 (hept, J = 25 Hz, C-2); 38.92 (C-17); 29.25 (C-18); 22.19 (C-19); 13.75 (C-20). Elem. Anal. Calcd. for $[C_{30}H_{19}F_6N_3O_5]_n$ $(615.49)_n$, C, 58.54%; H, 3.11%; N, 6.83%. Found: C, 57.98%; H, 2.99%; N, 6.27%.

PI-*t*-4: FT-IR (KBr, ν, cm^{-1}): 3409 (N–H, amide); 3090 (C–H, arom.); 2962, 2867 (C–H, aliph.); 1785 (C=O asym., imide); 1729 (C=O sym., imide); 1663 (C=O, amide); 1597, 1532, 1456 (C=C, arom. ring); 1355 (C–N, imide); 720 (imide ring deformation). ^{1}H NMR (600 MHz, DMSO-d_6, δ, ppm) 8.20 (d, J = 8.1 Hz Hz, 2H, H-9); 7.98 (m, 4H; H-4, H-10); 7.85 (broad peak, 1H, H-16); 7.77 (s, 2H, H-13); 7.70 (s, 1H, H-11); 1.36 (s, 9H, H-18). ^{13}C NMR (150 MHz, DMSO-d_6, δ, ppm) 166.88 (C-15); 166.25 (C-7); 165.98 (C-6); 138.06 (C-3); 136.41 (C-5); 136.50 (C-14); 133.86 (C-10); 133.32 (C-12); 133.19 (C-8); 129.18 (C-11); 126.73 (C-13); 124.89 (C-9); 124.40 (C-4); 123.89 (q, J = 286 Hz, C-1); 65.01 (hept, J = 25 Hz, C-2); 50.52 (C-17); 28.21 (C-18). Elem. Anal. Calcd. for $[C_{30}H_{19}F_6N_3O_5]_n$ $(615.49)_n$, C, 58.54%; H, 3.11%; N, 6.83%. Found: C, 58.03%; H, 3.19%; N, 6.73%.

PI-5: FT-IR (KBr, ν, cm^{-1}): 3403 (N–H, amide); 3087 (C–H, arom.); 2959, 2930, 2860 (C–H, aliph.); 1785 (C=O asym., imide); 1729 (C=O sym., imide); 1663 (C=O, amide); 1597, 1532, 1457 (C=C, arom. ring); 1355 (C–N, imide); 721 (imide ring deformation). ^{1}H NMR (600 MHz, DMSO-d_6, δ, ppm) 8.58 (broad peak, 1H, H-16); 8.20 (d, J = 7.9 Hz, 2H, H-9); 8.03 (s, 2H, H-4); 7.98 (broad peak, 2H, H-10); 7.77 (s, 2H, H-13); 7.74 (s, 1H, H-11); 3.26 (broad peak, 2H, H-17); 1.51 (broad peak, 2H, H-18); 1.28 (broad peak, 4H; H-19, H-20); 0.84 (broad peak, 3H, H-21). ^{13}C NMR (150 MHz, DMSO-d_6, δ, ppm) 166.87 (C-15); 166.30 (C-7); 166.15 (C-6); 138.02 (C-3); 136.48 (C-5); 136.48 (C-14); 133.79 (C-10); 133.35 (C-12); 133.21 (C-8); 129.17 (C-11); 126.97 (C-13); 124.96 (C-9); 124.21 (C-4); 123.84 (q, J = 286 Hz, C-1); 64.91 (hept, J = 25 Hz, C-2); 38.80 (C-17); 29.85 (C-19); 28.19 (C-18); 22.25 (C-20); 13.95 (C-21). Elem. Anal. Calcd. for [C$_{31}$H$_{21}$F$_6$N$_3$O$_5$]$_n$ (629.52)$_n$, C, 59.15%; H, 3.36%; N, 6.68%. Found: C, 58.99%; H, 3.03%; N, 6.52%.

PI-6: FT-IR (KBr, ν, cm^{-1}): 3407 (N–H, amide); 3088 (C–H, arom.); 2957, 2929, 2858 (C–H, aliph.); 1785 (C=O asym., imide); 1729 (C=O sym., imide); 1663 (C=O, amide); 1597, 1530, 1456 (C=C, arom. ring); 1354 (C–N, imide); 721 (imide ring deformation). ^{1}H NMR (600 MHz, DMSO-d_6, δ, ppm) 8.58 (broad peak, 1H, H-16); 8.20 (d, J = 7.8 Hz, 2H, H-9); 8.03 (s, 2H, H-4); 7.98 (broad peak, 2H, H-10); 7.77 (s, 2H, H-13); 7.74 (s, 1H, H-11); 3.26 (broad peak, 2H, H-17); 1.50 (m, 2H, H-18); 1.25 (m, 6H; H-19, H-20, H-21); 0.83 (broad peak, 3H, H-22). ^{13}C NMR (150 MHz, DMSO-d_6, δ, ppm) 166.86 (C-15); 166.19 (C-7); 166.09 (C-6); 138.02 (C-3); 136.49 (C-5); 136.46 (C-14); 133.80 (C-10); 133.35 (C-12); 133.21 (C-8); 129.18 (C-11); 126.99 (C-13); 124.97 (C-9); 124.22 (C-4); 123.85 (q, J = 286 Hz, C-1); 64.93 (hept, J = 25 Hz, C-2); 38.78 (C-17); 31.15 (C-20); 28.99 (C-18); 26.45 (C-19); 22.35 (C-21); 13.94 (C-22). Elem. Anal. Calcd. for [C$_{32}$H$_{23}$F$_6$N$_3$O$_5$]$_n$ (643.54)$_n$, C, 59.72%; H, 3.60%; N, 6.53%. Found: C, 59.21%; H, 3.43%; N, 6.38%.

[Structure of PI-6 with numbered atoms]

PI-6

2.4. Film Preparation

Poly(imide) films were prepared by solution casting using the following procedure: Each polymer (480 mg) was dissolved at room temperature in THF (12 mL) and filtered through a 200 µm Teflon syringe filter and poured onto an aluminium ring placed on a glass plate. Then, the solvent was evaporated at room temperature for 24 h. After that, the films were immersed in deionized water, stripped off the plate, and dried in a vacuum oven at 190 °C for 24 h. The thickness of the membranes ranged from 40 to 60 µm.

2.5. Instrumentation and Measurements

FT-IR spectra (KBr pellets) were recorded on a Nicolet 8700 Thermo Scientific FTIR spectrophotometer over the range of 4000–450 cm^{-1}. ^{1}H and ^{13}C NMR spectra for polymers were carried out on a 600 MHz instrument (Varian VNMRS) using DMSO-d_6 as solvent and TMS as internal standard, while ^{1}H and ^{13}C NMR spectra for monomers were recorded on a 400 MHz instrument (BRUKER AVANCE III HD-400). Viscosimetric measurements were made in an Ubbelohde viscosimeter number 50 at 30 °C (c = 0.5 g/dL). The size exclusion chromatography (SEC) measurements were performed on a GPC System 150cv (Santa Clara, CA, USA) at 20 °C equipped with a refractive index detector and a GPC KF-803 column (8.0 × 300 mm). The molecular weight (Mn and Mw) and polydispersity index were calculated according to the polyethylene glycol oxide standard and THF was used as a solvent. Soluble samples (c = 0.5 mg/mL) were filtered through micro-filters of 2 µm and then 100 µL were injected at 1 mL/min. Differential scanning calorimetry (DSC) was conducted on a TA Instruments Discovery DSC at a heating rate of 20 °C/min under nitrogen atmosphere. Thermogravimetric analysis (TGA) was performed using a thermogravimetric balance TGA-7 Perkin Elmer under a nitrogen atmosphere with a heating rate of 10 °C/min from 50 °C to 800 °C. Elemental analyses were made on a Fisons EA 1108-CHNS-O equipment. The mechanical properties of the films were measured with a Shimadzu AGS-X universal testing machine with a 100 N load cell at a strain rate of 1 mm/min, using strips of 5 mm wide, 50 mm long, and 40–60 µm thick, that were cut from polymer films. Wide-angle X-ray diffraction (XRD) was conducted on a Bruker D8 Advance diffractometer with CuKα radiation (wavelength λ_{Cu} = 1.542 Å), in range of 5° to 60° 2θ. The average d-spacing was calculated using Bragg's law:

$$d = \frac{n\lambda}{2\sin\theta}$$

where θ was assigned from the broad, amorphous peak maximum [20]. Poly(imide) density (ρ) was measured in a density gradient column (Techne Corporation, Minneapolis, MN, USA) with calcium nitrate solutions at 25 °C. Fractional free volume (FFV) was calculated by using the experimental density, and the theoretical volume occupied by the repeating unit of each polymer according to the following equation:

$$FFV = (V_{sp} - 1.3V_{w\text{-bondy}})/V_{sp}$$

where V_{sp} is the specific volume ($V_{sp} = \rho^{-1}$) and $V_{w\text{-bondy}}$ is the Van der Waals volume occupied by the repeating unit of the polymer, which was calculated with Bondi's group contribution method [21]. Pure gas permeability coefficients (P) were determined using a constant volume permeation cell of the type described elsewhere [22], according to the following equation:

$$P = \frac{273}{76} \frac{Vl}{ATp_0} \frac{dp}{dt}$$

where A and l are, the effective area and the thickness of the film, respectively. T is the temperature of the measurement (308.15 K), V is the constant volume of the permeation cell, p_0 is the pressure of the feed gas in the upstream, and dp/dt is the gas pressure increase with time under steady-state conditions measured in the permeation cell. P is expressed in Barrer [1 Barrer = 10^{-10} [cm^3·(STP)·cm cm^{-2}·s^{-1}·cmHg^{-1}]. The effective area of the film was 1.13 cm^2. Before each permeation test, the film was degassed for 24 h. The pure gases evaluated were He, H_2, O_2, N_2, CH_4, and CO_2, obtained from Praxair Corp. (San Salvador Xochimanca, Mexico) with purities > 99.99%. The measurements were carried out at 2 atm upstream pressure for each pure gas.

3. Results and Discussion

3.1. Monomers and Polymers: Synthesis and Characterization

The monomers synthesis started with the nucleophilic substitution reaction on the commercial acyl chloride (DNBC) using five primary amines as nucleophile agents. The 3,5-dinitro-N-alkylbenzamide derivatives were isolated as solids in good yields (90–95%). In the next step, the benzamide precursors were successfully reduced to the corresponding amide-diamine using a catalytic hydrogenolysis reaction (Yield: 87–93%). As it is known, one of the most critical factors to achieve high molecular weight polymers through polycondensation reactions is the purity of the monomers; therefore, all amide-diamine monomers and the commercial 6FDA were purified by sublimation technique before their use.

Monomers were structurally characterized by FT-IR and NMR techniques, as well as by elemental analysis (see Experimental Section). The FT-IR spectra of all monomers showed signals around 3400, 3330 and 1645 cm^{-1}, associated with the symmetric and asymmetric stretching of the amino (NH_2) and amide (NH and C=O) groups, respectively. Additionally, all the ^{1}H-NMR and ^{13}C-NMR signals were assigned for each monomer, where the most significant changes in NMR spectra were observed in the aliphatic region due to the different fragments [19].

The amide-diamine monomers were reacted with the commercial dianhydride 6FDA using anhydrous DMAc as solvent to obtain the aromatic PIs. In a first step, a viscous poly(amic acid) solution was obtained after 6 h of reaction, which underwent a chemical cyclization process with acetic anhydride and anhydrous pyridine as dehydrating agents. The fluorinated poly(imide)s were obtained in a 94–97% yield.

The success of the polymerization reactions was confirmed by FT-IR spectroscopy. The FT-IR spectra of the poly(imide)s are shown in Figure 1. The peaks at 1785 and 1729 cm^{-1} were attributed to the characteristic symmetric and asymmetric C=O stretching of the imide rings, respectively. The signal around 1663 cm^{-1} indicates the C=O stretching of the amide groups. The peaks around 1354 and 721 cm^{-1} correspond to the C-N stretching and the imide rings deformation, respectively. Furthermore, the absorption bands about 3000 cm^{-1} were ascribed to the C-H stretching of the phenyl and alkyl groups, and the broad signal around 3407 cm^{-1} was assigned to the N-H stretching of the imide groups [19]. In addition, no bands associated to the bi-functionality of the used monomers were observed (N-H stretching for diamine at 3388 cm^{-1} or one of the stretching bands for anhydride (1853 cm^{-1}).

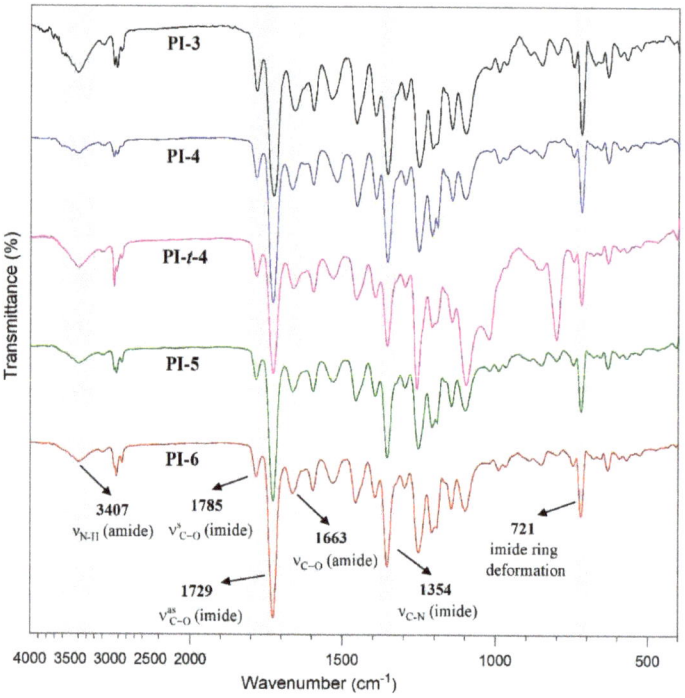

Figure 1. FT-IR spectra of the poly(imide)s.

The results of the NMR characterization also confirmed the chemical structure of each polymer [19,23]. Figure 2 shows the ^{1}H-NMR spectrum for each poly(imide). In the aromatic region, protons from the dianhydride (H-4, H-9, and H-10) and the diamine monomers (H-11, H-13) were observed, evidencing the success of the polymerization reaction. The main difference among PIs spectra was the chemical shift of the PI-*t*-4 amide proton (H-16), which was more shielded than their homologue signals. This is mainly due to the steric-hindrance exerted by the *tert*-butyl group attached to the nitrogen atom. Another difference was the chemical shift of the proton H-11 also in the PI-*t*-4 proton spectrum. The slightly higher positive inductive effect of the *tert*-butyl group, compared to the inductive effect of the linear chains, decreases the negative mesomeric effect of the *N*-R-carbamoyl group on the proton H-11 in the *para* position. Therefore, the proton H-11 is slightly more shielded in PI-*t*-4 than in similar PIs. Moreover, the chemical shift and the integrated signal intensities in the aliphatic region of the ^{1}H-NMR spectra were in accordance with the respective carbon chains of each poly(imide) [24].

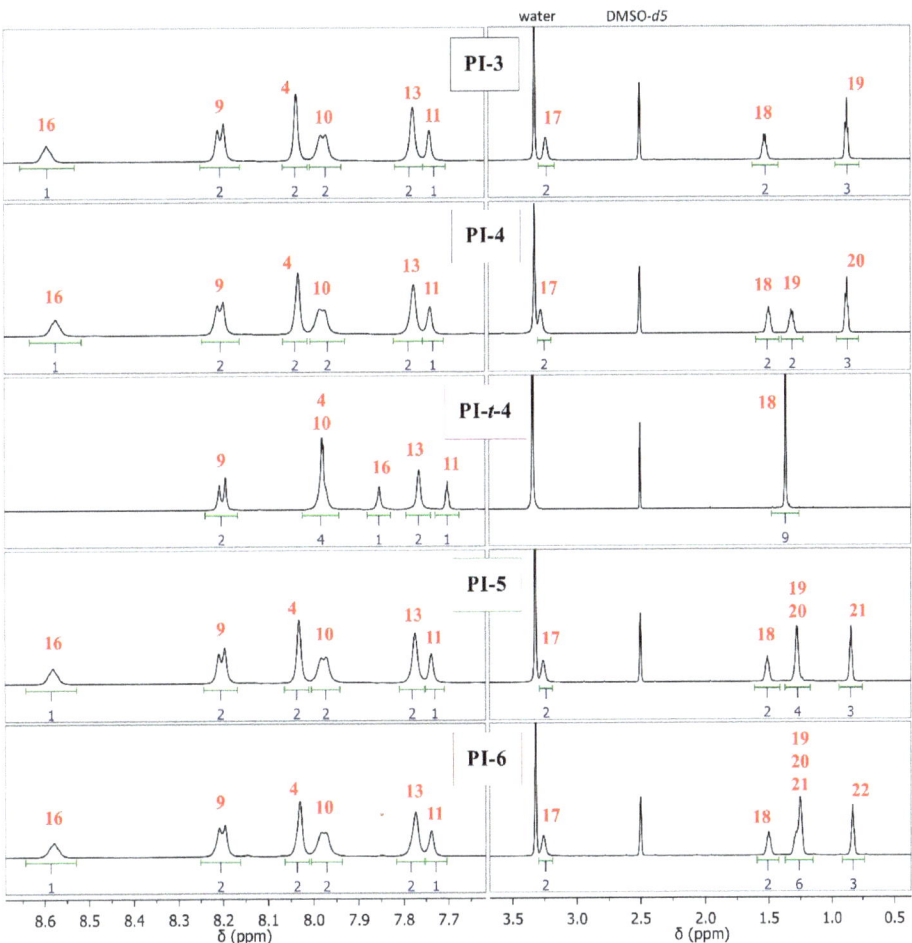

Figure 2. ^{1}H NMR spectra of all poly(imide)s.

3.2. Inherent Viscosity, Molecular Weight, and Solubility

Inherent viscosities were measured in a single point in NMP at 30 °C (c = 0.5 g/dL). The values ranged between 0.2 and 0.5 dL/g, indicating low to moderate molecular weights. Number (M_n) and weight (M_w) average molecular weights were measured by SEC. The values were in accordance with the results obtained for inherent viscosities, also indicating low to moderate molecular weights. Taking into account the molecular weights of their own repeating units, the chain ranged from 78–87 units long, except for PI-4, which was 45 units long. In fact, PI-4 exhibited the lowest inherent viscosity (0.2 dL/g) and molecular weight values in the series, leading to brittle thin film. It is not possible to offer a plausible explanation for the low molecular weight of this polymer, because M-4 were purified in the same way that other diamines, and the polymerization technique was the same one. Several attempts to synthetize PI-4, with special care in the monomers ratio (1:1) yielded the same result. Polydispersity indexes ranged from 1.3 to 1.5 and were within the range expected for condensation polymers.

Solubility is an important parameter for the processability of polymeric materials. In this sense, all poly(imide)s were soluble at 19 °C in a wide variety of aprotic polar organic solvents (Table 1) such as DMSO, NMP, DMF, and DMAc, as well as, THF, a relatively low boiling point solvent that favors their industrial processability. Additionally, PI-*t*-4 was soluble in chloroform. A possible interpretation of this outstanding solubility could be attributed to the bulky pendant groups (*tert*-butyl fragment) hanging along the polymer chain [15]. The *tert*-butyl groups increase the interchain distance, allowing the solvent to solvate the polymer chains more efficiently compared to the other PIs. As seen in Table 2, PI-*t*-4 had the highest fractional free volume (FFV) value, which corroborates the explanation for its excellent solubility.

Table 1. Inherent viscosity, molecular weight, and solubility.

PIs	η_{inh} [a] (dL/g)	Mn (×10^4 Da)	Mw (×10^4 Da)	DPI	Solubility at 19 °C					
					DMSO	DMF	DMAc	NMP	THF	CHCl$_3$
PI-3	0.3	4.728	6.321	1.3	+	+	+	+	+	-
PI-4	0.2	2.793	4.176	1.5	+	+	+	+	+	-
PI-*t*-4	0.4	5.385	7.364	1.4	+	+	+	+	+	+
PI-5	0.4	5.287	6.976	1.3	+	+	+	+	+	-
PI-6	0.5	5.897	8.131	1.4	+	+	+	+	+	-

[a] Measured in NMP (0.5 g/dL at 30 °C). DPI: polydispersity index. Solubility: +, Soluble; -, Insoluble.

Table 2. Density, fractional free volume, and *d*-spacing.

PIs	P [a] (g/cm^3)	V$_{w-bondy}$ [b] (cm^3/mol)	FFV [c]	Dsp [d] (Å)
PI-3	1.4423	278.23	0.132	5.0
PI-4	n.d.	n.d.	n.d.	5.1
PI-*t*-4	1.3806	288.45	0.158	5.1; 14.3
PI-5	1.3982	298.69	0.137	4.9
PI-6	1.3803	308.92	0.138	5.1; 10.9

[a] Density: measured at 25 °C. [b] Van der Waals volume calculated by Bondi's group contribution method [21]. [c] Fractional Free Volume. [d] *d*-spacing estimated by DRX.

3.3. Density, Fractional Free Volume, and Wide-Angle X-ray

The packing density of poly(imide)s was evaluated by determining the experimental density and theoretical FFV for each film (Table 2). The densities at 25 °C were in the range of 1.38–1.44 g/cm^3. PI-*t*-4 showed the lowest experimental density and the highest FFV in the series, mainly due to the bulky *tert*-butyl groups along the poly(imide) structure, which reduce the packing efficiency of the chains. PI-6 had the lowest density, but the volume occupied by the repeating unit (V$_{w-bondy}$) is the highest one, which reduces FFV.

Figure 3 shows the wide-angle X-ray diffraction profiles for poly(imide)s. A wide halo is observed in all spectra, which indicates that the samples are amorphous. PI-*t*-4 produced two peaks of 2θ angle at 6.2° and 17.8°, while PI-6 exhibited a little shoulder in 2θ at 8.1° and a maximum at 17.8°. The remaining poly(imide)s only showed one maximum. According to previous works, the interchain distance could be calculated through Bragg's equation, based on the position of the amorphous halo maximum [20]. PI-*t*-4 showed the highest interchain distance, with angles of 14.3 and 5.0 Å, respectively. This suggests that the introduction of *tert*-butyl moieties as pendant groups produces looser packing compared to the other *n*-alkyl pendant groups. PI-6, with six carbon atoms in the pendant chain, also showed two *d*-spacing values with an interchain distance of 10.9 and 5.1 Å.

In our previous work, with cycloalkyl pendant groups (cyclopentyl, cyclohexyl, cyclooctyl and adamantyl fragments) attached to the main chain, the *d*-spacing values were between 6.1 and 6.4 Å, which was indicative of moderate to high interchain distances [19]. However, none of those polymers showed two maximum halos in the X-ray pattern. Probably, PIs with acyclic alkyl pendant groups stack differently compared to cyclic side groups PIs in amorphous solid state, by virtue to their flexible nature. This packing would have created regions with greater interchain distance than others. The rigid nature of the cyclic fragments used in our previous work prevented this behaviour.

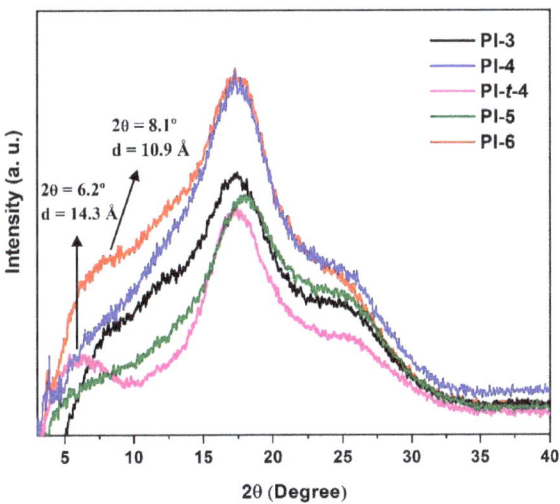

Figure 3. Wide-angle X-ray diffraction spectra of poly(imide)s.

3.4. Thermal and Mechanical Properties

The thermal properties of the poly(imide)s were determined by thermogravimetric analysis (TGA) and differential scanning calorimetry (DSC) under the nitrogen atmosphere. The TGA and DTGA curves of all poly(imide)s are shown in Figure 4, while the char yield and temperature at 5% and 10% weight loss values are summarized in Table 3.

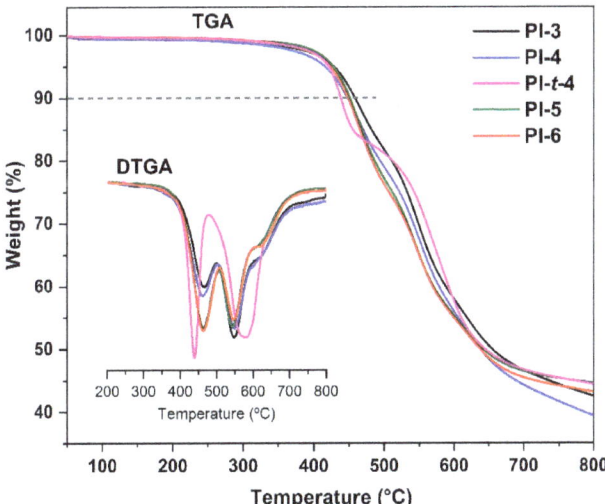

Figure 4. TGA and DTGA curves of poly(imide)s.

All poly(imide)s were highly thermally stable, with onset degradation temperature up to 415 °C. The thermograms had two well-defined stages of weight loss, as shown in the derivative curve (DTGA). The first stage, centered between 445 °C and 458 °C, is related to the decomposition of alkyl chains, while the second stage (550–580 °C) is attributed to the rupture of bonds in the poly(imide) backbone [18,25,26].

Table 3. Thermal properties of poly(imide)s.

PIs	T_5 (°C) [a]	T_{10} (°C) [b]	Char Yield (%)	T_g (°C) [c]
PI-3	428	458	42	270
PI-4	416	448	39	255
PI-t-4	421	455	44	265
PI-5	428	468	45	250
PI-6	423	448	43	240

[a,b] Thermal decomposition temperature at which 5% and 10% weight loss, respectively. [c] Glass transition temperature (taken from the second heating scan).

The DSC curves are presented in Figure 5, and the glass transition temperature (T_g) values are summarized in Table 3. T_g values of the poly(imide)s were in the range of 240–270 °C. In a polymer chain, the degrees of freedom increase with the number of carbon atoms, which leads to more flexible chains and, therefore, to T_gs [27–29]. In this sense, a decrease in T_g values was observed with the increase in the size lateral groups, from propyl to hexyl fragments. Because of its bulky nature, PI-t-4 does not follow this trend and gives a PI with higher rigidity and increased T_g.

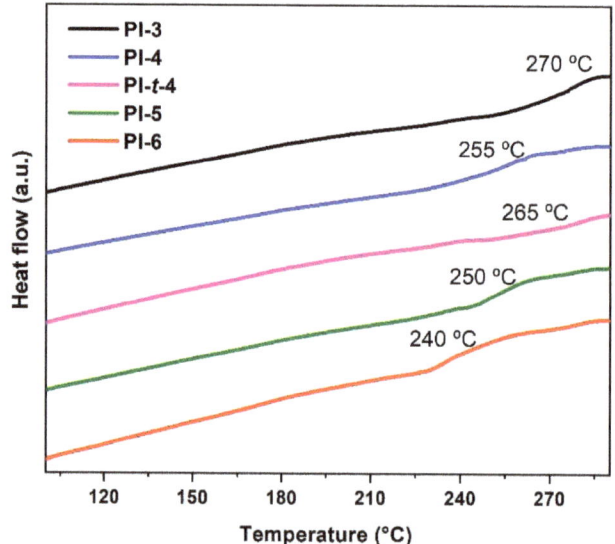

Figure 5. DSC PIcurves of poly(imide)s.

The mechanical properties values of the PI films are given in Table 4. The films had a tensile strength in the range of 45–87 MPa, elongation at break in the range of 3–11%, and tensile modulus in the range of 1.4–2.0 GPa. It was not possible to measure the mechanical properties of PI-4 film as it was brittle, which is attributed to its low molecular weight.

Table 4. Mechanical properties of PIs.

PIs	Tensile Strength (MPa)	Elongation at Break (%)	Young's Modulus (GPa)
PI-3	45	4.10	1.43
PI-t-4	53	3.2	2.19
PI-5	72	5.28	1.94
PI-6	87	11.98	1.74

Each sample was analyzed in triplicate. PI-4 film was too brittle to be measured.

It is expected that, for a same PIs backbone with different pendant acyclic alkyl chains, the Young's modulus increases with the length of these. The larger the pendant acyclic alkyl chains, the higher the cross-interactions between polymer backbones and the harder their mobility (increases the stiffness). This behaviour is observed when Young's Modulus of PI-3 is compared with those in PI-5 or PI-6. In the case of PI-t-4, the bulkiness of the *tert*-butyl moiety hinders the extension of mobility between PIs chains and therefore, a higher value is observed in the Young's modulus.

On the other hand, the observed increase in tensile strength from PI-3 to PI-6 can be attributed to the increase in molecular weight [30,31]. For similar structures, the higher the molecular weight, the stronger the intermolecular interactions and therefore the strength of the material. As can be noted, the elongation at breaks follows the same trend as in tensile strength except for PI-t-4, which showed a lower value (3.2%). This lower value could be explained considering that the mechanism of the plastic deformation is the sliding between the polymer chains in the material. The bulkiness of *tert*-butyl moiety in PI-t-4 do not favour the long-range interactions between the polymer chains as the linear alkyl moieties do. These long-range interactions are in charge of extending the elongation before the material breaks. Moreover, as side chains become lengthier, there are more possibilities of entangling the polymer chain, resulting in higher tensile strength. However, at the same time, the side chain becomes more flexible, which improves the elongation at break [32,33].

3.5. Gas Transport Properties

The effect of the acyclic alkyl-N-carbamoyl groups on the pure gas transport properties of poly(imide)s was evaluated, using a constant volume permeation cell at 2 atm and 35 °C. Table 5 shows the pure gas permeability coefficient for He, H_2, O_2, N_2, CH_4, and CO_2, gases and the ideal gas selectivity values for O_2/N_2, CO_2/CH_4, and the CO_2/N_2 gas pair. Since it was not possible to prepare an adequate film for PI-4, these properties could not be measured for this polymer. The permeability coefficient order for the remaining poly(imide)s was $PHe > PH_2 > PCO_2 > PO_2 > PN_2 > PCH_4$, which followed the same order as the gas kinetic diameter for these six gases (He 2.6 Å, H_2 2.89 Å, CO_2 3.3 Å, O_2 3.46 Å, N_2 3.64 Å, CH_4 3.8 Å) [34,35]. These results indicate that gas diffusion process plays a vital role in gas permeation through these polymers.

Table 5. Pure gas permeabilities and ideal selectivity for poly(imide)s.

PIs	Permeability (Barrer [a])						Ideal Selectivity (α)		
	He	H_2	O_2	N_2	CH_4	CO_2	O_2/N_2	CO_2/CH_4	CO_2/N_2
PI-3	52	46	5.4	1.2	1.1	23	4.5	21.3	19.9
PI-t-4	139	136	16.7	3.3	2.3	75	5.1	32.1	22.7
PI-5	44	42	5.9	1.4	1.2	27	4.2	21.9	19.8
PI-6	45	43	6.7	1.7	1.7	32	3.9	18.9	19.5
Matrimid *	-	-	2.1	0.3	0.3	10	6.5	35.7	31.2

[a] 1 Barrer = 10^{-10} [cm^3·(STP)·cm $cm^{-2} \cdot s^{-1} \cdot cmHg^{-1}$]. * Data reported in Ref. [3].

PI-t-4 showed the highest permeability and selectivity for all gases, which is attributed to the presence of the bulky *tert*-butyl pendant groups along the polymer chain [36]. Regarding the other poly(imide)s, the permeability for all gases decreased as follows: PI-6 > PI-5 > PI-3 (as the alkyl chains are shortened). In general, these results correlate with the calculated FFV and the obtained *d*-spacing values. Interestingly, permeability of PI-t-4 film was 2-3 times higher compared to PIs bearing acyclic alkyl groups films for all gases. PI-t-4 was actually more permeable than the polymers containing cyclic pendant groups, previously synthesized by our group [19]. Even PI-6 displayed a permeability for CO_2 (PCO_2 = 32 Barrer) similar to the homologous poly(imide) having an adamantyl pendant group (PCO_2 = 31.8 Barrer). The acyclic alkyl chains are probably less resistant to the pressure from gases, allowing them to diffuse more easily through the film. However, the selectivity of these poly(imide)s decreased as the gas permeability increased. Such behaviour has already been reported in previous works [19,37]. In Table 5, data available for Matrimid, a commercial poly(imide) membrane, was incorporated for comparison [3]. All PIs prepared in this work were more permeable than Matrimid, but Matrimid was more selective, as expected according to the trade-off between permeability and selectivity reported by Robeson [38,39].

4. Conclusions

Five new aromatic poly(imide)s with acyclic alkyl pendant groups were successfully synthesized by polycondensation reaction in solution, and structurally characterized. In general, PIs exhibited moderate molecular weights, were soluble in aprotic polar solvents, were thermally stable up to 415 °C, and had T_g values in the range of 240–270 °C. Mechanical properties were also measured, except for PI-4, giving values of tensile strength, elongation at break, and tensile modulus in the range of 45–87 MPa, 3–11%, and 1.4–2.0 GPa, respectively. The poly(imide) containing the most branched substituent (PI-*t*-4, *tert*-butyl fragment) was the most soluble and most permeable to gases. Those results are in agreement with the largest fractional free volume recorded for its film. The longest interchain distance in PI-*t*-4 was corroborated indirectly through density measurements, and directly by means of wide-range X-ray diffraction. The structure-property relationship for these new poly(imide)s is a crucial point to take into account because the results presented here demonstrated that the incorporation of *tert*-butyl-*N*-carbamoyl moieties along the polymer chain leads to films with increased permeability to different gases compared to previously reported PIs films bearing cyclic bulky groups.

Author Contributions: Conceptualization, C.A.T. and A.T.-C.; investigation, G.P.; data curation, F.E.R.-G.; writing—review and editing, I.J.; Project administration, L.H.T.; Resources, D.C., P.O. and M.A.-V.; Validation, R.M.-T.; Visualization, F.E.R.-G. and V.N. All authors have read and agreed to the published version of the manuscript.

Funding: This research was funded by FONDECYT Postdoctoral Fellowship, grant number 3160724 and FONDEQUIP EQM120021.

Acknowledgments: F.E. Rodríguez-González thanks the Agencia Nacional de Investigación y Desarrollo (ANID) for fellowship 21180253.

Conflicts of Interest: The authors declare no conflicts of interest.

References

1. Nunes, S.P.; Peinemann, K.V. *Membrane Technology: In the Chemical Industry*; Wiley-VCH: Hoboken, NJ, USA, 2006; ISBN 3527313168.
2. Baker, R.W.; Low, B.T. Gas separation membrane materials: A perspective. *Macromolecules* **2014**. [CrossRef]
3. Sanders, D.F.; Smith, Z.P.; Guo, R.; Robeson, L.M.; McGrath, J.E.; Paul, D.R.; Freeman, B.D. Energy-efficient polymeric gas separation membranes for a sustainable future: A review. *Polymer* **2013**, *54*, 729–4761. [CrossRef]
4. McKeown, N.B.; Budd, P.M. Polymers of intrinsic microporosity (PIMs): Organic materials for membrane separations, heterogeneous catalysis and hydrogen storage. *Chem. Soc. Rev.* **2006**, *35*, 675. [CrossRef] [PubMed]
5. Naiying, D.; Jingshe, S.; Robertson, G.P.; Pinnau, I.; Guiver, M.D. Linear high molecular weight ladder polymer via fast polycondensation of 5,5′,6,6′-tetrahydroxy-3,3,3′,3′- tetramethylspirobisindane with 1,4-dicyanotetrafluorobenzene. *Macromol. Rapid Commun.* **2008**. [CrossRef]
6. Park, H.B.; Han, S.H.; Jung, C.H.; Lee, Y.M.; Hill, A.J. Thermally rearranged (TR) polymer membranes for CO_2 separation. *J. Membr. Sci.* **2010**, *359*, 11–24. [CrossRef]
7. Yun, J.; Song, C.; Lee, H.; Park, H.; Jeong, Y.R.; Kim, J.W.; Jin, S.W.; Oh, S.Y.; Sun, L.; Zi, G.; et al. Stretchable array of high-performance micro-supercapacitors charged with solar cells for wireless powering of an integrated strain sensor. *Nano Energy* **2018**, *49*, 644–654. [CrossRef]
8. Jia, M.; Zhou, M.; Li, Y.; Lu, G.; Huang, X. Construction of semi-fluorinated polyimides with perfluorocyclobutyl aryl ether-based side chains. *Polym. Chem.* **2018**, *9*, 920–930. [CrossRef]
9. Wang, C.; Cao, S.; Chen, W.; Xu, C.; Zhao, X.; Li, J.; Ren, Q. Synthesis and properties of fluorinated polyimides with multi-bulky pendant groups. *RSC Adv.* **2017**. [CrossRef]

10. Nistor, C.; Shishatskiy, S.; Popa, M.; Nunes, S.P. Composite membranes with cross-linked matrimid selective layer for gas separation. *Environ. Eng. Manag. J.* **2008**. [CrossRef]
11. Wang, C.; Yu, B.; Jiang, C.; Zhao, X.; Li, J.; Ren, Q. Synthesis and characterization of an aromatic diamine and its polyimides containing asymmetric large side groups. *Polym. Bull.* **2020**. [CrossRef]
12. Liu, Y.; Guo, J.; Wang, J.; Zhu, X.; Qi, D.; Li, W.; Shen, K. A novel family of optically transparent fluorinated hyperbranched polyimides with long linear backbones and bulky substituents. *Eur. Polym. J.* **2020**. [CrossRef]
13. Wu, Q.; Ma, X.; Zheng, F.; Lu, X.; Lu, Q. High performance transparent polyimides by controlling steric hindrance of methyl side groups. *Eur. Polym. J.* **2019**. [CrossRef]
14. Tundidor-Camba, A.; Terraza, C.A.; Tagle, L.H.; Coll, D.; Ortiz, P.; De Abajo, J.; Maya, E.M. Novel aromatic polyimides derived from 2,8-di(3-aminophenyl)dibenzofuran. Synthesis, characterization and evaluation of properties. *RSC Adv.* **2015**. [CrossRef]
15. Yi, L.; Li, C.; Huang, W.; Yan, D. Soluble polyimides from 4,4′-diaminodiphenyl ether with one or two tert-butyl pedant groups. *Polymer* **2015**. [CrossRef]
16. Liu, Y.; Qian, X.; Shi, H.; Zhou, W.; Cai, Y.; Li, W.; Yao, H. New poly(amide-imide)s with trifluoromethyl and chloride substituents: Synthesis, thermal, dielectric, and optical properties. *Eur. Polym. J.* **2017**. [CrossRef]
17. Yen, H.J.; Wu, J.H.; Huang, Y.H.; Wang, W.C.; Lee, K.R.; Liou, G.S. Novel thermally stable and soluble triarylamine functionalized polyimides for gas separation. *Polym. Chem.* **2014**. [CrossRef]
18. Tundidor-Camba, A.; Terraza, C.A.; Tagle, L.H.; Coll, D.; Ortiz, P.; Pérez, G.; Jessop, I.A. Aromatic polyimides containing cyclopropylamide fragment as pendant group. A study of the balance between solubility and structural rigidity. *Macromol. Res.* **2017**. [CrossRef]
19. Pérez, G.; Terraza, C.A.; Coll, D.; Ortiz, P.; Aguilar-Vega, M.; González, D.M.; Tagle, L.H.; Tundidor-Camba, A. Synthesis and characterization of processable fluorinated aromatic poly(benzamide imide)s derived from cycloalkane substituted diamines, and their application in a computationally driven synthesis methodology. *Polymer* **2019**. [CrossRef]
20. Murthy, N.S. X-ray diffraction from polymers. In *Polymer Morphology: Principles, Characterization, and Processing*; Wiley: Hoboken, NJ, USA, 2016; ISBN 9781118892756.
21. Bondi, A. Van der waals volumes and radii. *J. Phys. Chem.* **1964**. [CrossRef]
22. Carrera-Figueiras, C.; Aguilar-Vega, M. Gas permeability and selectivity of hexafluoroisopropylidene aromatic isophthalic copolyamides. *J. Polym. Sci. Part B Polym. Phys.* **2005**. [CrossRef]
23. Cornelius, C.J.; Marand, E. Hybrid inorganic–organic materials based on a 6FDA–6FpDA–DABA polyimide and silica: Physical characterization studies. *Polymer* **2002**, *43*, 2385–2400. [CrossRef]
24. Jacobsen, N.E. *NMR Data Interpretation Explained: Understanding 1D and 2D NMR Spectra of Organic Compounds and Natural Products*, 1st ed; John Wiley & Sons, Inc.: Hoboken, NJ, USA, 2017; ISBN 978-1-118-37022-3.
25. Maya, E.M.; Lozano, A.E.; de Abajo, J.; de la Campa, J.G. Chemical modification of copolyimides with bulky pendent groups: Effect of modification on solubility and thermal stability. *Polym. Degrad. Stab.* **2007**, *92*, 2294–2299. [CrossRef]
26. Billmeyer, F.W. Textbook of Polymer Science. *Kobunshi* **1963**. [CrossRef]
27. Privalko, V.P.; Lipatov, Y.S. Glass Transition and Chain Flexibility of Linear Polymers. *J. Macromol. Sci. Part B* **1974**. [CrossRef]
28. Yang, S.; He, Y.; Liu, Y.; Leng, J. Shape-memory poly(arylene ether ketone)s with tunable transition temperatures and their composite actuators capable of electric-triggered deformation. *J. Mater. Chem. C* **2019**. [CrossRef]
29. Rivera Nicholls, A.; Kull, K.; Cerrato, C.; Craft, G.; Diry, J.B.; Renoir, E.; Perez, Y.; Harmon, J.P. Thermomechanical characterization of thermoplastic polyimides containing 4,4′-methylenebis(2,6-dimethylaniline) and polyetherdiamines. *Polym. Eng. Sci.* **2019**. [CrossRef]
30. Hasegawa, M.; Horiuchi, M.; Wada, Y. Polyimides containing trans-1,4-cyclohexane unit (II). Low-K and low-CTE semi- and wholly cycloaliphatic polyimides. *High Perform. Polym.* **2007**. [CrossRef]
31. Cheng, S.; Shen, D.; Zhu, X.; Tian, X.; Zhou, D.; Fan, L.J. Preparation of nonwoven polyimide/silica hybrid nanofiberous fabrics by combining electrospinning and controlled in situ sol-gel techniques. *Eur. Polym. J.* **2009**. [CrossRef]
32. Hasegawa, M.; Fujii, M.; Ishii, J.; Yamaguchi, S.; Takezawa, E.; Kagayama, T.; Ishikawa, A. Colorless polyimides derived from 1S,2S,4R,5R-cyclohexanetetracarboxylic dianhydride, self-orientation behavior during solution casting, and their optoelectronic applications. *Polymer* **2014**. [CrossRef]

33. Kim, G.; Byun, S.; Yang, Y.; Kim, S.; Kwon, S.; Jung, Y. Film shrinkage inducing strong chain entanglement in fluorinated polyimide. *Polymer* **2015**. [CrossRef]
34. Stevens, K.A.; Moon, J.D.; Borjigin, H.; Liu, R.; Joseph, R.M.; Riffle, J.S.; Freeman, B.D. Influence of temperature on gas transport properties of tetraaminodiphenylsulfone (TADPS) based polybenzimidazoles. *J. Membr. Sci.* **2020**. [CrossRef]
35. Ogieglo, W.; Puspasari, T.; Ma, X.; Pinnau, I. Sub-100 nm carbon molecular sieve membranes from a polymer of intrinsic microporosity precursor: Physical aging and near-equilibrium gas separation properties. *J. Membr. Sci.* **2020**. [CrossRef]
36. Hu, X.; He, Y.; Wang, Z.; Yan, J. Intrinsically microporous co-polyimides derived from ortho-substituted Tröger's Base diamine with a pendant tert-butyl-phenyl group and their gas separation performance. *Polymer* **2018**. [CrossRef]
37. Terraza, C.A.; Tagle, L.H.; Santiago-García, J.L.; Canto-Acosta, R.J.; Aguilar-Vega, M.; Hauyon, R.A.; Coll, D.; Ortiz, P.; Perez, G.; Herrán, L.; et al. Synthesis and properties of new aromatic polyimides containing spirocyclic structures. *Polymer* **2018**, *137*, 283–292. [CrossRef]
38. Robeson, L.M. Correlation of separation factor versus permeability for polymeric membranes. *J. Membr. Sci.* **1991**. [CrossRef]
39. Robeson, L.M. The upper bound revisited. *J. Membr. Sci.* **2008**. [CrossRef]

 © 2020 by the authors. Licensee MDPI, Basel, Switzerland. This article is an open access article distributed under the terms and conditions of the Creative Commons Attribution (CC BY) license (http://creativecommons.org/licenses/by/4.0/).

Article

Enhanced Selective Hydrogen Permeation through Graphdiyne Membrane: A Theoretical Study

Quan Liu [1,*], Long Cheng [2] and Gongping Liu [2,*]

1. Analytical and Testing Center, Anhui University of Science and Technology, Huainan 232001, China
2. State Key Laboratory of Materials-Oriented Chemical Engineering, College of Chemical Engineering, Nanjing Tech University, 30 Puzhu Road (S), Nanjing 211816, China; longcheng@njtech.edu.cn
* Correspondence: quanliu@aust.edu.cn (Q.L.); gpliu@njtech.edu.cn (G.L.)

Received: 29 September 2020; Accepted: 13 October 2020; Published: 15 October 2020

Abstract: Graphdiyne (GDY), with uniform pores and atomic thickness, is attracting widespread attention for application in H_2 separation in recent years. However, the challenge lies in the rational design of GDYs for fast and selective H_2 permeation. By MD and DFT calculations, several flexible GDYs were constructed to investigate the permeation properties of four pure gas (H_2, N_2, CO_2, and CH_4) and three equimolar binary mixtures (H_2/N_2, H_2/CO_2, and H_2/CH_4) in this study. When the pore size is smaller than 2.1 Å, the GDYs acted as an exceptional filter for H_2 with an approximately infinite H_2 selectivity. Beyond the size-sieving effect, in the separation process of binary mixtures, the blocking effect arising from the strong gas–membrane interaction was proven to greatly impede H_2 permeation. After understanding the mechanism, the H_2 permeance of the mixtures of H_2/CO_2 was further increased to 2.84×10^5 GPU by reducing the blocking effect with the addition of a tiny amount of surface charges, without sacrificing the selectivity. This theoretical study provides an additional atomic understanding of H_2 permeation crossing GDYs, indicating that the GDY membrane could be a potential candidate for H_2 purification.

Keywords: graphdiyne; molecular simulation; membrane separation; hydrogen purification

1. Introduction

As an attractive alternative fuel source, hydrogen (H_2) could eliminate the use of polluting fossil fuels in industry and transport in the future [1]. So, the H_2 energy is critical to reduce global greenhouse gases and promote sustainable development because of its natural abundance and high efficiency of combustion [2,3]. Today, in many industrial and drilling streams, H_2 is mainly produced from natural gas, hydro-electrolysis, and the combustion of hydrides [4,5]. However, these processes release several million tonnes of by-products (such as carbon dioxide, nitrogen, and methane) per year [6]. How to separate the target product of H_2 from the undesirable species is crucial to improve production efficiency and reduce the cost. Traditional separation techniques, such as pressure swing adsorption and cryogenic separation, would consume a considerable amount of energy to collect H_2 [7]. With the improved performances and lower operating conditions, the advanced membrane-based separation is seen as an alternative to significantly improve energy efficiency [8,9]. At room temperature and low transmembrane pressure, it is more instructive for experiments to develop membranes with both good H_2 permeance and selectivity. [10].

Recently, two-dimensional (2D) carbon-based membranes have sparked global attention due to their atomic thickness [11]. Among various membranes, graphene-based materials are, assuredly, one of the powerful candidates for membrane separation according to our previous experimental [12,13] and simulation studies [14,15]. According to the separation mechanism of the size sieving effect, controllable pores in the 2D-material membrane are imperative for gas separation. However, it is not cost-effective to drill uniform pores in graphene, and this process may introduce non-selective defects in the membrane [16]. Alternatively, newly emerged graphdiyne (GDY) allotropes that formed by the periodic combination of sp and sp^2 carbon atoms are proposed as alternative options for gas purification [16], because they process the well-distributed pores as well as the atomic membrane thickness [17]. Several studies have been reported to explore the gas permeation property through GDYs in recent years [18–20]. It is the comparable pores in GDYs that contribute to the outstanding gas separation properties, as Smith et al. first reported that the excellent H_2 separation performance was ascribed to the small triangular pores in the GDYs [21]. This later reports also showed that the various sized pores in diverse GDYs could be employed to distinguish the different sized molecules, such as He [18], O_2 [22], and CO_2 [23,24], concluding that the size sieving effect dominated the transport mechanism of GDYs for gas separation.

Although the pore size is an important factor for gas separation membranes, other factors, particularly the surface properties, should not be neglected. Sang et al. [25] and Smith et al. [26] found that the functionalized surface of GDYs shown a better separation performance than that of the pristine one. Moreover, the adsorption phenomenon in GDYs also affected the permeation of gases, such as H_2S [27] and CO_2 [24,28]. It means, beyond the well-understood size sieving effect, there could be a more comprehensive mechanism to better describe gas transport through GDYs [19]. However, the main challenge is how such a mechanism determines the gas permeation properties, especially for selective H_2 permeation through GDYs, and how it contributes to further improvement of the H_2 permeance and selectivity of GDYs. In addition, most previous works focused on the first-principles calculations density functional theory (DFT) to study the selectivity of H_2 over CO_2, CO, N_2, and CH_4 [25,26,29]. A few studies calculated the H_2 permeance by performing molecular dynamic (MD) simulations, which however were based on a rigid framework of GDYs [20,23,30] and might be too idealized for actual GDYs for gas separation. Moreover, it is unclear what the ultimate size of nanopores in GDYs is allowed to transport H_2 molecules. Therefore, it is necessary to further understand the underlying separation mechanism of H_2 purification through a carefully designed flexible GDY membrane that possesses both high H_2 permeance and selectivity and find the ultimate diameter in GDYs for H_2 permeation.

In this work, a series of 2D GDYs are computationally constructed to examine the permeation of four pure gases (H_2, N_2, CO_2 and CH_4) and their equimolar binary mixtures (H_2/N_2, H_2/CO_2 and H_2/CH_4). Assuming that the separation performance of H_2 is affected by pore structures and surface charges, we systematically investigated these two effects by both MD and DFT calculations. Following the introduction, the atomic models of GDYs, as well as the separation systems, are illustrated in Section 2. In Section 3, the H_2 permeance is calculated by the time evolution of the permeated molecules according to the MD results, and the ideal selectivity is evaluated by DFT calculations. This separation mechanism of H_2 from binary mixtures is revealed by analyzing the diffusion coefficient, density contour, and energy barrier of the permeation. After that, the H_2 permeance of the mixture of H_2/CO_2 is further improved by reducing the blocking effect. Finally, the concluding remarks are summarized in Section 4.

2. Models and Methods

The GDYs were constructed in the Material studio [31] and then subjected to geometry optimization in the Forcite module with 5000 iterations. As presented in Figure 1a–d, the dimensions of the membranes were 7.75×7.55 nm^2 in this setup, and the investigated pore diameters were varied from 1.5 to 2.5 Å, which were measured according to the formula: $D = 2\sqrt{A/\pi}$ by inserting a van der Waals sphere, where A is the open pore area, as depicted in Figure 1e–h. We noted that the membrane in Figure 1b coded as GDY_1.5Å_p7% has the porosity of 7%, which is larger than that of 3% in Figure 1a (GDY_1.5Å_p3%), although both of these two membranes have a similar pore diameter of 1.5 Å. This above-mentioned porosity was calculated by the formula $p = \frac{A_{pore}}{A_{mem}}$, where A_{pore} and A_{mem} denote to the areas of the totally unoccupied regions and the membrane surface, respectively.

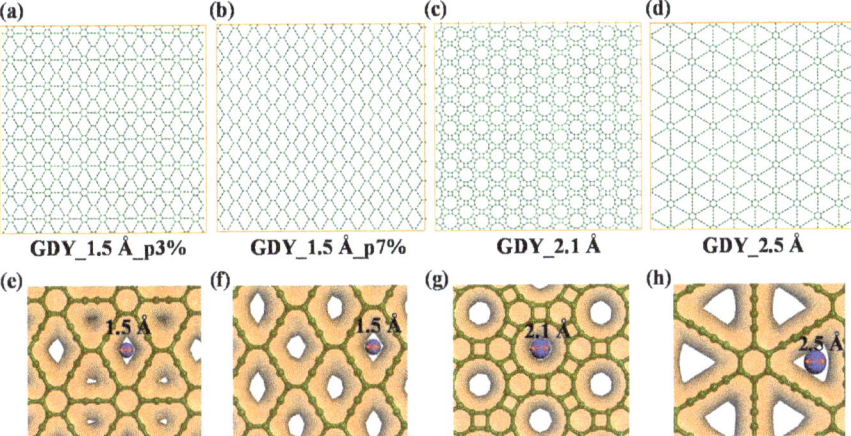

Figure 1. Membrane models. Atomic structures of graphdiyne (GDY) membranes with different pore structures: (**a**) GDY_1.5Å_p3%; (**b**) GDY_1.5Å_p7%; (**c**) GDY_2.1Å; and (**d**) GDY_2.5Å. This first two membranes have different porosities of 3% and 7%, respectively. (**e**–**h**) The criterion for the definition of the pore diameter in each GDY membranes.

Based on the optimized membranes, MD calculations were performed to simulate the gas separation. Figure 2 shows the simulation system, where two chambers were isolated by the GDY membrane. This left chamber was a gas reservoir, comprising 2000 molecules of pure gases (or binary mixtures with the mixing ratio of 1000:1000, in volume ratio). This right one treated as a vacuum is the permeate side, which is the most common setting in MD simulations for collecting the permeated molecules. [20,23,30] At both ends of each chamber, rigid graphene was placed to prevent molecules from roaming between the periodic boxes. Considering that carbon-based membranes usually have good flexibility in experiments [32], all pores in GDYs were treated as flexible in our present work so that the atoms around pores could perform small displacements. Thus, the aperture would be enlarged to allow the passage of H_2 even though the molecular size is larger than the pore size. On the contrary, the atoms on the edge of membranes were imposed with a position restriction to ensure the uniform lattice of GDYs and decrease the impact on gas permeation [33]. Furthermore, no collapse appeared in our system, suggesting good stability of the flexible structures. This framework of GDYs was described by the all-atom optimized potential. This four gases were represented by Lennard–Jones (LJ) and electrostatic potentials originating from our previous work [14]. Between dissimilar atoms, the interactions were assembled by the Lorentz–Berthelot combination rule [34].

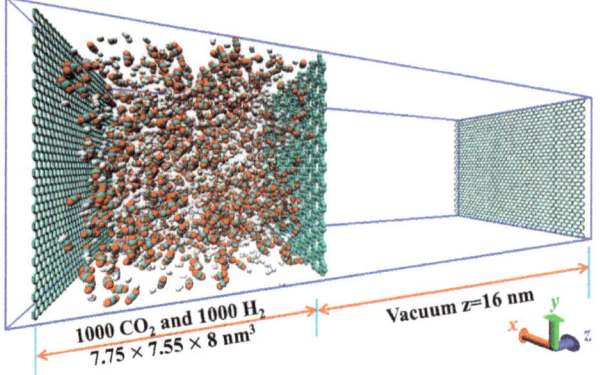

Figure 2. Simulation system of the equimolar binary mixture of H_2/CO_2 permeating through the GDY_2.1Å membrane. Atom: C (Cyan); O (red); H (white).

In this study, MD simulations were all carried out using the GROMACS package (version 4.5.5) [35]. A static minimization with the steepest descent method was first carried out to remove the unreasonable contacts among each atom. Then, the separation system was pre-equilibrated in the isobaric–isothermal ensemble (constant temperature and constant pressure ensemble, named as NPT) for 2 ns only in the z-direction. Following this, another 20 ns MD simulations were performed in the canonical ensemble (constant temperature and constant volume ensemble, named as NVT) for data collection and further analysis, and the trajectories were stored every 1 ps. During the simulations, the classical equations of motion were integrated with a time step of 1 fs by using the leapfrog algorithm. This temperature was maintained at 300 K by coupling with the velocity rescaling [36] thermostats. This long-range electrostatic interactions were computed by using the method of particle mesh Ewald [37]. While the short-range van der Waals interactions were truncated at a cut-off distance of 1.2 nm, the periodic boundary conditions were implemented in all three directions. With the partial pressure gradient as the driving force [38], the gases would permeate through the GDYs. For a better understanding of the separation process, an animation is provided in the Supplementary Materials (Video S1).

3. Results and Discussion

3.1. Gas Permeation Behavior

The relationship between gas flux (J, mol·m^{-2}·s^{-1}) and permeance (S, mol·m^{-2}·s^{-1}·Pa^{-1}) was described as Equation (1):

$$J = \frac{dN}{A_{mem}N_A dt} = \Delta P S \quad (1)$$

where N refers to the number of permeated molecules, N_A is the Avogadro constant, and ΔP is the transmembrane gas partial pressure estimated by Equation (2):

$$\Delta P = \frac{(N_0 - N_{ad} - N)kT}{V_l} - \frac{NkT}{V_r} \quad (2)$$

in which N_{ad} refers to the number of gases adsorbed on the membrane surface, V_l and V_r are the volumes of the left and right chambers, respectively, and k represents the Boltzmann constant. Therefore, the time evolution of the number of permeated molecules is integrated as Equation (3), where R is the gas constant,

$$N = \frac{(N_0 - N_{ad})L_r}{(L_l + L_r)}\left(1 - e^{-\frac{RTS(L_l + L_r)}{L_l L_r}t}\right) = \frac{2(N_0 - N_{ad})}{3}\left(1 - e^{-467.7St}\right). \quad (3)$$

As presented in Figure 3a, the number of permeated H_2 is increased exponentially with the simulation time, which agrees well with the above mathematical analysis. This permeance of pure H_2 is shown in Figure 3b. Obviously, it is remarkably enhanced with the increase of porosity and pore size. With the smallest pore of 1.5 Å, the porosity of 0.03 and 0.07 have the H_2 permeance of 1.34×10^5 and 2.55×10^5 GPU (1 GPU = 3.35×10^{-10} mol·m^{-2}·s^{-1}·Pa^{-1}), respectively. This permeation of other gases (i.e., CO_2, N_2 and CH_4) through the GDYs was also simulated. It was shown that only the biggest pore with 2.5 Å allowed the passage of N_2, CO_2, and CH_4, as presented in Figure 3c. This incompatibility of the kinetic diameter of gases (i.e., H_2, 2.89 Å; N_2, 3.64 Å; CO_2, 3.30 Å; CH_4, 3.8 Å) [39], and the pore size is ascribed to the flexible structures, implying that the membrane of GDY_2.5Å is not suitable for H_2 purification. Moreover, the comparable pore with 2.1 Å diameter can not only completely block the other three gases but also process the good H_2 permeance of 5.94×10^5 GPU, which implies that the selectivities of H_2 over CO_2, N_2, and CH_4 can be extremely high in the membrane of GDY_2.1Å. We noted that the molecules of N_2, CO_2, and CH_4 can not be detected in the permeation side as long as the pore diameter was smaller than 2.1 Å, indicating the infinitely low permeance of N_2, CO_2, and CH_4.

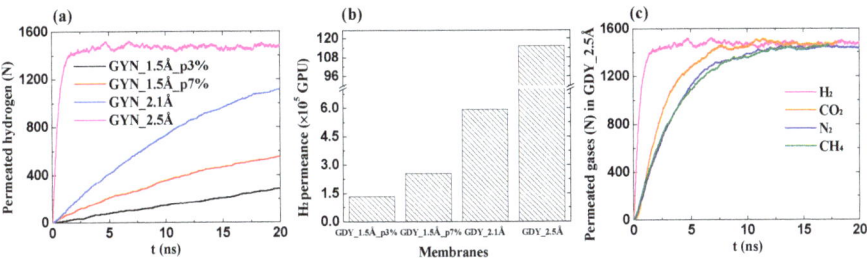

Figure 3. Pure gas permeation. (**a**) The time evolution of permeated H_2 molecules; (**b**) The permeance of H_2 through different GDYs; (**c**) The permeation of four gases in GDY_2.5Å.

To evaluate the ideal selectivities of H_2 over other three gases in GDY_2.1 Å, the DFT calculations were performed to calculate permeation barriers as per our previous study [15]. Figure 4a illustrates the minimum energy pathway (MEP) of four gases crossing the membrane. This inset configurations are the energetically stable states of CO_2 permeation at different locations. By searching the saddle point in MEPs, the energy barriers of permeation can be calculated. Evidently, the permeation barrier of H_2 crossing the flexible GDY_2.1 Å is drastically reduced to 1.55 kJ/mol (Figure 4a, inset), which results in extraordinary H_2 permeance. On the contrary, the permeation barriers of CO_2, N_2, and CH_4 are all extremely high. In Figure 4b, the temperature-dependent H_2 selectivity has an inverse correlation with the temperature according to the Arrhenius Equation (4), whereas P_i is the permeation rate, A is the permeation prefactor that can be assumed as 10^{11} s^{-1} [38], and T is the temperature. For H_2/CO_2 and H_2/N_2, the ideal selectivities of H_2 can be up to 10^{17} at 300 K, which is the same order with the modified GDYs [25]. It remains very high (>10^8) even at 600 K, further suggesting the extraordinary H_2 separation performance through GDY_2.1 Å.

$$S_{i/j} = \frac{P_i}{P_j} = \frac{A_i exp\left(-\frac{E_{barrier,\ i}}{RT}\right)}{A_j exp\left(-\frac{E_{barrier,\ j}}{RT}\right)} \qquad (4)$$

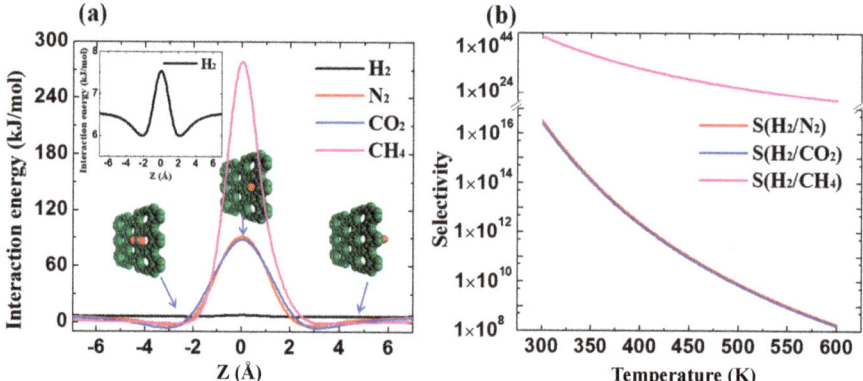

Figure 4. DFT calculations of gases crossing the membrane of GDY_2.1Å. (**a**) Minimum energy pathways of four gases. (**b**) Ideal selectivities of H_2 over N_2, CO_2, and CH_4 as functions of temperature.

3.2. Transport Mechanism: Blocking Effect

For binary gas mixtures, the MD calculations were also performed to simulate the separation process. To visualize H_2 purification properties through the membrane of GDY_2.1Å, the equilibrium configurations of final frames are presented in Figure 5a–c. As seen, the GDY_2.1Å membrane acts as an effective filter for H_2 separation while completely blocking the passage of CO_2, N_2, and CH_4, so that the H_2 is largely gathering in the vacuum chamber. This corresponding H_2 permeance of the three binary mixtures is presented in Figure 5d. Interestingly, the mixture of H_2/N_2 exhibits the highest H_2 permeance of 4.71×10^5 GPU, which is a little higher than that of H_2/CH_4 and almost twice as many as the mixture of H_2/CO_2. This primary reason is that on the membrane surface, the strongly interacting gas (CO_2) preferentially adsorbs, blocking the transport pores of H_2 molecules, thus resulting in a relatively low H_2 permeance.

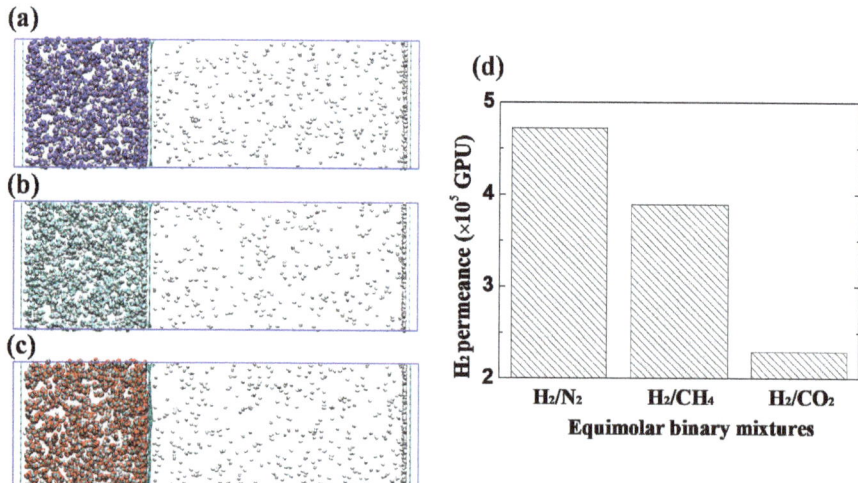

Figure 5. H_2 permeation of three equimolar binary mixtures crossing the GDY_2.1Å. This final snapshots: (**a**) H_2/N_2, (**b**) H_2/CH_4, and (**c**) H_2/CO_2. Blue: N; cyan: C; red: O; white: H. (**d**) The H_2 permeance of different binary mixtures.

Further analyses are carried out to understand this blocking effect. This radial distribution function (RDF) was calculated to present the affinity of GDY_2.1Å to four gases by using Equation (5):

$$g_{ij}(r) = \frac{N_{ij}(r, r+\Delta r)V}{4\pi r^2 \Delta r N_i N_j} \qquad (5)$$

where $N_{ij}(r, r+\Delta r)$ is the number of species j around i within a shell from r to $r + \Delta r$, r is the distance between two species, and N_i and N_j refer to the numbers of atom types i and j, respectively. As presented in Figure 6a, there is an increasing trend of the gas–membrane interaction as $H_2 < N_2 < CH_4 < CO_2$. This weakest interaction together with the smallest molecular size endows H_2 with exceptional permeance. Meanwhile, the strong interaction promotes the gases, particularly CO_2 to preferentially adsorb on the membrane surface. To offer more intuitionistic information, the density contours of the distribution of N_2, CH_4, and CO_2 were plotted on the GDY_2.1Å surface. This general distribution behavior of the three gases is similar as shown in Figure 6b–d, where the pores that are largely clogged by these gases are larger than the pore diameters. Nevertheless, the intensity of gas accumulation is quite different. As seen in Figure 6d, the pores are the most clogged, with the number of adsorbed CO_2 molecules exceeding 6.5 N_w/uc in every pore. Similar to the affinity analysis in Figure 6a, the extent of the blocking effect follows the increasing trend of $N_2 < CH_4 < CO_2$, decreasing the placeholders of H_2 on the membrane surface in Figure 7a.

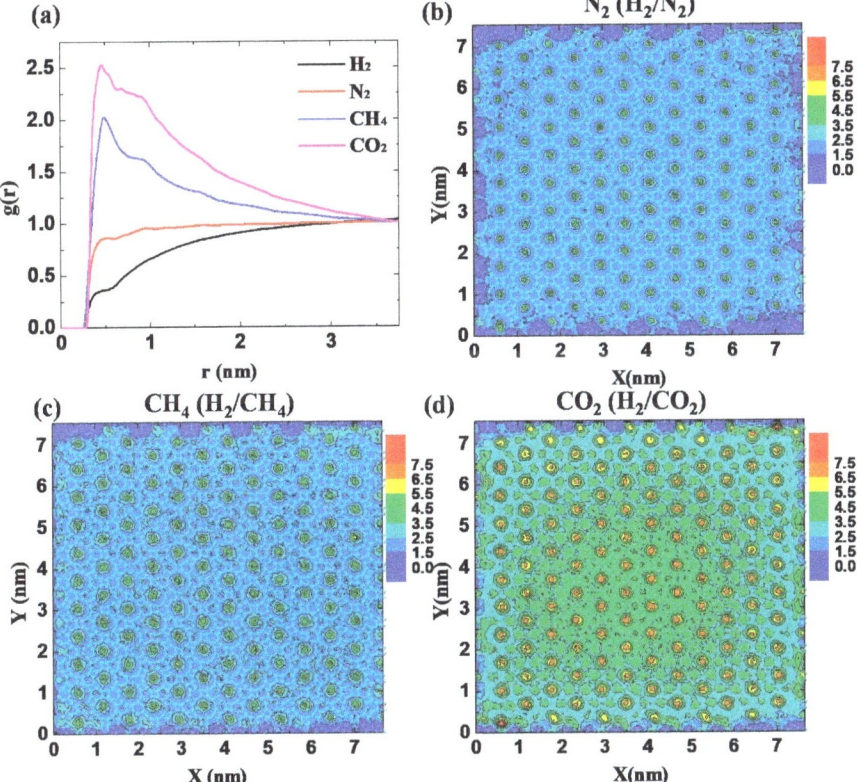

Figure 6. Blocking effect on the surface of GDY_2.1Å. (a) Radial distribution function (RDF) of gases around the membrane. Density contours of the impermeable gases on the surface: (b) N_2 (H_2/N_2); (c) CH_4 (H_2/CH_4); (d) CO_2 (H_2/CO_2). This unit of density (N_w/uc) is 1/(1.25Å^3).

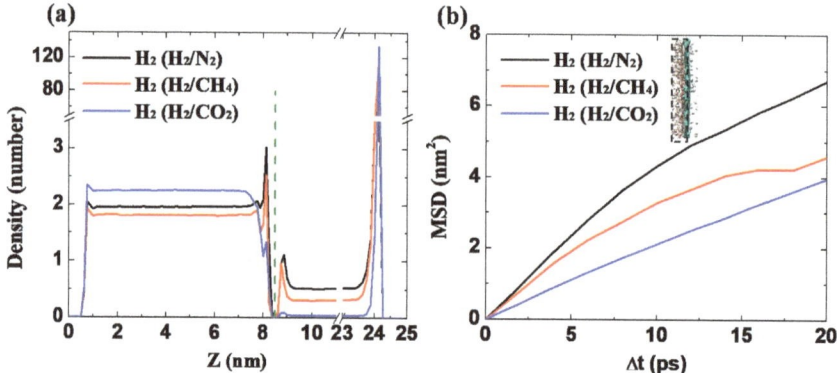

Figure 7. Permeation behavior of H_2 in binary mixtures. (**a**) The number distributions along the z-direction in the last 10 ns. This membrane is located at the green dotted line; (**b**) The mean square displacement (MSD) curves of H_2 crossing the GDY_2.1Å membrane. (Inset) The region within 0.5 nm of the membrane surface for MSD calculation.

According to the first peak in Figure 6a, the mean square displacement (MSD) of H_2 molecules was analyzed within 0.5 nm of the membrane surface (Figure 7b, inset) from

$$MSD(t) = \frac{1}{N}\left\langle \sum_{i=1}^{N}(r_i(t)^2 - r_i(t_0)^2) \right\rangle \quad (6)$$

$$D = \frac{1}{6} \lim_{t \to \infty} \frac{dMSD(t)}{dt} \quad (7)$$

where $r_i(t)^2 - r_i(t_0)^2$ refers to the distance traveled by the atom (*i*) over the time interval of $t - t_0$. This diffusion coefficient (*D*) is determined by the linear slope of the MSD curve with Equation (7). This diffusion behavior in this confined region cannot sustain a normal movement with a large time interval, and it is sufficient to calculate the diffusion coefficient within 20 ps. This greater placeholders of H_2, ascribing to the lower blocking effect of N_2, promote the faster movement of H_2, as presented in Figure 7b. H_2 in three binary mixtures of H_2/N_2, H_2/CH_4, and H_2/CO_2 exhibits the diffusion coefficients of 0.57×10^{-2}, 0.44×10^{-2}, and 0.33×10^{-2} cm^2/s, respectively. That is, the H_2 permeation is indeed impeded by the preferentially adsorbed CO_2 molecules. In other words, the GDY_2.1Å can separate H_2 faster and more selectively from such a kind of binary mixture that the other species has a weak interaction toward GDY surfaces, such as H_2/N_2.

3.3. Surface Charge Effect

Two important mechanisms dominated the gas separation. This first one is the size sieving effect, which favors the permeation of small-sized molecules of H_2. While beyond the size sieving effect, in the separation process of binary mixtures, the blocking effect greatly impedes H_2 permeation. This above understanding of the separation mechanism is beneficial to further guide the improvement of H_2 permeance particularly for the mixture of H_2/CO_2 without sacrificing its selectivity. An effective way proposed here is to reduce the most severe blocking effect of CO_2 by surface charge modification. As depicted in the inset figure in Figure 8a, the positive and negative charges are uniformly imposed on the network of GDY_2.1Å from ±0.00001 to ±0.35 e/atom, whereas the net charge of the whole membrane is zero. This CO_2 molecules still cannot cross the membrane regardless of surface charges. Compared to the pristine GDY membrane, the increased H_2 permeance of the binary mixture of H_2/CO_2 was observed when surface charges are lower than ±0.1 e/atom in Figure 8b. It can be up to 2.84×10^5 GPU by imposing a tiny amount of surface charge of ±0.035–0.050 e/atom. This exceptionally high permeance is several orders of magnitude greater than the existed experiments [40–42], which is ascribed to the ultimate pore size and atomic thickness of GDYs. This following decreasing trend in Figure 8b was ascribed to the surface overcharge, where the generated strong electrostatic repels not only CO_2 but also H_2 from approaching. Thus, achieving the perfect balance between these two mechanisms, a tiny amount of surface charges is qualified to maximize the H_2 permeance by reducing the blocking effect of strong interlaced molecules of CO_2, meanwhile ensuring the placeholders of small-sized molecules of H_2 as well.

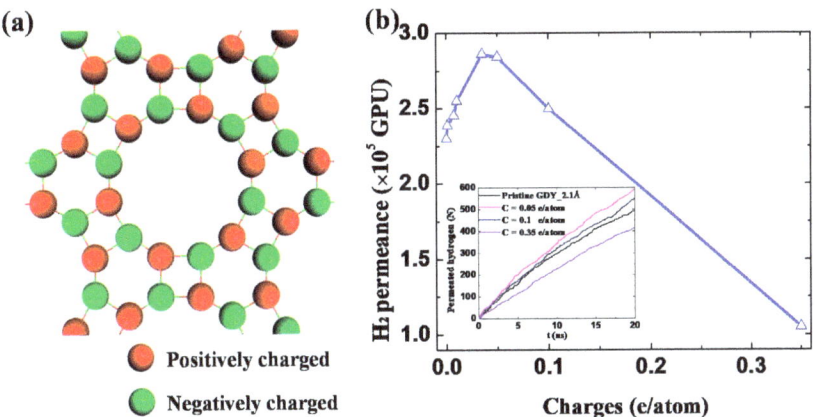

Figure 8. The effect of the surface charges on gas permeation. (**a**) The distribution of charges on the network of GDY_2.1Å; (**b**) H_2 permeance of the binary mixture of H_2/CO_2 as a function of surface charges. (Inset) The time evolution of permeated H_2 molecules through the charged GDYs.

4. Conclusions

In summary, a multiscale study combining MD and DFT calculations was performed to investigate the gas permeation through carefully designed flexible GDY membranes with different pore structures and surface charges. Four single gases and three equimolar binary mixtures (H_2/other gas) were simulated to study the selective gas permeation through GDYs. Approximately infinite selectivities of H_2 over N_2, CO_2, and CH_4 in GDY_2.1Å membranes were demonstrated by DFT calculations. This underlying mechanism indicated that the blocking effect impeded H_2 permeation and the GDY_2.1Å is prone to separate H_2 from such binary mixtures in which the other species has a weak gas–membrane interaction. Moreover, by imposing a tiny amount of surface charges, the H_2 permeance of the binary H_2/CO_2 mixture was further enhanced up to 2.84×10^5 GPU without sacrificing the selectivity. These excellent transport properties make the GDYs a promising candidate for efficient H_2 purification. Although the present work is a theoretical study, it is believable that the GDYs can be elaborately designed for realistic separations with the improvement of a well-controlled synthesis strategy.

Supplementary Materials: The following are available online at http://www.mdpi.com/2077-0375/10/10/286/s1, Video S1: The separation process of the binary mixture of H_2/CO_2 through GDY_2.1Å.

Author Contributions: Conceptualization, Q.L. and G.L.; Data curation, Q.L.; Formal analysis, Q.L.; Funding acquisition, G.L.; Investigation, Q.L.; Methodology, Q.L.; Resources, G.L.; Writing—original draft, Q.L.; Writing—review & editing, L.C. and G.L. All authors have read and agreed to the published version of the manuscript.

Funding: This work was financially supported by the National Natural Science Foundation of China (grant numbers 21922805, 21776125). We are grateful to the High-Performance Computing Center of Nanjing Tech University for supporting the computational resources.

Conflicts of Interest: The authors declare no conflict of interest.

References

1. Corredor, J.; Perez-Pena, E.; Rivero, M.J.; Ortiz, I. Performance of rGO/TiO$_2$ photocatalytic membranes for hydrogen production. *Membranes* **2020**, *10*, 218. [CrossRef] [PubMed]
2. Staffell, I.; Scamman, D.; Velazquez Abad, A.; Balcombe, P.; Dodds, P.E.; Ekins, P.; Shah, N.; Ward, K.R. This role of hydrogen and fuel cells in the global energy system. *Energy Environ. Sci.* **2019**, *12*, 463–491. [CrossRef]
3. Parvasi, P.; Mohammad Jokar, S.; Basile, A.; Iulianelli, A. An on-board pure H$_2$ supply system based on a membrane reactor for a fuel cell vehicle: A theoretical study. *Membranes* **2020**, *10*, 159. [CrossRef]
4. Zhang, Y.; Xu, P.; Liang, S.; Liu, B.; Shuai, Y.; Li, B. Exergy analysis of hydrogen production from steam gasification of biomass: A review. *Int. J. Hydrog. Energy* **2019**, *44*, 14290–14302. [CrossRef]
5. Franchi, G.; Capocelli, M.; De Falco, M.; Piemonte, V.; Barba, D. Hydrogen production via steam reforming: A critical analysis of MR and RMM technologies. *Membranes* **2020**, *10*, 10. [CrossRef]
6. Zhang, Y.; Lee, W.H.; Seong, J.G.; Bae, J.Y.; Zhuang, Y.; Feng, S.; Wan, Y.; Lee, Y.M. Alicyclic segments upgrade hydrogen separation performance of intrinsically microporous polyimide membranes. *J. Membr. Sci.* **2020**, *611*, 118363. [CrossRef]
7. Ockwig, N.W.; Nenoff, T.M. Membranes for hydrogen separation. *Chem. Rev.* **2007**, *107*, 4078–4110. [CrossRef]
8. Casadei, R.; Giacinti Baschetti, M.; Yoo, M.J.; Park, H.B.; Giorgini, L. Pebax@2533/graphene oxide nanocomposite membranes for carbon capture. *Membranes* **2020**, *10*, 188. [CrossRef]
9. Castel, C.; Favre, E. Membrane separations and energy efficiency. *J. Membr. Sci.* **2018**, *548*, 345–357. [CrossRef]
10. Richard, W. Baker. Overview of membrane science and technology. In *Membrane Technology and Applications*; John Wiley & Sons, Ltd.: Chichester, UK, 2012; pp. 1–14.
11. Wang, S.; Yang, L.; He, G.; Shi, B.; Li, Y.; Wu, H.; Zhang, R.; Nunes, S.; Jiang, Z. Two-dimensional nanochannel membranes for molecular and ionic separations. *Chem. Soc. Rev.* **2020**, *49*, 1071–1089. [CrossRef]
12. Liu, G.; Jin, W.; Xu, N. Graphene-based membranes. *Chem. Soc. Rev.* **2015**, *44*, 5016–5030. [CrossRef] [PubMed]
13. Liu, G.; Jin, W.; Xu, N. Two-Dimensional-Material Membranes: A new family of high-performance separation membranes. *Angew. Chem.* **2016**, *55*, 13384–13397. [CrossRef] [PubMed]
14. Liu, Q.; Gupta, K.M.; Xu, Q.; Liu, G.; Jin, W. Gas permeation through double-layer graphene oxide membranes: The role of interlayer distance and pore offset. *Sep. Purif. Technol.* **2019**, *209*, 419–425. [CrossRef]
15. Liu, Q.; Liu, Y.; Liu, G. Simulation of cations separation through charged porous graphene membrane. *Chem. Phys. Lett.* **2020**, *753*, 137606. [CrossRef]
16. James, A.; John, C.; Owais, C.; Myakala, S.N.; Chandra Shekar, S.; Choudhuri, J.R.; Swathi, R.S. Graphynes: Indispensable nanoporous architectures in carbon flatland. *RSC Adv.* **2018**, *8*, 22998–23018. [CrossRef]
17. Gao, X.; Liu, H.; Wang, D.; Zhang, J. Graphdiyne: Synthesis, properties, and applications. *Chem. Soc. Rev.* **2019**, *48*, 908–936. [CrossRef]
18. Rezaee, P.; Naeij, H.R. Graphenylene–1 membrane: An excellent candidate for hydrogen purification and helium separation. *Carbon* **2020**, *157*, 779–787. [CrossRef]
19. Qiu, H.; Xue, M.; Shen, C.; Zhang, Z.; Guo, W. Graphynes for water desalination and gas separation. *Adv. Mater* **2019**, *31*, e1803772. [CrossRef]
20. Mahdizadeh, S.J.; Goharshadi, E.K. Multicomponent gas separation and purification using advanced 2D carbonaceous nanomaterials. *RSC Adv.* **2020**, *10*, 24255–24264. [CrossRef]
21. Jiao, Y.; Du, A.; Hankel, M.; Zhu, Z.; Rudolph, V.; Smith, S.C. Graphdiyne: A versatile nanomaterial for electronics and hydrogen purification. *Chem. Commun.* **2011**, *47*, 11843–11845. [CrossRef]
22. Meng, Z.; Zhang, X.; Zhang, Y.; Gao, H.; Wang, Y.; Shi, Q.; Rao, D.; Liu, Y.; Deng, K.; Lu, R. Graphdiyne as a high-efficiency membrane for separating oxygen from harmful gases: A first-principles study. *ACS Appl. Mater. Interfaces* **2016**, *8*, 28166–28170. [CrossRef] [PubMed]
23. Zhao, L.; Sang, P.; Guo, S.; Liu, X.; Li, J.; Zhu, H.; Guo, W. Promising monolayer membranes for $CO_2/N_2/CH_4$ separation: Graphdiynes modified respectively with hydrogen, fluorine, and oxygen atoms. *Appl. Surf. Sci.* **2017**, *405*, 455–464. [CrossRef]

24. Apriliyanto, Y.B.; Lago, N.F.; Lombardi, A.; Evangelisti, S.; Bartolomei, M.; Leininger, T.; Pirani, F. Nanostructure selectivity for molecular adsorption and separation: The case of graphyne layers. *J. Phys. Chem. C* **2018**, *122*, 16195–16208. [CrossRef]
25. Sang, P.; Zhao, L.; Xu, J.; Shi, Z.; Guo, S.; Yu, Y.; Zhu, H.; Yan, Z.; Guo, W. Excellent membranes for hydrogen purification: Dumbbell-shaped porous γ-graphynes. *Int. J. Hydrog. Energy* **2017**, *42*, 5168–5176. [CrossRef]
26. Jiao, Y.; Du, A.; Smith, S.C.; Zhu, Z.; Qiao, S.Z. H_2 purification by functionalized graphdiyne–role of nitrogen doping. *J. Mater. Chem. A* **2015**, *3*, 6767–6771. [CrossRef]
27. Lei, G.; Liu, C.; Li, Q.; Xu, X. Graphyne nanostructure as a potential adsorbent for separation of H_2S/CH_4 mixture: Combining grand canonical Monte Carlo simulations with ideal adsorbed solution theory. *Fuel* **2016**, *182*, 210–219. [CrossRef]
28. Bartolomei, M.; Giorgi, G. A nnovel nanoporous graphite based on graphynes: First-principles structure and carbon dioxide preferential physisorption. *ACS Appl. Mater. Interfaces* **2016**, *8*, 27996–28003. [CrossRef]
29. Zhang, H.; He, X.; Zhao, M.; Zhang, M.; Zhao, L.; Feng, X.; Luo, Y. Tunable hydrogen separation in sp-sp^2 hybridized carbon membranes: A first-principles prediction. *J. Phys. Chem. C* **2012**, *116*, 16634–16638. [CrossRef]
30. Bartolomei, M.; Carmona-Novillo, E.; Hernández, M.I.; Campos-Martínez, J.; Pirani, F.; Giorgi, G. Graphdiyne pores: "Ad Hoc" openings for helium separation applications. *J. Phys. Chem. C* **2014**, *118*, 29966–29972. [CrossRef]
31. Delley, B. An all-electron numerical method for solving the local density functional for polyatomic molecules. *J. Chem. Phys.* **1990**, *92*, 508–517. [CrossRef]
32. Lee, S.-M.; Kim, J.-H.; Ahn, J.-H. Graphene as a flexible electronic material: Mechanical limitations by defect formation and efforts to overcome. *Mater. Today* **2015**, *18*, 336–344. [CrossRef]
33. Zhou, K.; Xu, Z. Renormalization of ionic solvation shells in nanochannels. *ACS Appl. Mater. Interfaces* **2018**, *10*, 27801–27809. [CrossRef]
34. Wennberg, C.L.; Murtola, T.; Páll, S.; Abraham, M.J.; Hess, B.; Lindahl, E. Direct-space corrections enable fast and accurate Lorentz–Berthelot combination rule Lennard-Jones lattice summation. *J. Chem. Theory Comput.* **2015**, *11*, 5737–5746. [CrossRef] [PubMed]
35. Hess, B.; Kutzner, C.; van der Spoel, D.; Lindahl, E. GROMACS 4: Algorithms for highly efficient, load-balanced, and scalable molecular simulation. *J. Chem. Theory Comput.* **2008**, *4*, 435–447. [CrossRef] [PubMed]
36. Bussi, G.; Donadio, D.; Parrinello, M. Canonical sampling through velocity rescaling. *J. Chem. Phys.* **2007**, *126*, 014101. [CrossRef]
37. Essmann, U.; Perera, L.; Berkowitz, M.L.; Darden, T.; Lee, H.; Pedersen, L.G. A smooth particle mesh Ewald method. *J. Chem. Phys.* **1995**, *103*, 8577–8593. [CrossRef]
38. Xu, J.; Zhou, S.; Sang, P.; Li, J.; Zhao, L. Inorganic graphenylene as a promising novel boron nitrogen membrane for hydrogen purification: A computational study. *J. Mater. Sci.* **2017**, *52*, 10285–10293. [CrossRef]
39. Van Reis, R.; Zydney, A. Bioprocess membrane technology. *J. Membr. Sci.* **2007**, *297*, 16–50. [CrossRef]
40. Omidvar, M.; Nguyen, H.; Liang, H.; Doherty, C.M.; Hill, A.J.; Stafford, C.M.; Feng, X.; Swihart, M.T.; Lin, H. Unexpectedly strong size-sieving ability in carbonized polybenzimidazole for membrane H_2/CO_2 separation. *ACS Appl. Mater. Interfaces* **2019**, *11*, 47365–47372. [CrossRef]
41. Sun, Y.; Song, C.; Guo, X.; Liu, Y. Concurrent Manipulation of out-of-plane and regional in-plane orientations of NH_2-UiO-66 membranes with significantly reduced anisotropic grain boundary and superior H_2/CO_2 separation performance. *ACS Appl. Mater. Interfaces* **2020**, *12*, 4494–4500. [CrossRef]
42. Robeson, L.M. This upper bound revisited. *J. Membr. Sci.* **2008**, *320*, 390–400. [CrossRef]

Publisher's Note: MDPI stays neutral with regard to jurisdictional claims in published maps and institutional affiliations.

© 2020 by the authors. Licensee MDPI, Basel, Switzerland. This article is an open access article distributed under the terms and conditions of the Creative Commons Attribution (CC BY) license (http://creativecommons.org/licenses/by/4.0/).

Article

Hyper Cross-Linked Polymers as Additives for Preventing Aging of PIM-1 Membranes

Federico Begni [1], Elsa Lasseuguette [2], Geo Paul [1], Chiara Bisio [1,3], Leonardo Marchese [1], Giorgio Gatti [1,*] and Maria-Chiara Ferrari [2,*]

[1] Dipartimento di Scienze e Innovazione Tecnologica, Università degli Studi del Piemonte Orientale "Amedeo Avogadro", Viale Teresa Michel 11, 15121 Alessandria, Italy; federico.begni@uniupo.it (F.B.); geo.paul@uniupo.it (G.P.); chiara.bisio@uniupo.it (C.B.); leonardo.marchese@uniupo.it (L.M.)

[2] School of Engineering, University of Edinburgh, Robert Stevenson Road, Edinburgh EH9 3FB, UK; e.lasseuguette@ed.ac.uk

[3] CNR-SCITEC Instituto di Scienze e Tecnologie Chimiche "G. Natta", Via C. Golgi 19, 20133 Milano, Italy

* Correspondence: giorgio.gatti@uniupo.it (G.G.); m.ferrari@ed.ac.uk (M.-C.F.)

Citation: Begni, F.; Lasseuguette, E.; Paul, G.; Bisio, C.; Marchese, L.; Gatti, G.; Ferrari, M.-C. Hyper Cross-Linked Polymers as Additives for Preventing Aging of PIM-1 Membranes. *Membranes* 2021, *11*, 463. https://doi.org/10.3390/membranes11070463

Academic Editor: Maria Grazia De Angelis

Received: 21 May 2021
Accepted: 17 June 2021
Published: 23 June 2021

Publisher's Note: MDPI stays neutral with regard to jurisdictional claims in published maps and institutional affiliations.

Copyright: © 2021 by the authors. Licensee MDPI, Basel, Switzerland. This article is an open access article distributed under the terms and conditions of the Creative Commons Attribution (CC BY) license (https://creativecommons.org/licenses/by/4.0/).

Abstract: Mixed-matrix membranes (MMMs) are membranes that are composed of polymers embedded with inorganic particles. By combining the polymers with the inorganic fillers, improvements can be made to the permeability compared to the pure polymer membranes due to new pathways for gas transport. However, the fillers, such as hyper cross-linked polymers (HCP), can also help to reduce the physical aging of the MMMs composed of a glassy polymer matrix. Here we report the synthesis of two novel HCP fillers, based on the Friedel–Crafts reaction between a tetraphenyl methane monomer and a bromomethyl benzene monomer. According to the temperature and the solvent used during the reaction (dichloromethane (DCM) or dichloroethane (DCE)), two different particle sizes have been obtained, 498 nm with DCM and 120 nm with DCE. The change in the reaction process also induces a change in the surface area and pore volumes. Several MMMs have been developed with PIM-1 as matrix and HCPs as fillers at 3% and 10wt % loading. Their permeation performances have been studied over the course of two years in order to explore physical aging effects over time. Without filler, PIM-1 exhibits the classical aging behavior of polymers of intrinsic microporosity, namely, a progressive decline in gas permeation, up to 90% for CO_2 permeability. On the contrary, with HCPs, the physical aging at longer terms in PIM-1 is moderated with a decrease of 60% for CO_2 permeability. ^{13}C spin-lattice relaxation times (T1) indicates that this slowdown is related to the interactions between HCPs and PIM-1.

Keywords: mixed-matrix membranes; physical aging; hyper cross-linked polymer; gas permeation; ^{13}C spin-lattice relaxation times; SS-NMR spectroscopy

1. Introduction

Membrane-based materials have played an important role in the field of gas separation [1]. The relatively low energy requirement of membrane processes [2,3] makes them an attractive alternative to more intensive separation processes, such as pressure swing adsorption, chemical absorption and cryogenic [4–7]. A promising area of development for gas separation is the use of hybrid membranes or mixed-matrix membranes (MMMs). MMMs are composite membranes made by combining a filler (dispersed phase) and a polymer matrix (the continuous phase). By using two materials with different transport properties, these membranes have the potential to synergistically combine the easy processability of polymers and the superior gas-separation performance of filler materials, and therefore provide separation properties surpassing the Robeson upper bound [8]. In particular, enhancements have been noticed with nanostructured and highly porous additives, such as metal–organic frameworks, zeolites, carbon nanotubes or hyper crosslinked polymers (HCP) [9–15]. The addition of porous fillers in the continuous phase plays an important role in the transport properties of MMMs, providing new pathways for gas transport, hence increasing gas permeability [16–21]. Furthermore, the incorporation of fillers might be a solution to tackle the physical aging of glassy polymers [22–27].

HCPs are a class of high surface area amorphous polymers, which can be obtained either from the post-crosslinking of polystyrene-type precursors in their swollen state, or from the cross-linking of small building blocks, following Friedel–Crafts chemistry [28,29]. The resulting polymers consist of aromatic rings joined together through aliphatic bridges. The formation of a highly interconnected network results in a stable nanoporous structure. HCPs present several advantages, they are stable, tunable, inexpensive, scalable [30] and present high surface areas (1200–2000 $m^2 g^{-1}$) [29], which allow high gas uptakes [31]. For these reasons HCPs are attractive porous solids to be used as fillers within mixed-matrix membranes in order to improve separation performances. Numerous studies show their ability to provide high permeability membranes which can surpass the classical Robeson upper bound. Hou et al. [32] developed MMMs with Poly(1-trimethylsilyl-1-propyne) (PTMSP) matrix and 10 wt % HCP as fillers, based on α,α'-dichloro-p-xylene. Considerable selectivity improvement was achieved thanks to the efficient nanoparticle dispersion and sufficient interaction between the matrix and the filler. The enhancement of gas permeability of MMMs was attributed to the pore channels added by the highly crosslinked fillers which provided additional pathways for gas transport. Lau et al. [33] studied the physical aging of these MMMs. Over time, the CO_2/N_2 separation performance increased whilst CO_2 permeability remained stable. This selective-aging effect is explained by the retention of fractional free volume (FFV) content by HCP particles. The ^{13}C solid-state NMR spectroscopy showed that the bulky part of the PTMSP chains was immobilized by the HCP. Mitra et al. [34] developed MMMs composed of PIM1 and HCP based on poly(vinyl benzyl chloride). The incorporation of fillers induced an increase of gas permeability due to the higher Brunauer–Emmett–Teller (BET) surface area of HCP (1700 $m^2 g^{-1}$) compared to PIM1 (750 $m^2 g^{-1}$). On the contrary, the CO_2/N_2 selectivity decreased with the presence of HCP, especially with high amounts of filler (>15 wt %). Moreover, the presence of HCP retarded the physical aging of the MMM with a lower decrease in CO_2 permeability as compared to PIM-1 alone.

In this paper, we report on the synthesis of new HCPs based on the Friedel–Crafts reaction between a tetraphenyl methane monomer and a bromomethyl benzene monomer. Depending on the solvent used during the reaction, i.e., dichloroethane (DCE) or dichloromethane (DCM), HCPs with different particle sizes were obtained. To the best of our knowledge, the effect of the synthetic conditions on the HCP particle sizes has not been covered in the scientific literature. The synthesized materials have been characterized in terms of their structure and properties. The effect on the membrane gas separation performance over time was investigated when HCPs are used as additives in PIM-1 membranes.

2. Materials and Methods

2.1. Materials Preparation

The synthesis of PIM-1 was performed following the procedure reported in literature [35], by mixing anhydrous K_2CO_3 (11.05 g, 80 mmol), 5,5′,6,6′-tetrahydroxy- 3,3,3′,3′-tetramethyl-1,1′-spirobisindane (3.4 g, 10 mmol), and 2,3,5,6-tetrafluoroterephthalonitrile (2.0 g, 10 mmol) in anhydrous dimethylformamide (65 mL) at 338 K for 72 h under an N_2 atmosphere. On cooling, the mixture was added to water (500 mL), and the crude product collected by filtration. Repeated precipitations from methanol gave 4.96 g (92% yield) of fluorescent yellow polymer (PIM-1) with a Mw ~ 87,000 g mol^{-1}. 3,3,3′,3′-Tetramethyl-1,1′-spirobiindane-5,5′,6,6′-tetraol and 2,3,5,6- tetrafluorophthalonitrile were purified before use by recrystallization in methanol and ethanol, respectively. Ethanol and methanol were purchased from Fisher Chemicals. Anhydrous dimethylformamide (99.9%) and 5,5′,6,6′-tetrahydroxy- 3,3,3′,3′-tetramethyl-1,1′-spirobisindane (97%) were purchased from Alfa

Aesar. Potassium carbonate (99.5%) and 2,3,5,6-tetrafluoroterephthalonitrile (98%) were purchased from Fluorochem.

HCPs were named ABT01 and ABT02, AB from aluminum bromide and T from tetraphenyl methane, while 01 is used for the material obtained by using dichloromethane in the synthetic procedure and 02 for the material obtained using dichloroethane.

Here the synthesis of ABT01 is reported, however the same procedure was employed for the material named ABT02, except that dichloroethane and a temperature of 80 °C were used instead of dichloromethane at 35 °C.

ABT01 synthetic procedure was carried out in a 250 mL three-necked bottom flask by adding 1 g ($3.12 \cdot 10^{-3}$ mol) of tetraphenyl methane (TPM, Capot Chemical Company (97%)) to 90 mL of dichloromethane (DCM) (Sigma-Aldrich $\geq$ 99,8%) at room temperature. After approximately 10 min the addition of 10.02 g (2.8×10^{-2} mol) of the cross-linker 1,3,5-tris(bromomethyl)benzene (Sigma-Aldrich, >97%) was carried out which after 10 min, which was followed by the addition of 7.46 g (2.8×10^{-2} mol) of the catalyst aluminum (III) bromide (Sigma-Aldrich, $\geq$98%). The reaction mixture was left under stirring for approximately 20 min and then heated under reflux over night at 308 K. A molar ratio of 1:9:9 between the monomer, the cross-linker and the catalyst was used for the synthesis of both materials. After approximately 22 h the resulting mixture appeared as a dark brown gel. The reaction was quenched by addition of diluted HCl with deionized water (v/v 2:1), then the material was washed with ethanol and deionized water. After the washing step, the material was put in an oven at 343 K for 24 h. ABT01 appears as a brown powder.

The reaction scheme associated with the synthesis of ABT materials is reported in Figure 1.

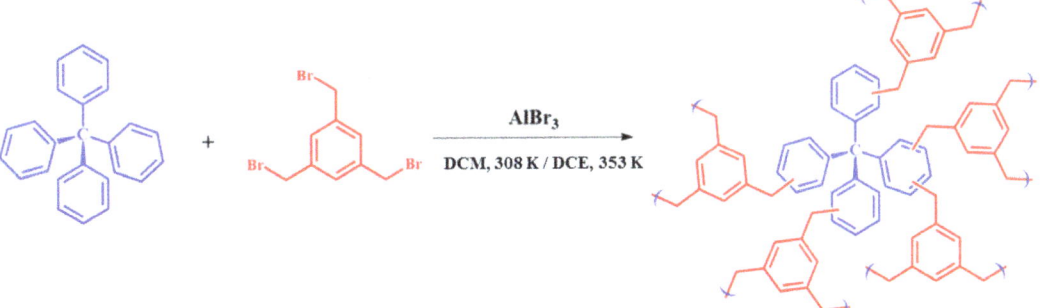

Figure 1. Reaction scheme for the synthesis of ABT hyper crosslinked polymers.

2.2. Membranes Preparation

Solution casting [35] at ambient conditions was used to fabricate dense film membranes with a filler content of 3 wt % and 10 wt % (with respect to PIM-1 weight). For the preparation of the membranes a suspension of filler (6 mg for 3 wt % or 20 mg for 10 wt %) in 5 mL of $CHCl_3$ was sonicated with an ultrasound probe (Fisher Scientific, Model CL18, 120 W) for 1 h by using a water bath to maintain the flask at room temperature. Meanwhile, 200 mg of PIM-1 was dissolved in 5 mL of $CHCl_3$. After complete dissolution, the PIM-1 solution was added to the additive suspension with other 5 mL of $CHCl_3$. The mixture was then sonicated again for 2 h at room temperature. The resulting solution was poured into a 5 cm glass Petri dish. The membrane was allowed to form by slow solvent evaporation for 24–36 h under a fume cupboard. Five membranes were obtained, namely, a pure PIM-1 membrane (PIM-1), two MMMs composed of PIM-1 and ABT01 as a filler at 3 and 10 wt % named, respectively, PIM1-ABT01-3% and PIM1-ABT01-10% and two MMMs composed of PIM-1 and ABT02 as a filler at 3 and 10 wt %, namely, PIM1-ABT02-3% and PIM1-ABT02-10%. After the drying steps, the thickness of the membranes was determined with a digital micrometer (Mitutoyo). Before performing permeability measurements, the membranes were immersed in methanol for 2 h, followed by a drying step under a fume cupboard for 1 h and then under vacuum at room temperature overnight. It is well known that the permeability performances of PIM-1 membranes may vary based on the casting conditions (for example, the choice of solvents) and the history of the sample. Alcohol washes away residual casting solvent and provides a comparable starting point for evaluation of different membranes [35].

2.3. Membranes Characterizations

Permeation measurements—The permeation properties of the MMMs were tested using the constant volume-variable pressure method in an in-house built time-lag apparatus of which a schematic can be seen in Figure 2.

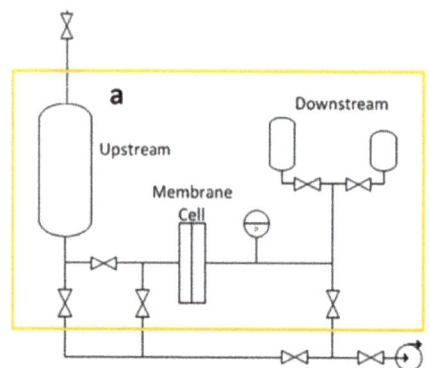

Figure 2. Permeation rig: (**a**) schematic; (**b**) rig with the cell.

The permeation cell consists of three main parts, namely, upstream, permeation cell and downstream. The upstream, or feed side, of the permeation cell consists of a controlling valve, a pressure gauge and a 2000 cm^3 volume gas reservoir. The sample is positioned in the gas permeation cell and sealed with two rubber O-rings. The downstream volume is fixed and a pressure transducer is used to detect pressure changes.

The permeability is obtained from the evolution of pressure of the downstream side (Brooks Transducer 1000 mBar, CMC Series). The permeability coefficient, P, was determined from the slope of the pressure vs. time curve under steady state condition (Equation (1)).

$$P(G) = \frac{l}{A} \frac{V_{down}}{P_{up}RT} \left(\frac{dP_{down}}{dt} \right)_{ss} \tag{1}$$

where l is the membrane thickness, A is the membrane area, V_{down} is the downstream volume, P_{up} is the upstream pressure (P_{up} = 1.1 bar), P_{down} is the downstream pressure, T is the temperature recorded during analysis and R is the gas constant.

Before each experiment, the apparatus is vacuum-degassed and a leak rate is determined from the pressure increase in the downstream part.

The ideal selectivity between two gas species i and j is the ratio of the two single-gas permeabilities (Equation (2)).

$$\alpha_{ij} = \frac{P(i)}{P(j)} \tag{2}$$

For the aging tests, the initial measurement is performed right after the methanol treatment. The membrane is then stored in a sealed plastic bag at ambient temperature and tested over time.

Scanning Electron Microscopy (SEM)—The morphological properties of the membranes were examined with a Quanta 200 FEI (Hillsboro, Oregon) operating at 20 kV and equipped with an EDAX (Mahwah, New Jersey) EDS attachment. Before SEM analysis, the samples were prepared by sputtering with a 20-nm layer of gold to form a conductive surface.

N_2 physisorption analysis—The specific surface area (SSA) was measured by means of nitrogen adsorption at liquid nitrogen temperature (77 K) in the pressure range of 1×10^{-6} Torr to $1 \, P/P_0$ by using an Autosorb-1-MP (Quantachrome Instruments). Prior to adsorption, the samples were outgassed for 3 h at 423 K, (final pressure lower than 10^{-6} Torr). The SSA of the samples was determined by the Brunauer–Emmett–Teller (BET) equation, in a pressure range $0.05–0.15 \, P/P_0$ selected to maximize the correlation coefficient of the fitted linear equation. The pore size distribution was calculated by means of the nonlocal density functional theory (NLDFT) method for slit pores.

FT-IR spectroscopy—Infrared spectra were collected on a Thermo Electron Corporation FT Nicolet 5700 spectrometer (resolution 4 cm^{-1}). Pellets were prepared by mixing the prepared materials with KBr (1:10 weight ratio) and placed into an IR cell with KBr windows permanently connected to a vacuum line (residual pressure: 1×10^{-4} mbar), allowing all treatments to be performed in situ. Samples were degassed for 3 h, using an oil-free apparatus and grease-free vacuum line.

SS-NMR spectroscopy—Solid-state NMR spectra were acquired on a Bruker Avance III 500 spectrometer and a wide bore 11.7 Tesla magnet with operational frequencies for ^1H and ^{13}C of 500.13 and 125.77 MHz, respectively. A 4 mm triple resonance probe, in double resonance mode, with magic angle spinning (MAS) was employed in all the experiments. The as-cast polymer membranes were cut into small pieces so that they could be packed in a 4 mm Zirconia rotor and were spun at a MAS rate of 12 kHz. For the ^{13}C cross-polarization (CP) MAS experiments, the proton radio frequencies (RF) of 55 and 28 kHz were used for initial excitation and decoupling, respectively. During the CP period the ^1H RF field was ramped using 100 increments, whereas the ^{13}C RF field was maintained at a constant level. During the acquisition, the protons were decoupled from the carbons by using a Spinal-64 decoupling scheme. A moderate ramped RF field of 55 kHz was used for spin locking, while the carbon RF field was matched to obtain optimal signal (40 kHz). T1 measurements were performed with a CPXT1 pulse sequence using a 10 ms spin-lock of 55 kHz and 40 kHz for ^1H and ^{13}C, respectively, immediately followed by $\pi/2 - \tau - \pi/2$ sequence on ^{13}C with variable delay (τ) ranging from 0.1 to 45 s. Spectra were recorded with a spectral width of 42 kHz and 256 transients were accumulated at 298 K. A line broadening of 50 Hz and zero filling to 2048 points were used. All chemical shifts are reported using δ scale and are externally referenced to TMS at 0 ppm. Data analyses were performed using Bruker software Dynamics Center, version 2.5.6 and T1 curves were obtained by plotting the intensity of the carbon signals versus time. A single exponential decay was used to fit the data using the following equation (Equation (3)):

$$I_t = I_0 e^{\left(\frac{-t}{T_1}\right)} \tag{3}$$

3. Results
3.1. Characterization of the HCPs
3.1.1. SEM

In Figure 3, SEM images of ABT materials are reported.

ABT01 and ABT02 appear as aggregates of round-shaped particles. Particle dimensions for ABT01 are 498 ± 39 nm while for ABT02 they are 120 ± 23 nm. From the Energy Dispersive X-Ray elemental analysis (EDX) reported in Table S1, it is seen that ABT materials are mainly composed of carbon, which is expected. The presence of aluminum and bromine is to be attributed to unreacted catalyst species, which is present in higher quantities in ABT01. Bromine could also be present within the polymeric framework as partially unreacted crosslinker species as well as side reaction products.

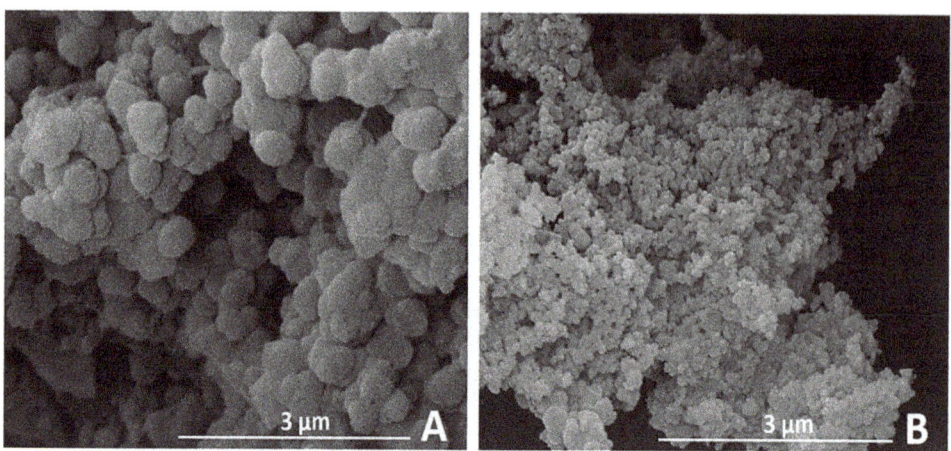

Figure 3. SEM images at 50,000× magnification of ABT01 (**A**) and ABT02 (**B**).

3.1.2. N_2 Physisorption Analysis

The results of the N_2 physisorption analysis are reported in Figure 4.

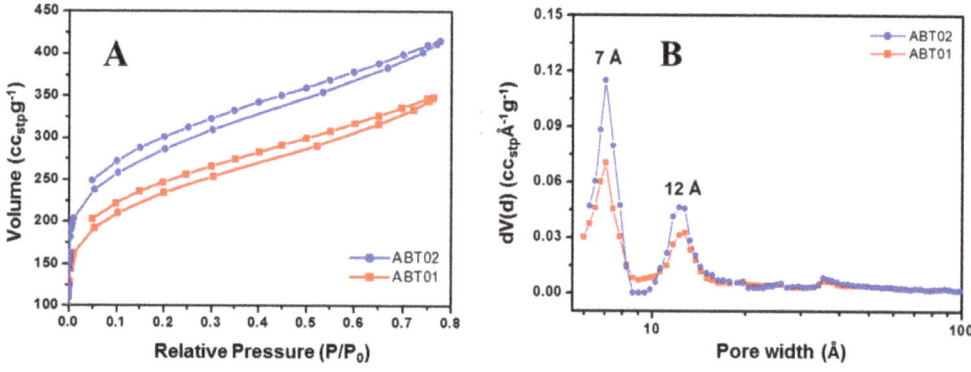

Figure 4. N_2 physisorption isotherms at 77 K (**A**) and pore size distribution (**B**) of ABT01 (■) and ABT02 (■).

As it can be seen in Figure 4, ABT isotherms (with relative upper pressure of only 0.76–0.77) do not show a horizontal plateau after the initial filling of the micropores, therefore a clear classification of the isotherms cannot be made. In particular, a distinction between type I and II is difficult. A fair amount of gas (>200 cm^3/g) is adsorbed at low relative pressures (up to 0.1 P/P$_0$), indicating permanent micro porosity for both materials. Towards higher values of P/P$_0$ a gradual increase in the amount of adsorbed nitrogen is observed indicating the filling of mesopores in the range between 0.45 and 0.8 P/P$_0$. Open hysteresis loops for both materials are observed for the whole desorption branch, which is consistent with swelling effects of the polymeric network, due to gas sorption [36,37]. The non-reversible desorption at low relative pressures indicates that N_2 could either be trapped in pockets or free volume elements with a size comparable to that of the N_2 molecule, or swelling of ABT is locking some of the pockets or free volume elements on the time scale of the experiments, or in combination. From the data reported in Table 1 it can be seen that ABT02 possess higher surface area with respect to ABT01, namely, 990 m^2/g versus 823 m^2/g; in addition, pore volume values associated with both micro- and mesoporosities are slightly higher in ABT02, which is probably a solvent-induced effect since dichloroethane is known to be a particularly suited solvent for the development of high porosity degree in hyper crosslinked polymers [38].

Table 1. Textural properties of ABT materials, assessed via N_2 physisorption analysis performed at 77 K.

Sample	SSA_{BET} (m²/g)	V_{Tot} (cc/g)	V_{micro} (cc/g)			V_{meso} (cc/g)
			Total	<7Å	7 < Å < 20	20 < Å < 100
ABT01	823	0.52	0.28	0.08	0.20	0.24
ABT02	990	0.61	0.35	0.10	0.25	0.26

3.1.3. FT-IR Spectroscopy

IR spectra of ABT materials are reported in Figure 5 with the assignments of the IR absorption bands in Table 2.

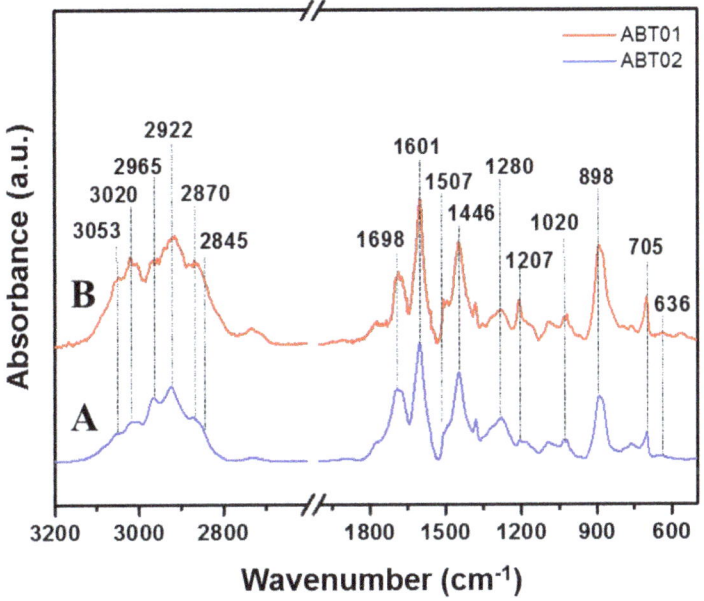

Figure 5. Infrared spectra of ABT01 — (**A**) and ABT02 — (**B**) acquired under vacuum conditions (minimum pressure below 10^{-4} mbar) and beam temperature. Prior to analysis, the samples were treated at 423 K under vacuum (10^{-4} mbar) for three hours.

The IR spectra of ABT materials present signals in the ranges 3200–2800 cm^{-1} and 1750–400 cm^{-1}. In the high wavenumber region, broad bands are found centered at 3053, 3020, 2965, 2922, 2870 and 2845 cm^{-1}. The signals at 3053 and 3020 cm^{-1} are respectively assigned to the asymmetric and symmetric stretching of aromatic C-H groups [39,40]. Between 3000 and 2800 cm^{-1}, the signals of aliphatic C-H stretching vibrations are found. In particular, the band at 2965 cm^{-1} is assigned to the C-H asymmetric stretching mode of methyl groups [37,39,40] while the band at 2922 cm^{-1} is assigned to the asymmetric stretching mode of the -CH_2- group [39,40]. The corresponding symmetric stretching vibrations are found at 2870 cm^{-1} for the methyl group and at 2845 cm^{-1} for the methylene group [37,39,40]. The ratio between the intensity of the aliphatic over aromatic C-H stretching modes bands is higher for the ABT02 with respect to ABT01, which could be an indication of a more crosslinked network. In the low frequency region, a series of signals is found, and assignments have been made for the main absorption bands, namely, the bands at 1698, 1601, 1507, 1446, 1270, 1020, 898 and 705 cm^{-1}. Between 1700 and approximately 1200 cm^{-1}, signals associated with stretching modes of -C=C- bonds are found [39,40]. The band at 1698 cm^{-1} is associated with hindered vibrations of the aromatic rings, probably due to high crosslinking degree of the polymeric framework [41]. The relative intensity of this signal slightly increases for ABT02 with respect to ABT01. This can be linked to higher interconnectivity of the polymeric network of ABT02, probably as a consequence of the higher temperature adopted during the synthetic procedure which accelerated the crosslinking reaction. Additional evidence of the higher crosslinking degree for ABT02 is also the lower intensity of the signal at 1207 cm^{-1} assigned to skeletal C-C bond vibrations [39,40]. The absence of a sharp peak around 1500 cm^{-1} and the presence of the intense signal at 1601 cm^{-1} are indications of possible predominant meta substitution of the aromatic rings, which is expected when AlBr$_3$ is used in the synthetic procedure [39,40,42].

Table 2. Assignments of the main IR absorption bands of ABT materials.

Band Positions (cm^{-1})	Assignments [39,40]
3053	ν_{As} Aromatic C-H
3020	ν_S Aromatic C-H
2965	ν_{As} Aliphatic C-H (-CH$_3$)
2922	ν_{As} Aliphatic C-H (-CH$_2$-)
2870	ν_S Aliphatic C-H (-CH$_3$)
2845	ν_S Aliphatic C-H (-CH$_2$-)
1280	ν skeletal -C-C-
1700–1210	Collective stretching vibrations of poly-substituted benzene rings
900–700	Collective bending vibrations of poly-substituted benzene rings
636	ν aliphatic C-Br (–CH$_2$Br)

In-plane bending modes of aromatic C-H groups are found between 1225 and 950 cm^{-1} [40] while out-of-plane bending modes are found between 1020 and 700 cm^{-1} [40]. The presence of the intense signal centered at 898 cm^{-1} may be interpreted as a sign of 1,3 substitution of the aromatic ring [40].

Between 700 and 600 cm^{-1} the stretching vibrations of the C-Br bond are found. For ABT01, a weak signal at 636 cm^{-1} is found, which indicates that a small amount of bromine is directly linked to the polymeric network, probably as –CH$_2$Br groups, since the aromatic C-Br bond stretching vibrations are found around 680 cm^{-1} [40]. For ABT02, only very weak broad bands are observed in the region of the C-Br stretching modes.

3.1.4. SS-NMR Spectroscopy

In Figure 6, ^{13}C CPMAS NMR spectra of the two samples are reported.

Peaks associated with aromatic carbons are found in the region between 150 and 120 ppm. Around 142 ppm the signal is associated with both the carbon directly attached to the central quaternary carbon and the carbon directly attached to the methylene group of the crosslinker [43].

Aromatic carbons associated with C-H groups are found at 134 and 128 ppm [43]. The central quaternary carbon of the monomer unit is found at 58 ppm while around 35 ppm the carbons associated with the methylene group of the crosslinker are found [43]. In the region around 40–35 ppm the signals associated with -CH$_2$Br can also be found [44]. Carbons associated with methyl groups are found around 17 and 12 ppm [37,43]. It is interesting to observe the presence of these functional groups since they are not present in either the monomer or crosslinker. This finding is also confirmed by the FT-IR analysis (vide supra).

As a possible explanation, the presence of ethyl groups can be a consequence of a small fraction of dichloroethane molecules reacting with the catalyst while methyl groups could be explained via mechanisms involving the carbocationic sites of the crosslinker. This type of functional group was previously found in hyper crosslinked aromatic polymers and derives from secondary reactions that can occur in the presence of dichloroethane solvent [43].

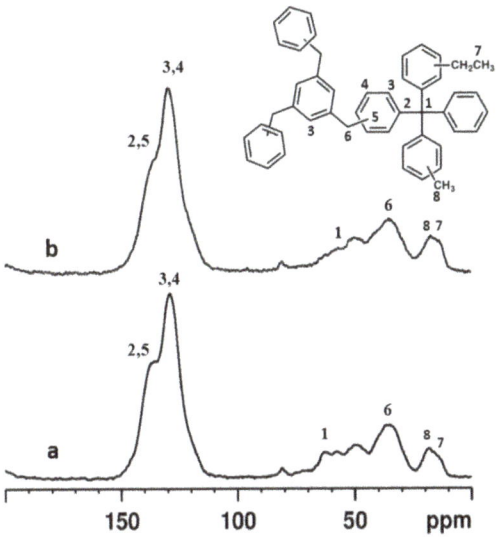

Figure 6. ^{13}C CPMAS NMR spectra of ABT01 (**a**) and ABT02 (**b**).

3.2. Membranes Characterization

Membrane photographs of PIM-1 and PIM-1-ABT0 membranes are presented in Figure 7.

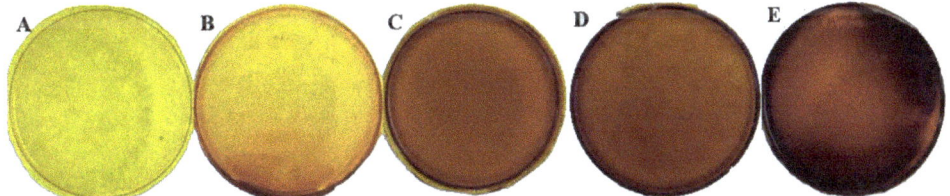

Figure 7. Pristine PIM-1 membrane (**A**), PIM1-ABT01 (3 wt %) (**B**), PIM-ABT02 (3 wt %) (**C**), PIM1-ABT01 (10 wt %) (**D**) and PIM1-ABT02 (10 wt %) (**E**).

As Figure 7 shows, good dispersion of the filler within the membrane has been obtained. No aggregates were visible.

3.2.1. Permeability Measurements

Freshly cast membranes

Table 3 shows the separation performances of PIM-1 and the four MMMs after MeOH treatment, at t_0.

The permeation data of PIM-1 are in the range of those reported in the literature [35]. PIM-1 is highly permeable to CO_2 and presents a reasonable CO_2/N_2 selectivity as well. By adding 3 wt % of the filler ABT02, a decrease of 35.2% in CO_2 permeability with the

CO$_2$/N$_2$ selectivity almost unaffected is observed. Addition of 10 wt % of ABT02 results in a drop of 44% in CO$_2$ permeability while an increase in selectivity from 15 to 18 is also observed. The decline in permeability could have multiple sources and it might be due to densification of the PIM-1 polymeric chain, filling of fractional free volumes by the HCP particles or a partial blockage of the HCP pores by polymer chains [45]. Another possible explanation could be that non-covalent interactions between the fillers and the PIM-1 matrix do not allow for the complete regeneration of the MMMs via methanol treatment. For the MMM sample prepared with the addition of 10 wt % of ABT01, we have a similar separation performance decline of 34% in CO$_2$ permeability, but also a drop of 13% in selectivity. This drop might be explained as for ABT02 with a possible rigidification of the polymeric chain due to the filler, by occupation of fractional free volume within the PIM-1 matrix by addition of the fillers or by a partial blockage of the ABT01 pores by PIM-1 chains. The addition of ABT01 at low content stands out as it induces an increase in CO$_2$ permeability and selectivity. The addition of ABT01 with a 10 wt % loading causes a drop in the selectivity while the addition of ABT02 with the same loading causes the CO$_2$/N$_2$ selectivity to increase. This may be ascribed to particle dimension effects, meaning that the larger particles associated with ABT01 could partially disrupt chain packing in the PIM-1 matrix resulting in the formation of non-selective gas diffusion pathways.

Table 3. Permeation data for PIM-1, PIM-ABT01 (3 and 10% wt), PIM-ABT02 (3% and 10% wt) at t_0 (MeOH treatment). (1 Barrer = 10^{-10} cm^3 (STP)·cm·cm^{-2}·s^{-1}·cmHg^{-1}).

Filler	% wt	Permeability CO$_2$ (Barrer) (±5%)	Selectivity CO$_2$/N$_2$
ABT01	0	13,400	15
	3	14,700	18
	10	8800	13
ABT02	0	13,400	15
	3	8690	14
	10	7500	18

Aging behavior—Permeability measurements were conducted over the course of approximately two years. Measurements were performed after leaving the sample under vacuum over-night at room temperature.

Figure 8 summarizes the aging characteristics of the pristine PIM-1 and MMMs between t_0 and t_f. The mixed matrix containing 3% of ABT01 was only measured up to 250 days.

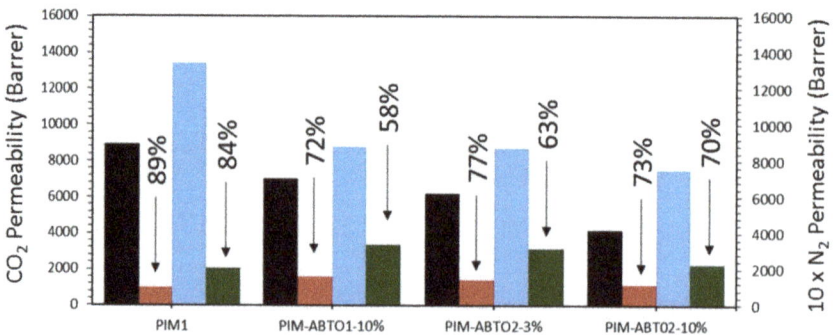

Figure 8. Permeability values and loss of permeability in % of mixed-matrix membranes (MMMs). (■) N$_2$ permeability on t_0; (■) N$_2$ permeability on t_f; (■) CO$_2$ permeability on t_0; (■) CO$_2$ permeability on t_f (850 days).

In Figure 9 the initial drop in the normalized permeation data is reported as a function of time over approximatively the first year of aging.

All the samples present a drop in permeability with time. Aging effects on all samples are clearly visible in the first few days after methanol treatment regardless of the size of the particles (i.e., ABT01 or ABT02) or the concentration of the filler (i.e., 3 or 10 wt %). PIM-1 displays the classical aging behavior exhibited by polymers of intrinsic microporosity [46,47], namely, a progressive drop in the permeation capacity associated with both CO_2 and N_2 caused by the collapse of free volume [46]. After approximately one year of aging, CO_2 permeability of pure PIM-1 decreases by 85% with respect to t_0 value, while for N_2 it reduces by 90% of the initial permeability, hence an increase in selectivity over time towards CO_2 is observed.

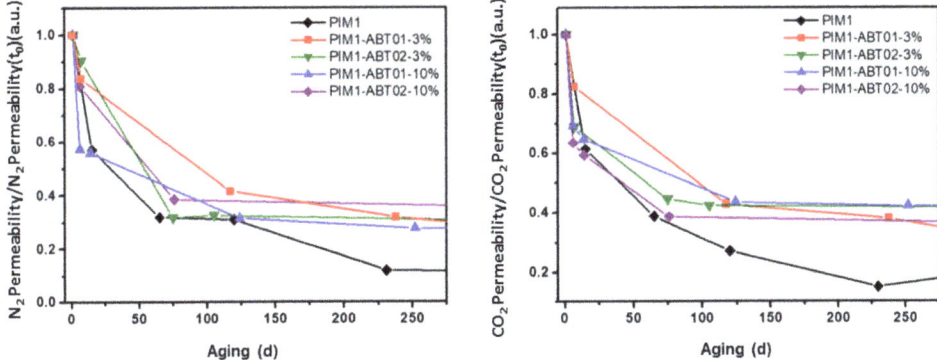

Figure 9. Normalized permeability data with respect to t_0 values of PIM-1 (◆), PIM1-ABT01-3% (■), PIM1-ABT01-10% (▲), PIM1-ABT02-3% (▼), PIM1-ABT02-10% (◆), as function of time over 1 year. Lines are drawn to guide eye.

As shown in Figure 9, the nanocomposite membranes show a better resistance to the aging than the pristine membrane at 200 days where the permeability drop appears to have stopped. The addition of the ABT compounds leads to an arrest in polymer aging and permeability loss. The particles incorporated within the polymer prevent the collapse of the free volume [22,25,27,33,34]. The HCPs possess a rigid network due to the high level of crosslinks which prevents the collapse of the structure. The PIM1-ABT01-3% membrane shows a different behavior compared to the other samples, with a slower rate of aging more similar to PIM-1 with aging progressing even when the other samples have stabilized. In Table 4 the final permeability at ~850 days is reported for the mixed matrices that had stopped aging, confirming that the insertion of the HCPs has prevented further aging compared to the one-year mark and with no significant difference between the filler produced with a different solvent (i.e., of different particle sizes) or the concentration of the filler. Only 3% of the filler ABT02 is sufficient to inhibit aging compared to the pristine PIM-1.

Table 4. CO_2 and N_2 permeability coefficients and selectivity data of PIM-1, PIM1-ABT01-3%, PIM1-ABT01-10%, PIM1-ABT02-3%, PIM1-ABT02-10% at t_0 and t_f.

Membrane	P_{CO_2} (Barrer)		P_{N_2} (Barrer)		Selectivity CO_2/N_2	
	t_0	t_f	t_0	t_f	t_0	t_f
PIM-1	13,400	2040	890	100	15	21
PIM1-ABT01-10%	8800	3390	700	160	13	21
PIM1-ABT02-3%	8690	3170	620	140	14	22
PIM1-ABT02-10%	7500	2270	420	110	18	20

3.2.2. SS-NMR ^{13}C T1 Measurements

The typical ^{13}C CPMAS NMR spectra for PIM-1 and PIM1-ABT-based as-cast membranes are shown in Figure 10. The only visible difference between these spectra is that in the MMMs we see an additional broad peak due to aromatic carbons (highlighted by red arrow) belonging to the ABT fraction. The ^{13}C resonances originating from the PIM-1 backbone have enough peak resolution and intensities such that the ^{13}C spin-lattice relaxation time (T_1) measurements have been carried out. The influence of additives on the relaxation behavior of individual carbons in PIM-1 can reveal the molecular level dynamics in the polymer membranes.

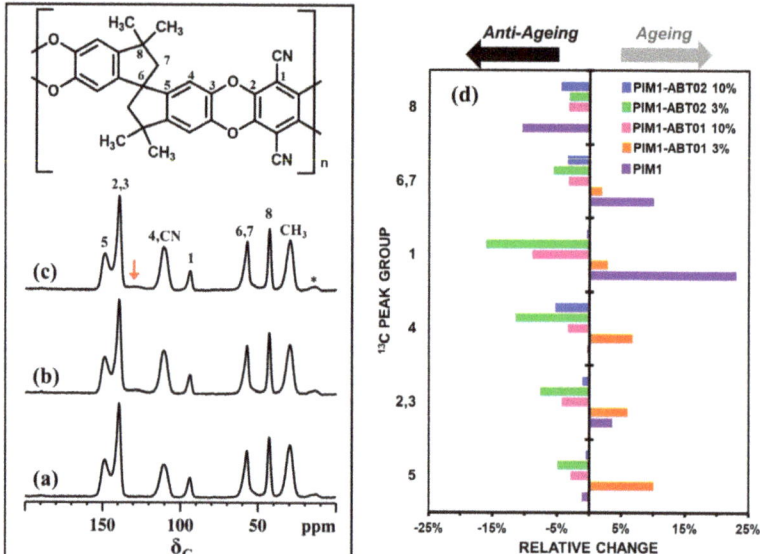

Figure 10. ^{13}C CPMAS NMR spectra of various PIM-1 based membranes: PIM-1 (**a**), PIM1-ABT01 10% (**b**) and PIM1-ABT02 10% (**c**). The spectra were recorded using a magic angle spinning (MAS) rate of 12 kHz and a CP contact time of 10 ms. * Denotes spinning sidebands. Bar chart (**d**) representing percentage changes in T_1 relaxation times of chemical groups assigned in (**c**) between t_{12} (1-year aging) and t_0 (MeOH treatment).

The T_1 values are related to the mobility/rigidity of specific carbon atoms within the polymer and the ^{13}C relaxation studies allow one to estimate the aging in polymer membranes. We have recorded the ^{13}C T_1 values, at ambient temperature, on the freshly cast (t_0) as well as on the one-year aged (t_{12}) membranes. The relative changes (($t_{12} - t_0$)/t_0) in the ^{13}C spin–lattice relaxation time values, which have been evaluated to estimate the aging behavior in PIM1 based membranes, are shown in Figure 10d.

As the aging starts, an increase in the T_1 values for the polymer carbons is expected due to the rigidification of the polymer backbone and the gradual reduction in its excess free volume [22,25]. This is the case in the PIM-1 membrane where the backbone carbons (C1, C2, C3, C6 and C7) have longer T_1 values at t_{12}. Similarly, uniformly higher spin-lattice relaxation times at t_{12} were detected for all carbons in the PIM1-ABT01-3% aged composite membrane. These results show that as the polymer membrane ages, the relaxation times of polymer carbons increases, and the chain mobility decreases due to densification. These results confirm the similar trend in permeability over time for the PIM1-ABT01-3% and pristine PIM-1 membranes.

As far as the other composite membranes (namely, PIM1-ABT01-10%, PIM1-ABT02-3%, and PIM1-ABT-02 10%) are concerned, significant reductions in the T_1 values at t_{12} for all carbons have been noted. In particular, carbons C1–C4 show the significant decrease in ^{13}C T_1 values between t_0 and t_{12}, evidencing the preferable influence of additives on those aromatic carbons. These reductions in the spin-lattice relaxation times on the aged membranes are due to greater molecular mobility of the polymer backbone carbons because of their preferential interactions with ABT components. Previous studies [25,33] have shown that Porous Aromatic Frameworks (PAF) based additives have moderated the aging in PIM-1-based MMMs. Here ABT-based additives should make non-covalent interactions with the PIM-1 backbone, thus preventing the compactification of the polymer chains. On average, weak non–covalent interactions are expected to be at play between the filler and the PIM-1 matrix. Shorter T_1 values could be observed, as is in the case of the present study, if these interactions could result in an enhancement of the mobility of chain backbones, which would also cause a freeing of some fractional free volume, in turn increasing the segmental motion further. As the excess free volume gradually increases, physical aging is diminished in the longer terms in the remaining three ABT-based composite membranes (PIM1-ABT01-10%, PIM1-ABT02-3%, and PIM1-ABT02-10%).

In summary, although the carbon atoms in PIM-1 can be categorized into flexible and bendable units or bulky units or contortion points, their effective packing can lead to diminution of chain motions. On the contrary, when a PIM-1 membrane is cast with the addition of ABT as a filler, new pathways for the gas transport could be generated. In addition, the intimate mixing at the molecular level introduces non-covalent interactions which lead to the retention of fractional free volume in the MMMs. Since the relaxation behavior of aromatic carbons in PIM-1 are mostly influenced, π-π stacking-based interactions can be envisaged as the ABT additive belongs to a hyper-crosslinked aromatic polymer system.

The difference in aging behavior between ABT01 and ABT02 MMMs could be ascribed in part to the difference in textural properties between the two fillers, however the difference in particles size between the two fillers could lead to higher contact surface for ABT02, being the fillers with the smaller particles (see SEM characterization).

The interactions induce anti-aging properties into PIM-1 membranes revealing the chemical compatibility between the ABT polymer and PIM-1. Consequently, physical aging at longer terms in PIM-1 is moderated by the presence of ABT-based additives.

4. Conclusions

Novel hyper cross-linked polymers (HCPs) named ABT01 and ABT02, based on tetraphenyl methane and 1,3,5-tris(bromomethyl)benzene, were successfully synthesized and characterized via FT-IR and ^{13}C-SS-NMR spectroscopy, via scanning electron microscopy (SEM) and via N_2 physisorption analysis at 77 K. By changing the solvent and the temperature of the reaction mixture, control over the particle dimension could be exerted. Dichloroethane and a higher temperature resulted in higher specific surface area and pore volumes.

ABT materials were tested as fillers for the production of PIM-1 based mixed-matrix membranes for CO_2 and N_2 gas separation applications. Four MMMs were obtained by adding ABT01 and ABT02 with a 3 and 10 wt % loading with respect to PIM-1. The addition of the fillers causes a reduction in permeability performances with respect to pure PIM-1 towards both CO_2 and N_2 at t_0, right after methanol treatment. In terms of CO_2/N_2 selectivity, only PIM1-ABT01-3% and PIM1-ABT02-10% showed higher values with respect to pure PIM-1 while the other samples showed a slight decrease in selectivity.

With aging, all the membranes showed a reduction in gas permeability. However, while pure PIM-1 showed a reduction of 85 and 90% for CO_2 and N_2 permeability, respectively, MMMs exhibited a visible slowdown of the aging rate after two months from t_0. After almost three years of aging, MMMs retained approximately 40% of the initial CO_2 permeability and approximately 25% of the N_2 permeability with respect to t_0 values, hence an increase in CO_2/N_2 selectivity was also observed.

^{13}C spin-lattice relaxation times (T_1) allowed monitoring of molecular-level dynamics and degree of flexibility of polymers in membranes. It was found that for the pure PIM-1 membrane and the MMM containing 3% ABT01, T_1 values increased over time with respect to T_1 values measured at t0. Longer T_1 values are a sign of reduction of fractional free volumes within the PIM-1 matrix, due to physical aging. However, for the other MMMs the opposite trend was noticed and shorter T_1 values were observed with respect to those measured for the freshly cast membranes. This indicates that interactions between the PIM-1 polymer matrix and the ABT fillers provide a way to slow down physical aging for PIM-1-based gas separation membranes.

Supplementary Materials: The following are available online at https://www.mdpi.com/article/10.3390/membranes11070463/s1, Table S1: Elemental analysis performed via EDX spectroscopy on ABT01 and ABT02 materials. Five measurements were performed on each sample and the mean values are here reported.

Author Contributions: Conceptualization, M.-C.F., L.M., C.B. and G.G.; formal analysis, F.B., G.P., E.L.; investigation, F.B., G.P. and E.L.; writing—original draft preparation, F.B., G.P. and E.L.; writing—review and editing, M.-C.F., G.G., C.B. and L.M.; supervision, M.-C.F., G.G. and C.B. All authors have read and agreed to the published version of the manuscript.

Funding: This research received no external funding.

Institutional Review Board Statement: Not applicable.

Informed Consent Statement: Not applicable.

Data Availability Statement: The data presented in this study are available on request from the corresponding author. The data are not publicly available due to privacy.

Conflicts of Interest: The authors declare no conflict of interest.

References

1. Credence Research. *Gas Separation Membranes Market by Type, By Application—Growth Future Prospects and Competitive Analysis, 2016–2024*; Credence Research: San Jose, CA, USA, 2017.
2. Abanades, J.; Arias, B.; Lyngfelt, A.; Mattisson, T.; Wiley, D.; Li, H.; Ho, M.; Mangano, E.; Brandani, S. Emerging CO_2 capture systems. *Int. J. Greenh. Gas Control.* **2015**, *40*, 126–166. [CrossRef]
3. Ho, M.T.; Allinson, G.W.; Wiley, D.E. Reducing the Cost of CO_2 Capture from Flue Gases Using Membrane Technology. *Ind. Eng. Chem. Res.* **2008**, *47*, 1562–1568. [CrossRef]
4. Duong, S. White Paper: Cost Effective Alternative to Distillation for Olefin Purification and Extraction. Available online: https://www.imtexmembranes.com/cost-effective-alternative-to-distillation-for-olefin-purification-and-extraction (accessed on 1 January 2019).
5. Puri, P.S. Chapter Commercial Applications of Membranes in Gas Separations. In *Membrane Engineering for the Treatment of Gases: Gas-Separation Problems with Membranes*; Royal Society of Chemistry (RSC): Cambridge, UK, 2011; pp. 215–244.
6. Sanders, D.F.; Smith, Z.P.; Guo, R.; Robeson, L.M.; McGrath, J.E.; Paul, D.R.; Freeman, B.D. Energy-efficient polymeric gas separation membranes for a sustainable future: A review. *Polymer* **2013**, *54*, 4729–4761. [CrossRef]
7. Robinson, S.; Jubin, R.; Choate, B. *Materials for Separation Technology: Energy and Emission Reduction Opportunities*; U.S. Department of Energy: Washington, DC, USA, 2005.
8. Robeson, L.M. The upper bound revisited. *J. Membr. Sci.* **2008**, *320*, 390–400. [CrossRef]
9. Gür, T.M. Permselectivity of zeolite filled polysulfone gas separation membranes. *J. Membr. Sci.* **1994**, *93*, 283–289. [CrossRef]
10. Süer, M.G.; Baç, N.; Yilmaz, L. Gas permeation characteristics of polymer-zeolite mixed matrix membranes. *J. Membr. Sci.* **1994**, *91*, 77–86. [CrossRef]
11. Hussain, M.; König, A. Mixed-Matrix Membrane for Gas Separation: Polydimethylsiloxane Filled with Zeolite. *Chem. Eng. Technol.* **2012**, *35*, 561–569. [CrossRef]
12. Zarshenas, K.; Raisi, A.; Aroujalian, A. Mixed matrix membrane of nano-zeolite NaX/poly (ether-blockamide) for gas separation applications. *J. Membr. Sci.* **2016**, *510*, 270–283. [CrossRef]
13. Dechnik, J.; Gascon, J.; Doonan, C.; Janiak, C.; Sumby, C. Mixed-Matrix Membranes. *Angew. Chem.* **2017**, *56*, 9292–9310. [CrossRef] [PubMed]
14. Nejad, M.; Asghari, M.; Afsari, M. Investigation of Carbon Nanotubes in Mixed Matrix Membranes for Gas Separation: A Review. *ChemBioEng Rev.* **2016**, *3*, 276–298. [CrossRef]
15. Khan, M.M.; Filiz, V.; Bengtson, G.; Shishatskiy, S.; Rahman, M.M.; Lillepaerg, J.; Abetz, V. Enhanced gas permeability by fabricating mixed matrix membranes of functionalized multiwalled carbon nanotubes and polymers of intrinsic microporosity (PIM). *J. Membr. Sci.* **2013**, *436*, 109–120. [CrossRef]

16. Sutrisna, P.D.; Hou, J.; Li, H.; Zhang, Y.; Chen, V. Improved operational stability of Pebax-based gas separation membranes with ZIF-8: A comparative study of flat sheet and composite hollow fibre membranes. *J. Membr. Sci.* **2017**, *524*, 266–279. [CrossRef]
17. Li, T.; Pan, Y.; Peinemann, K.V.; Lai, Z. Carbon dioxide selective mixed matrix composite membrane containing ZIF-7 nanofillers. *J. Membr. Sci.* **2013**, *425*, 235–242. [CrossRef]
18. Bushell, A.F.; Budd, P.M.; Attfield, M.P.; Jones, J.T.; Hasell, T.; Cooper, A.I.; Bernanrdo, P.; Bazzarelli, F.; Clarizia, G.; Jansen, J. Nanoporous organic polymer/cage composite membranes. *Angew. Chem.* **2013**, *52*, 1253–1256. [CrossRef]
19. Şen, D.; Kalıpçılar, H.; Yilmaz, L. Development of polycarbonate based zeolite 4A filled mixed matrix gas separation membranes. *J. Membr. Sci.* **2007**, *303*, 194–199. [CrossRef]
20. Zhao, Z.; Jiang, J. POC/PIM-1 mixed-matrix membranes for water desalination: A molecular simulation study. *J. Membr. Sci.* **2020**, *608*, 118173. [CrossRef]
21. Yu, G.; Li, Y.; Wang, Z.; Liu, T.X.; Zhu, G.; Zou, X. Mixed matrix membranes derived from nanoscale porous organic frameworks for permeable and selective CO_2 separation. *J. Membr. Sci.* **2019**, *591*, 117343. [CrossRef]
22. Begni, F.; Paul, G.; Lasseuguette, E.; Mangano, E.; Bisio, C.; Ferrari, M.C.; Gatti, G. Synthetic Saponite Clays as Additives for Reducing Aging Effects in PIM1 Membranes. *ACS Appl. Polym. Mater.* **2020**, *2*, 3481–3490. [CrossRef]
23. Lau, C.H.; Konstas, K.; Doherty, C.M.; Smith, S.J.; Hou, R.; Wang, H.; Carta, M.; Yoon, H.; Park, J.; Freeman, R.-M.; et al. Tailoring molecular interactions between microporous polymers in high performance mixed matrix membranes for gas separations. *Nanoscale* **2020**, *12*, 17405–17410. [CrossRef] [PubMed]
24. Smith, S.J.D.; Lau, C.H.; Mardel, J.I.; Kitchin, M.; Konstas, K.; Ladewig, B.P.; Hill, M.R. Physical aging in glassy mixed matrix membranes; tuning particle interaction for mechanically robust nanocomposite films. *J. Mater. Chem. A* **2016**, *4*, 10627–10634. [CrossRef]
25. Lau, C.H.; Konstas, K.; Thornton, A.W.; Liu, A.C.; Mudie, S.; Kennedy, D.F.; Howard, S.C.; Hill, A.J.; Hill, M. Gas Separation Mem-branes Loaded with Porous Aromatic Frameworks that Improve with Age. *Angew. Chem.* **2015**, *54*, 2669–2673. [CrossRef]
26. Khdhayyer, M.; Bushell, A.F.; Budd, P.M.; Attfield, M.P.; Jiang, D.; Burrows, A.D.; Esposito, E.; Bernardo, E.; Monteleone, M.; Fuoco, A.; et al. Mixed matrix membranes based on MIL-101 metal−organic frameworks in polymer of intrinsic microporosity PIM-1. *Sep. Purif. Technol.* **2019**, *212*, 545–554. [CrossRef]
27. Alberto, M.; Bhavsar, R.; Luque-Alled, J.M.; Vijayaraghavan, A.; Budd, P.M.; Gorgojo, P. Impeded physical aging in PIM-1 membranes containing graphene-like fillers. *J. Membr. Sci.* **2018**, *563*, 513–520. [CrossRef]
28. Huang, J.; Turner, S.R. Hypercrosslinked Polymers: A Review. *Polym. Rev.* **2017**, *58*, 1–41. [CrossRef]
29. Tsyurupa, M.; Davankov, V. Hypercrosslinked polymers: Basic principle of preparing the new class of polymeric materials. *React. Funct. Polym.* **2002**, *53*, 193–203. [CrossRef]
30. Dawson, R.; Cooper, A.I.; Adams, D.J. Nanoporous organic polymer networks. *Prog. Polym. Sci.* **2012**, *37*, 530–563. [CrossRef]
31. Yang, Y.; Tan, B.; Wood, C.D. Solution-processable hypercrosslinked polymers by low cost strategies: A promising platform for gas storage and separation. *J. Mater. Chem. A* **2016**, *4*, 15072–15080. [CrossRef]
32. Hou, R.; O'Loughlin, R.; Ackroyd, J.; Liu, Q.; Doherty, C.M.; Wang, H.; Hill, M.R.; Smith, S.J.D. Greatly Enhanced Gas Selectivity in Mixed-Matrix Membranes through Size-Controlled Hyper-cross-linked Polymer Additives. *Ind. Eng. Chem. Res.* **2020**, *59*, 13773–13782. [CrossRef]
33. Lau, C.H.; Mulet, X.; Konstas, K.; Doherty, C.M.; Sani, M.-A.; Separovic, F.; Hill, M.R.; Wood, C.D. Hypercrosslinked Additives for Ageless Gas-Separation Membranes. *Angew. Chem. Int. Ed.* **2016**, *55*, 1998–2001. [CrossRef]
34. Mitra, T.; Bhavsar, R.S.; Adams, D.; Budd, P.M.; Cooper, A.I. PIM-1 mixed matrix membranes for gas separations using cost-effective hypercrosslinked nanoparticle fillers. *Chem. Commun.* **2016**, *52*, 5581–5584. [CrossRef] [PubMed]
35. Budd, P.M.; McKeown, N.B.; Ghanem, B.S.; Msayib, K.J.; Fritsch, D.; Starannikova, L.; Belov, N.; Sanfirova, O.; Yampolskii, Y.; Shantarovich, V. Gas permeation parameters and other physicochemical properties of a polymer of intrinsic microporosity: Polybenzodioxane PIM-1. *J. Membr. Sci.* **2008**, *325*, 851–860. [CrossRef]
36. Thommes, M.; Kaneko, K.; Neimark, A.V.; Olivier, J.P.; Rodriguez-Reinoso, F.; Rouquerol, J.; Sing, K.S.W. Physisorption of Gases, With Special Reference to the Evaluation of Surface Area and Pore Size Distribution (IUPAC Technical Report). *Pure Appl. Chem.* **2015**, *87*, 1051. [CrossRef]
37. Paul, G.; Begni, F.; Melicchio, A.; Golemme, G.; Bisio, C.; Marchi, D.; Cossi, M.; Marchese, L.; Gatti, G. Hy-per-Cross-Linked Polymers for the Capture of Aromatic Volatile Compounds. *ACS Appl. Polym. Mater.* **2020**, *2*, 647–658. [CrossRef]
38. Liu, Y.; Fan, X.; Jia, X.; Zhang, B.; Zhang, H.; Zhang, A.; Zhang, Q. Hypercrosslinked polymers: Controlled preparation and effective adsorption of aniline. *J. Mater. Sci.* **2016**, *51*, 8579–8592. [CrossRef]
39. Socrate, G. *Infrared and Raman Characteristic Group Frequencies: Tables and Charts*, 3rd ed.; Wiley & Sons Ltd.: West Sussex, UK, 2004.
40. Coates, J. Interpretation of infrared spectra, a practical approach. In *Encyclopedia of Analytical Chemistry: Applications, Theory and Instrumentation*; Meyers, R.A., Ed.; John Wiley & Sons Ltd.: Hoboken, NJ, USA, 2000.
41. Tsyurupa, M.P.; Blinnikova, Z.K.; Davidovich, Y.A.; Lyubimov, S.E.; Naumkin, A.V.; Davankov, V.A. On the nature of "func-tional groups" in non-functionalized hypercrosslinked polystyrenes. *React. Funct. Polym.* **2012**, *72*, 973–982. [CrossRef]
42. Olah, G.A.; Kobayashi, S.; Tashiro, M. Aromatic Substitution. XXX.1 23Friedel-Crafts Benzylation of Benzene and Toluene with Benzyl and Substituted Benzyl Halides. *J. Am. Chem. Soc.* **1972**, *21*, 94.

43. Errahali, M.; Gatti, G.; Tei, L.; Paul, G.; Rolla, G.A.; Canti, L.; Fraccarollo, A.; Cossi, M.; Comotti, A.; Sozzani, P.; et al. Microporous Hyper Cross-Linked Aromatic Polymers Designed for Methane and Carbon Dioxide Adsorption. *J. Phys. Chem. C* **2014**, *118*, 28699–28710. [CrossRef]
44. Van der Made, A.W.; Van der Made, R.H. A Convenient Procedure for Bromomethylation of Aromatic Compounds. Selective Mono-, Bis-, or Trisbromomethylation. *J. Org. Chem.* **1993**, *58*, 1262–1263. [CrossRef]
45. Bastani, D.; Esmaeili, N.; Asadollahi, M. Polymeric mixed matrix membranes containing zeolites as a filler for gas separation applications: A review. *J. Ind. Eng. Chem.* **2013**, *19*, 375–393. [CrossRef]
46. Swaidan, R.; Ghanem, B.; Pinnau, I. Fine-Tuned Intrinsically Ultramicroporous Polymers Redefine the Permeability/Selectivity Upper Bounds of Membrane-Based Air and Hydrogen Separations. *ACS Macro Lett.* **2015**, *4*, 947–951. [CrossRef]
47. Harms, S.; Rätzke, K.; Faupel, F.; Chaukura, N.; Budd, P.M.; Egger, W.; Ravelli, L. Aging and Free Volume in a Polymer of Intrinsic Microporosity (PIM-1). *J. Adhes.* **2012**, *88*, 608–619. [CrossRef]

Article

Effect of Water and Organic Pollutant in CO₂/CH₄ Separation Using Hydrophilic and Hydrophobic Composite Membranes

Clara Casado-Coterillo *, Aurora Garea and Ángel Irabien

Department of Chemical and Biomolecular Engineering, Universidad de Cantabria, Av. Los Castros s/n, 39005 Santander, Spain; gareaa@unican.es (A.G.); irabienj@unican.es (Á.I.)
* Correspondence: casadoc@unican.es; Tel.: +34-942-20-6777

Received: 21 November 2020; Accepted: 4 December 2020; Published: 8 December 2020

Abstract: Membrane technology is a simple and energy-conservative separation option that is considered to be a green alternative for CO_2 capture processes. However, commercially available membranes still face challenges regarding water and chemical resistance. In this study, the effect of water and organic contaminants in the feed stream on the CO_2/CH_4 separation performance is evaluated as a function of the hydrophilic and permselective features of the top layer of the membrane. The membranes were a commercial hydrophobic membrane with a polydimethylsiloxane (PDMS) top layer (Sulzer Chemtech) and a hydrophilic flat composite membrane with a hydrophilic [emim][ac] ionic liquid–chitosan (IL–CS) thin layer on a commercial polyethersulfone (PES) support developed in our laboratory. Both membranes were immersed in NaOH 1M solutions and washed thoroughly before characterization. The CO_2 permeance was similar for both NaOH-treated membranes in the whole range of feed concentration (up to 250 GPU). The presence of water vapor and organic impurities of the feed gas largely affects the gas permeance through the hydrophobic PDMS membrane, while the behavior of the hydrophilic IL–CS/PES membranes is scarcely affected. The effects of the interaction of the contaminants in the membrane selective layer are being further evaluated.

Keywords: composite membranes; CO_2/CH_4 separation; water and organic pollutants; hydrophilic/hydrophobic character; biogas upgrading; sustainable energy

---

1. Introduction

Membrane gas separation of CO_2 from synthesis gas, natural gas or biogas is a potential energy-efficient alternative to other separation techniques with potential for biogas upgrading as an alternative energy source from anaerobic digestion and landfills [1,2]. The availability of organic biomass could facilitate countries in meeting sustainable development goals related with creating and providing sustainable energy sources [3]. The typical components of biogas are mainly high added value as combustible methane and the major non-combustible component CO_2, together with traces of impurities, such as H_2S, water vapor, aromatics, chlorinated hydrocarbons, volatile organic compounds and siloxanes [4–7].

Separation of CO_2 from methane is also important to use biogas as a natural gas substitute [3]. CO_2/CH_4 separation is particularly challenging due to the variability of the biogas composition, which depends on the source and seasonal conditions [7]. Both feed flow rate and feed concentrations change during the CO_2/CH_4 separation when the feed concentration of the valuable component is at least 65 mol % at a pressure in the range between 3.1 and 4.1 bar [8].

Biogas upgrading, thus, does not require high pressure and temperatures as other CO_2/CH_4 separations, such as natural gas separation. In 1989, a two-stage process involving an adsorption step to remove the impurities and a membrane unit for CO_2 separation was first proposed for biogas's upgrading [9]. Since then, several small landfills in Europe have set up pilot membrane biogas upgrading plants, but they are still not commercially available [3]. The correlation of theoretical and technological studies should emphasize feasibility based on laboratory-scale experiments for further development of biogas plants to large-scale projects, by improving stability and efficiency [10]. In this light, there remains significant scope for the development of better performing CO_2 selective membranes to improve the separation power and the durability of materials [11].

There are different commercial polymeric membranes for CO_2 separation processes made of different materials, basically polymers: Polaris® ES Polaris™ EN LUGAR DE Polaris®? (Membrane Technology Research, Inc. (Newark, CA, USA), MTR) [12], and Dupont/Air Liquid, Air Products & Chemicals Inc. (Allentown, PA, USA) [13], General Electric (Boston, MA, USA) [14], Honeywell (Charlotte, NC, USA) [15] and Evonik (Essen, Germany) [16] are also other major players in membrane technology. Their main advantages are their commercial availability or scale, low cost and easy reproducibility. The main drawbacks are the thermal, mechanical and chemical resistance, as well as the uncertainty of their performance regarding the presence of impurities such as water and organic vapors. A major challenge for developing effective gas separation membranes is overcoming the well-known permeability–selectivity trade-off for light bases in polymeric materials. Since popularized by Robeson in 2008, [17] this upper bound constituted the reference for characterizing advances in highly permselective membrane materials. Recently, deviations of the upper bound have been reported in mixed CO_2/CH_4 gas permeation experiments, for CO_2/CH_4 separation, for some of the mostly studied membranes in biogas upgrading, like cellulose acetate (CA) and polyimides [18], mixed matrix membranes (MMMs) [19] and new polymers with increased intrinsic microporosity [20].

However, the acceptance of membrane technology continues being hindered by the materials' stability under operating conditions. In order to account for the uncertainty of the membrane behavior in the presence of impurities [21], Chenar et al. [22] compared two commercial polymer hollow fiber membranes with different hydrophilic (Cardo polyimide) and hydrophobic (PPO) selective layer in dry and wet conditions. They observed that the CO_2 and CH_4 permeances were differently affected by the hydrophobic or hydrophilic character of the membranes in CO_2/CH_4 separation in the presence of water vapor in the feed.

Simcik et al. observed that the water affinity has a key role for the separation of CO_2/CH_4 as representative biogas binary mixture [23]. Likewise, the permeance and selectivity of glassy polyimide membranes have been affected by the presence of wet non-methane hydrocarbons in natural gas applications. For example, the CO_2 permeance and CO_2/CH_4 selectivity of 6FDA-based dense membranes and CA thin-film composite membranes were observed to decrease in the presence of n-hexane or pentane in the feed gas, due to the plasticization of the polymer [6,24]. On the other hand, the presence of toluene caused a decrease of CO_2 permeance of cross-linkable polyimide membranes [25,26] attributed to competitive sorption of toluene impurity. Cerveira et al. [27] compared two different polymer membranes, cellulose acetate (CA) and polydimethylsiloxane (PDMS), in the separation of CO_2 from CH_4 at 22 °C and 50 psig, observing that CO_2 caused plasticization, resulting in lower separation factors in the mixed-gas mixture separation tests that were accentuated by the presence of impurities other than water vapor in the feed.

Hydrophilicity, as the other membrane properties, like mechanical resistance, pore size and morphology, can be tuned up by modifying membrane material composition [28]. The hydrophobic or hydrophilic character of a membrane may help resolve mass transfer limitations and the removal of impurities in the biogas feed that prevent the integration of membrane technology in biogas production plants [8,29]. Hydrophilicity can influence CO_2 facilitated transport in mixed matrix membranes by modification of "Janus" materials with Ag^+ ions [30]. In a recent work, we observed that the hydrophobicity of PTMSP can be changed by adding hydrophilic Zeolite 4A to polymer matrix, obtaining a new mixed matrix membrane material with increasing CO_2 permeance and increasing or constant selectivity with increasing relative humidity in the feed up to 50% [31].

However, when reducing the thickness of the membranes to prepare thin-film composite membranes, the high selectivity of this MMM was significantly decreased [32] as the probability of defects grew, as usual in MMM synthesis [33]. This was not observed when the chitosan biopolymer (CS) matrix hybridized with 1-ethyl-3-methylimidazolium acetate ([emim][acetate]) ionic liquid (IL) as filler was coated on the polyethersulfone (PES) support, because of the good compatibility between IL and CS [34]. Surface modification of robust supports is a common way of tuning up the membrane separation properties and correcting defects [35].

The aim of this study was to investigate the performance of two types of highly CO_2 high permeable thin-film composite membranes whose selective layer possesses different hydrophobic and hydrophilic character, in the separation of CO_2/CH_4, in the presence and absence of water vapor and organic pollutants. These membranes were previously studied in the separation of CO_2/N_2 gas mixtures [32,36], so the purpose of the present work is to estimate the potential of membranes for biogas upgrading.

2. Materials and Methods

2.1. Materials and Modules

The membranes used for this study are flat-sheet composite Membranes, both PDMS PERVAP 4060 (Sulzer Chemtech GmbH, Alschwill, Switzerland), with a 1–1.5 µm thick PDMS top layer and a total thickness of 180 µm, and the IL–CS/PES composite membrane fabricated in our laboratory, with similar selective layer thickness of about 1.5 µm as the commercial hydrophobic membrane to facilitate comparison [32].

2.2. Gas Permeation Experiments

The flat composite membranes studied in this work, with an effective membrane area of 15.6 cm^2, were placed in a stainless-steel module and tested in the separation of CO_2/CH_4 as a function of feed composition in a homemade separation plant represented in Figure 1, which enables working with different module geometry configuration in dry and humid feed conditions [37]. The feed gas concentration was set by mass flow controllers (KOFLOC 8500, Sequopro S.L., Madrid, Spain), while the retentate and permeate flow rates were measured by a bubble flow meter at the exit of the permeate. Flow control was achieved by means of backpressure regulators on the retentate stream. Permeate was kept at atmospheric pressure, and N_2 was used as sweep carrier gas at a flow rate 10 mL/min.

Once the membrane performance reached the steady state, the permeate was measured by using a bubble flow meter at the end of the system, at least 3 times, for about 1 h, to confirm the membrane stability at a given operating condition. The composition of the permeate stream is determined by a gas analyzer (BIOGAS5000, Geotech, Tamarac, FL, USA).

As reported in our previous work with composite hollow fiber membranes [37], the humid gas experiments were carried out at approximately 50% relative humidity in the feed by passing half of the feed gas stream through a tank filled with water before entering the module. Stop-valves prevented the entrance of liquid water into the membrane modules. Once the performance of the membranes has been characterized in dry and humid conditions, 2 mL toluene (Sigma Aldrich, St. Louis, MI, USA) was added, as a model organic contaminant [26] to the water tank, and the humid experiments were repeated in the presence of this organic contaminant, for both types of membranes.

The permeance of gas i, $(P/t)_i$, in GPU (1 GPU = 10^{-6} cm^3(STP) cm^{-2} s^{-1} cmHg^{-1}), is defined as the pressure-normalized flux of a gas through a membrane:

$$\left(\frac{P}{t}\right)_i = \frac{Q_p y_i}{(p_r x_i - p_p y_i)A} \times 10^6 \qquad (1)$$

where P is the intrinsic permeability of the selective membrane layer, in Barrer (1 Barrer = 10^{-10} cm^3(STP) cms^{-1} cmHg^{-1}); p_p is the retentate and permeate pressure, in cmHg, respectively; A the effective area of the membrane; t is the selective layer thickness for the separation; and Q_p is the permeate flow rate (cm^3 (STP)/s) at measurement pressure and temperature conditions.

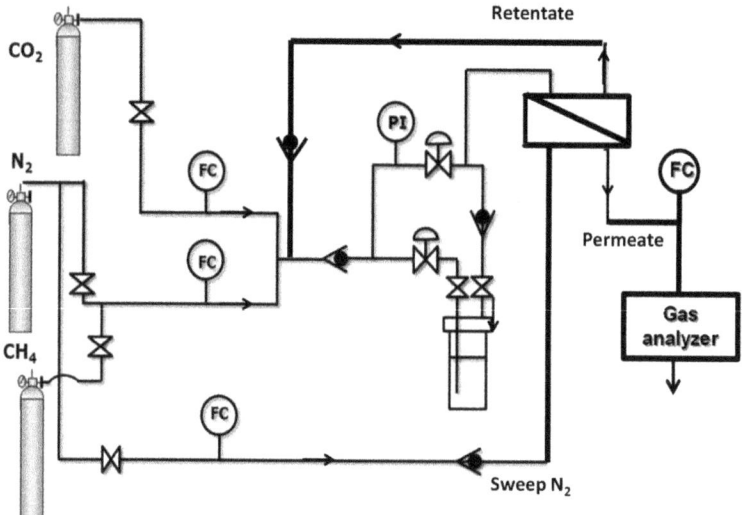

Figure 1. Diagram of the gas separation plant used in the experiments.

The separation factor of gas i over gas j, α_{ij}, is defined as the ratio of the concentration of gas i in the permeate and the retentate relative to the same ratio for gas j:

$$\alpha_{ij} = \frac{(y_i/y_j)}{(x_i/x_j)} \quad (2)$$

The selectivity is calculated as the ratio between the permeance of the fast and slow gas components in a gas pair, in the case of this work, CO$_2$ and CH$_4$, respectively.

3. Results

3.1. Membrane Performance in Terms of the Robeson's Upper Bound

The most popular means of comparing membrane performances for gas pair separations acknowledged worldwide is the Robeson's plot, usually on the development of new membrane materials [17]. Although initially settled from pure gas permeances and selectivities, this upper bound was also employed to analyze the effect of mixed gas feed mixtures in the separation performance of available or new membranes. This upper bound has been redefined by Professor Robeson and other authors, in order to take new developments on membrane materials and configurations into account [20].

The upper bound developed by Robeson and the modifications for new polymers upgraded [20] are also plotted for comparison with the pure gas upper bound Robeson's plot in Figure 2 [17]. This upper bound is an useful screening tool for the development of new membrane materials, usually as self-standing single-layer membranes, but it has limitations to establish the state-of-the-art of thin-film composite membranes as those commercially available, since the trade-off plots described are for pure-gas permeation [38].

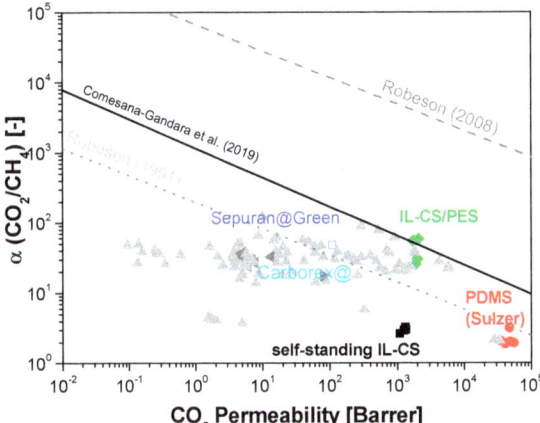

Figure 2. Effect of morphology on the CO_2/CH_4 separation performance of the fresh hydrophobic PDMS and hydrophilic ionic liquid–chitosan (IL–CS) composite membranes, against the CO_2/CH_4 upper bound.

In this light, for instance, Lin and Yavari highlighted that the upper bound for CO_2/CH_4 mixed-gas separation in the presence of non-methane hydrocarbon impurities could reduce the upper bound up to 37% for 20%CO_2/80%CH_4 mixtures [21]. They developed a model based on the free volume theory, to define the variations observed for composite membranes and process operation variables (such as pressure or the presence of contaminants) when compared with the upper bound generated from single-gas permeation depicted in Figure 2. The upper bound has been revised to include high free-volume polymers with intrinsic porosity [39], which intrinsically give much higher permeabilities and lower selectivities than other polymers, thus making the comparison difficult. These high free-volume polymers are also susceptible to suffer from phenomenological effects such as plasticization and physical aging, in detrition of their CO_2 separation performance. Specifically, Comesaña-Gandara et al. revised the CO_2/CH_4 upper bound, in order to include novel polymers whose intrinsic microporosity provided high permeability and selectivity not contemplated in previous upper bounds [20]. This upper bound is utilized in Figure 2, to compare the influence of morphology when comparing dense- and thin-film composite IL–CS/PES membranes reported elsewhere for CO_2/N_2, regarding their CO_2/CH_4 separation performance. The performance is observed to be highly dependent on the active layer thickness, increasing up to the latter upper bound mentioned. Those authors studied polyimide membranes with the same thickness and flat-sheet configuration we used in this work. In this work, we compare the performance of this hydrophilic IL–CS/PES and hydrophobic commercial PDMS membrane in the separation of CO_2/CH_4 mixtures, in the presence of humidity and organic pollutants.

3.2. Comparison of Membrane Performance in the Presence of Impurities

The permeability and selectivity in Figure 2 were calculated from single-gas permeation experiments as a function of feed pressure, between 2 and 5 bar. The feed pressure in biogas pilot plant experiments is not usually very high, so the gas-mixture separation experiments were performed at a feed pressure of about 4.5 bar, focusing on the influence of other components in the mixture that could affect the gas-separation performance by the presence of impurities in the feed, upon feed concentration and flux decline by membrane plasticization, as well as decrease of glass transition upon addition of organic pollutants [40].

The experimental results of the influence of water vapor in the feed on the CO_2 and CH_4 permeances obtained in this work are represented in Figure 3 for a flat hydrophobic and a hydrophilic composite membrane, respectively. The CO_2 and CH_4 dry permeances through the IL–CS composite membranes are of the same order of magnitude as those reported for a commercial CA membrane [41]. The results shown in Figure 3 for the CO_2:CH_4 separation in the presence of humidity affect the hydrophilic IL–CS/PES composite membrane differently than the PDMS membrane. The presence of toluene as model organic pollutant in the humid feed stream enhances these differences. The permeance of the hydrophobic membrane is affected by the presence of damp impurities, while the permeance and selectivity of the hydrophilic membrane are almost invariable in the presence of humid streams.

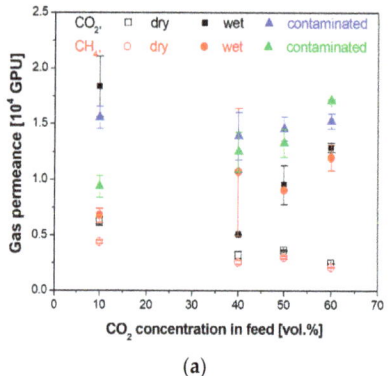
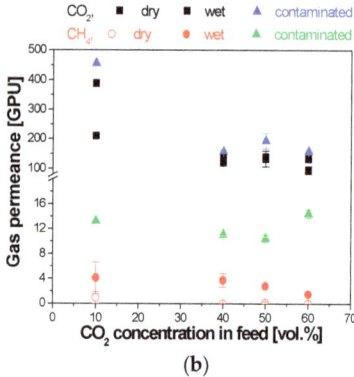

Figure 3. Influence of dry (void symbols) and humid (full symbols) conditions of the feed in the CO_2 (black squares) and CH_4 (red circles) permeance through the hydrophobic PDMS (**a**) and the hydrophilic IL–CS flat composite membrane (**b**). Data obtained after contamination with the toluene solution are drawn by the full triangles (blue for CO_2, and green for CH_4).

As observed in composite hollow fiber membrane geometry elsewhere, the presence of water vapor in the feed affects most significantly the CO_2 permeance through the hydrophobic PDMS than the hydrophilic IL–CS/PES membrane [22,37]. The permeances through the hydrophobic membrane, in Figure 3a, are not as affected by the CO_2 concentration in the feed in the whole range of feed gas concentration as by the humidity or the presence of organic pollutant, while the behavior of the hydrophilic membrane (Figure 3b) is mostly affected by CO_2 concentration in the feed rather than the presence of impurities. The CO_2 permeances decrease with increasing CO_2 concentration in the feed in a more remarkable way for the hydrophobic PDMS composite membrane than for the hydrophilic IL–CS-based composite membrane, which may be attributed to the water-facilitated transport through the hydrophilic membrane [14].

NaOH treatment is used in IL–CS-based membranes to neutralize the acetate groups of the CS polymer matrix and enhance the affinity towards acid gas molecules such as CO_2. This treatment has also been used to increase CO_2 separation properties from other gases in ZrO_2 ceramic membranes that were not so selective as-made [42]. NaOH immersion in MMMs has also been reported to enhance the hydrophilicity and facilitated transport through Ag^+-particles MMMs in CO_2/CH_4 separation [30]. In chitosan-based membranes, the same NaOH treatment is used to neutralize the functional groups of the IL–CS matrix to attract CO_2 preferentially [43]. After immersion in NaOH 1M and removal of excess NaOH solution from the surface of the membrane, the membrane is dried at 50 °C prior to characterization. The selectivity of the hydrophobic PDMS commercial membrane increases both in single-gas permeation and gas-mixture separation experiments after this treatment. Table 1 collects these characterization results for the case of dry feed gas measurements. Although the CO_2 permeance of the PDMS membrane decreases in the order of magnitude 86–95% upon NaOH treatment, the permeance of the NaOH-treated PDMS membrane is 169 ± 9.9 GPU, 165 ± 10 GPU and 320 ± 39 GPU for 60/40, 50/50 and 40/60 CO_2/CH_4 (vol %) gas mixture concentrations in the feed. These values are in the order of magnitude reported in the literature [26]. The CH$_4$ permeance decreases more than 99% and determines the enhancement of the CO$_2$/CH$_4$ selectivity of the membrane summarized for the dry membranes in Table 1.

Table 1. Theoretical and mixed gas selectivity results for hydrophobic PDMS and hydrophilic IL–CS/polyethersulfone (PES) membranes.

Membrane	Theoretical [1] α(CO$_2$/CH$_4$)	Mixed Gas CO$_2$/CH$_4$ Selectivity		
		60/40 (vol %)	50/50 (vol %)	40/60 (vol %)
PDMS	3.13	1.15	1.21	1.26
IL–CS/PES	2.66	48.47 ± 8.6	39.59 ± 7.0	36.49 ± 2.6
PDMS–NaOH	10.65	16.52 ± 2.0	15.0 ± 0.3	12.11 ± 1.42

[1] From single gas-permeation experiments.

The relationship between the permeable gas and the gas pair selectivity obtained for the NaOH-treated membranes follows the same trend that was reported by Lokhandwala et al. [41] for N$_2$/CH$_4$ separation.

The separation factor through the membranes is plotted in Figure 4. The hydrophobic PDMS membrane shows lower CO$_2$/CH$_4$ selectivity than the hydrophilic IL–CS composite membrane (Figure 4). The CO$_2$/CH$_4$ of the untreated hydrophobic PDMS membrane is not affected by feed-gas-mixture concentration. This agrees with the study of Chenar et al., when they compared the performance of Cardo polyimide with PPO hollow fiber membranes in dry and humid CO$_2$/CH$_4$ gas mixture separation [22]. The CO$_2$/CH$_4$ separation factor of the PDMS membrane was increased by the NaOH-treatment for the whole range of CO$_2$ concentration in the feed mixture. As expected, the CO$_2$/CH$_4$ separation of the hydrophilic membranes was not diminished by the presence of water vapor, as it was also previouslyobserved in Figure 3b.

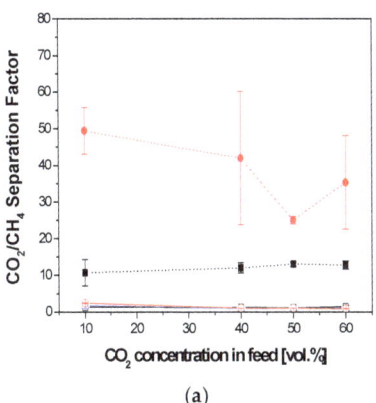
(a)

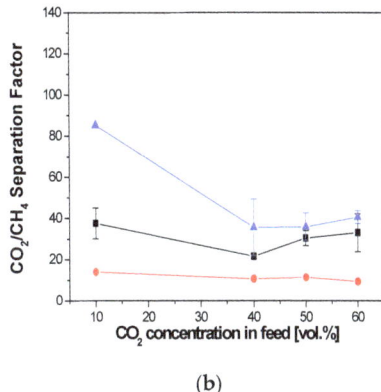
(b)

Figure 4. Influence of feed conditions on the CO$_2$/CH$_4$ separation factor of the commercial hydrophobic PDMS membrane (**a**) and the hydrophilic IL–CS/PES (**b**), in the absence (squares) and presence of water vapor (blue) and toluene (red) impurities. Void symbols in Figure 4a correspond to the experimental values measured with the un-treated commercial PDMS membrane for comparison purposes. Lines are a guide to the eye.

The red full circles in Figure 4 represent the influence of the presence of toluene contaminating the feed to the membrane module. This corresponds to a toluene concentration of 3460 ppm, which is high enough to accelerate the observation of any contamination effect affecting the aging of the membrane-separation operation in a single experiment [25]. The decrease of CH$_4$ permeance through the rubbery hydrophobic membranes with increasing CO$_2$ concentration in the feed was not significant due to the competition between sorption and plasticization of the CO_2 in the PDMS layer. In the case of the hydrophilic IL–CS/PES membrane, there was a favorable competition between plasticization and water preferential solubility for CO_2 [19]. This agrees with the favorable competition on CO_2 and CH_4 permeances reported by Jusoh et al. for polyamide membranes in the presence and absence of humid pentane [6].

On the other hand, Figure 4 shows how the presence of the model organic pollutant, toluene, in the feed mixture increased the separation factor of the rubbery hydrophobic PDMS membrane, but decreased that of the hydrophilic IL–CS/PES membrane. This may be compared to the antiplasticizing effect of toluene in SSZ-13-PDMC mixed matrix membranes (hydrophobic) and un-crosslinked Matrimid membranes (more hydrophilic) observed by professor Koros' group [25,26]. The comparatively large increase in selectivity of the NaOH-treated PDMS membrane may be due to the organic pollutant blocking the pass of the CO_2 and CH_4 molecules through the membrane [25]. Since this was particularly remarkable for the NaOH-treated PDMS membrane, one may point out to some changes in the chemical composition of the membrane surface altering the interaction of the membrane surface with the gas feed mixture and thus the CO_2/CH_4 separation performance.

3.3. Surface Characterization

ATR–FTIR has been used to characterize the surface chemistry of fresh membranes before their separation performance [44]. In this work, the ATR–FTIR spectra in Figure 5a reveal that the NaOH treatment increases the hydrophilicity of the PDMS hydrophobic layer, as derived by the band at 3500 cm^{-1}, typical of the OH stretching vibration, appearing at the NaOH-treated PDMS surface, measured on the dry membrane. This accounts for the different performance of the PDMS membrane after an NaOH treatment. The PDMS characteristic peaks at 2964–2950 cm^{-1} are still present after NaOH treatment, so the main polymer backbone of the top layer of the commercial membrane should not have been damaged after the experimental runs in the presence of toluene in the feed gas mixture, although the magnitude of the OH band of the freshly NaOH-treated PDMS membrane makes it difficult to discern these peaks in the clean diagram [45].

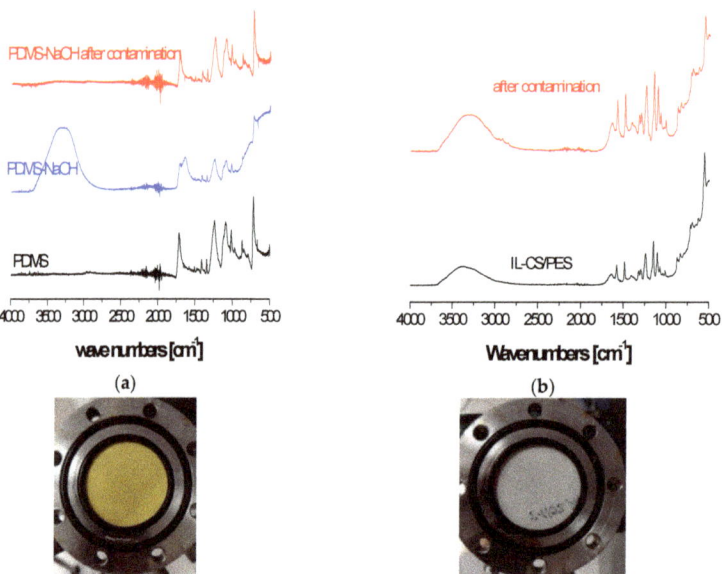

Figure 5. ATR–FTIR spectra of the PDMS (**a**) and IL–CS/PES (**b**) membranes. Pictures of the top surface of the composite membranes after 10 h experimental runs are presented below.

Regarding the effect of the organic contaminant, the 1024 and 1070 cm^{-1} bands of in-plane C–H stretching of toluene are appreciated in the contaminated IL–CS/PES membrane in Figure 5b. These bands are blurred in the PDMS-based membranes spectra in Figure 4a before contamination, so no conclusion can be extracted at this point. No other apparent changes are appreciated in the IL–CS IR peaks [46] in the IL–CS/PES membrane. This implies that the membrane is not deteriorated by the presence of toluene contaminant, which agrees with the different coloring of the membrane surface of the hydrophobic (left) and hydrophilic (right) membranes included in the pictures below the spectra. The PDMS-based membrane from Sulzer was also reported to turn yellowish brown, as well after CO_2 separation from flue gas in pilot plant measurements [47].

4. Conclusions

Biogas upgrading development is one of the key features to attain circular economy. The main drawback of membrane technology to be actually seen as the potential technical and environmentally friendly alternatives on the treatment of residual gas effluents is the uncertainty of available membranes' behavior in the presence of impurities in practical waste feed stream mixtures. This work aims at analyzing the behavior of membranes in the presence and absence of water and model organic compounds as a function of the membrane's top-layer characteristics. The perm-selectivity of commercial PDMS membranes can be altered by an alkaline treatment commonly used to stabilize polyelectrolyte membranes commonly studied as ion-exchange membranes in several applications, including acid gas purification.

This NaOH treatment facilitating the transport properties, i.e. the CO_2 permeance and CO_2/CH_4 separation characteristics of the hydrophilic IL-CS/PES membrane even in the presence of impurities, also maintains the transport properties, of the hydrophobic PDMS membrane regardless the CO_2 concentration in the feed.

The tuning up of the hydrophilic/hydrophobic character of the membrane surface can be an effective way of improving facilitated transport properties and improving membrane performance in CO_2 capture applications. The outcomes of this study are being validated by gas separation models, in order to analyze the correlation between the presence of impurities and the membrane performance as an approach to overcome the uncertainty of membrane technology in the treatment of residual gas streams.

Author Contributions: All authors contributed to the evaluation of the conceptualization, methodology, validation and final writing of the manuscript. All authors have read and agreed to the published version of the manuscript.

Funding: This research was funded by the Spanish Ministry of Science and Innovation; project CTQ2016-76231-C2-(AEI/FEDER, UE) and project PID2019-108136RB-C31).

Acknowledgments: Teresa Jagiello (TU Graz, Austria) is gratefully thanked for helping with the conduction of several of the experimental gas separation tests during her Erasmus stay at the University of Cantabria.

Conflicts of Interest: The authors declare no conflict of interest. The funders had no role in the design of the study; in the collection, analyses, or interpretation of data; in the writing of the manuscript; or in the decision to publish the results.

References

1. Harasimowicz, M.; Orluk, P.; Zakrzewskatrznadel, G.; Chmielewski, A. Application of polyimide membranes for biogas purification and enrichment. *J. Hazard. Mater.* **2007**, *144*, 698–702. [CrossRef] [PubMed]
2. Scholz, M.; Melin, T.; Wessling, M. Transforming biogas into biomethane using membrane technology. *Renew. Sustain. Energy Rev.* **2013**, *17*, 199–212. [CrossRef]
3. Esposito, E.; Dellamuzia, L.; Moretti, U.; Fuoco, A.; Giorno, L.; Jansen, J.C. Simultaneous production of biomethane and food grade CO_2 from biogas: An industrial case study. *Energy Environ. Sci.* **2019**, *12*, 281–289. [CrossRef]
4. Ajhar, M.; Bannwarth, S.; Stollenwerk, K.-H.; Spalding, G.; Yüce, S.; Wessling, M.; Melin, T. Siloxane removal using silicone–rubber membranes. *Sep. Purif. Technol.* **2012**, *89*, 234–244. [CrossRef]

5. Friess, K.; Lanč, M.; Pilnáček, K.; Fíla, V.; Vopička, O.; Sedláková, Z.; Cowan, M.G.; McDanel, W.M.; Noble, R.D.; Gin, D.L.; et al. CO_2/CH_4 separation performance of ionic-liquid-based epoxy-amine ion gel membranes under mixed feed conditions relevant to biogas processing. *J. Membr. Sci.* **2017**, *528*, 64–71. [CrossRef]
6. Jusoh, N.; Lau, K.K.; Shariff, A.; Yeong, Y. Capture of bulk CO_2 from methane with the presence of heavy hydrocarbon using membrane process. *Int. J. Greenh. Gas Control.* **2014**, *22*, 213–222. [CrossRef]
7. Rasi, S.; Veijanen, A.; Rintala, J. Trace compounds of biogas from different biogas production plants. *Energy* **2007**, *32*, 1375–1380. [CrossRef]
8. Stern, S.; Krishnakumar, B.; Charati, S.; Amato, W.; Friedman, A.; Fuess, D. Performance of a bench-scale membrane pilot plant for the upgrading of biogas in a wastewater treatment plant. *J. Membr. Sci.* **1998**, *151*, 63–74. [CrossRef]
9. Rautenbach, R.; Welsch, K. Treatment of landfill gas by gas permeation—Pilot plant results and comparison to alternatives. *J. Membr. Sci.* **1994**, *87*, 107–118. [CrossRef]
10. Mao, C.; Feng, Y.; Wang, X.; Ren, G. Review on research achievements of biogas from anaerobic digestion. *Renew. Sustain. Energy Rev.* **2015**, *45*, 540–555. [CrossRef]
11. Rufford, T.E.; Smart, S.; Watson, G.C.Y.; Graham, B.F.; Boxall, J.; Da Costa, J.D.; May, E.F. The removal of CO_2 and N_2 from natural gas: A review of conventional and emerging process technologies. *J. Pet. Sci. Eng.* **2012**, *94–95*, 123–154. [CrossRef]
12. Merkel, T.C.; Lin, H.; Wei, X.; Baker, R. Power plant post-combustion carbon dioxide capture: An opportunity for membranes. *J. Membr. Sci.* **2010**, *359*, 126–139. [CrossRef]
13. Zhai, H.; Rubin, E.S. Techno-economic assessment of polymer membrane systems for postcombustion carbon capture at coal-fired power plants. *Environ. Sci. Technol.* **2013**, *47*, 3006–3014. [CrossRef] [PubMed]
14. Kai, T.; Kouketsu, T.; Duan, S.; Kazama, S.; Yamada, K. Development of commercial-sized dendrimer composite membrane modules for CO_2 removal from flue gas. *Sep. Purif. Technol.* **2008**, *63*, 524–530. [CrossRef]
15. Ferrari, M.-C.; Bocciardo, D.; Brandani, S. Integration of multi-stage membrane carbon capture processes to coal-fired power plants using highly permeable polymers. *Green Energy Environ.* **2016**, *1*, 211–221. [CrossRef]
16. Evonik Fibres GmbH, SEPURAN®Green Membrane Technology for Upgrading BIOGAS efficiently. 2016. Available online: http://www.sepuran.com/sites/lists/RE/DocumentsHP/SEPURAN-green-for-upgrading-biogas-EN.pdf (accessed on 4 July 2020).
17. Robeson, L.M. The upper bound revisited. *J. Membr. Sci.* **2008**, *320*, 390–400. [CrossRef]
18. Scholes, C.A.; Stevens, G.W.; Kentish, S.E. Membrane gas separation applications in natural gas processing. *Fuel* **2012**, *96*, 15–28. [CrossRef]
19. Öztürk, B.; Demirciyeva, F. Comparison of biogas upgrading performances of different mixed matrix membranes. *Chem. Eng. J.* **2013**, *222*, 209–217. [CrossRef]
20. Comesaña-Gándara, B.; Chen, J.; Bezzu, C.G.; Carta, M.; Rose, I.; Ferrari, M.-C.; Esposito, E.; Fuoco, A.; Jansen, J.C.; McKeown, N.B. Redefining the Robeson upper bounds for CO_2/CH_4 and CO_2/N_2 separations using a series of ultrapermeable benzotriptycene-based polymers of intrinsic microporosity. *Energy Environ. Sci.* **2019**, *12*, 2733–2740. [CrossRef]
21. Lin, H.; Yavari, M. Upper bound of polymeric membranes for mixed-gas CO_2/CH_4 separations. *J. Membr. Sci.* **2015**, *475*, 101–109. [CrossRef]
22. Chenar, M.P.; Soltanieh, M.; Matsuura, T.; Tabe-Mohammadi, A.; Khulbe, K. The effect of water vapor on the performance of commercial polyphenylene oxide and Cardo-type polyimide hollow fiber membranes in CO2/CH4 separation applications. *J. Membr. Sci.* **2006**, *285*, 265–271. [CrossRef]
23. Simcik, M.; Ruzicka, M.; Karaszova, M.; Sedláková, Z.; Vejražka, J.; Veselý, M.; Capek, P.; Friess, K.; Izak, P. Polyamide thin-film composite membranes for potential raw biogas purification: Experiments and modeling. *Sep. Purif. Technol.* **2016**, *167*, 163–173. [CrossRef]
24. White, L.S.; Blinka, T.A.; Kloczewski, H.A.; Wang, I.-F. Properties of a polyimide gas separation membrane in natural gas streams. *J. Membr. Sci.* **1995**, *103*, 73–82. [CrossRef]
25. Omole, I.C.; Bhandari, D.A.; Miller, S.J.; Koros, W.J. Toluene impurity effects on CO_2 separation using a hollow fiber membrane for natural gas. *J. Membr. Sci.* **2011**, *369*, 490–498. [CrossRef]

26. Ward, J.K.; Koros, W.J. Crosslinkable mixed matrix membranes with surface modified molecular sieves for natural gas purification: II. Performance characterization under contaminated feed conditions. *J. Membr. Sci.* **2011**, *377*, 82–88. [CrossRef]
27. Cerveira, G.S.; Borges, C.P.; Kronemberger, F.D.A. Gas permeation applied to biogas upgrading using cellulose acetate and polydimethylsiloxane membranes. *J. Clean. Prod.* **2018**, *187*, 830–838. [CrossRef]
28. Saedi, S.; Madaeni, S.; Hassanzadeh, K.; Shamsabadi, A.A.; Laki, S. The effect of polyurethane on the structure and performance of PES membrane for separation of carbon dioxide from methane. *J. Ind. Eng. Chem.* **2014**, *20*, 1916–1929. [CrossRef]
29. Iovane, P.; Nanna, F.; Ding, Y.; Bikson, B.; Molino, A. Experimental test with polymeric membrane for the biogas purification from CO_2 and H_2S. *Fuel* **2014**, *135*, 352–358. [CrossRef]
30. Zhou, T.; Luo, L.; Hu, S.; Wang, S.; Zhang, R.; Wu, H.; Jiang, Z.; Wang, B.Y.; Yang, J. Janus composite nanoparticle-incorporated mixed matrix membranes for CO_2 separation. *J. Membr. Sci.* **2015**, *489*, 1–10. [CrossRef]
31. Fernández-Barquín, A.; Rea, R.; Venturi, D.; Giacinti-Baschetti, M.; De Angelis, M.G.; Casado, C.; Irabien, A. Effect of relative humidity on the gas transport properties of zeolite A/PTMSP mixed matrix membranes. *RSC Adv.* **2018**, *8*, 3536–3546. [CrossRef]
32. Fernández-Barquín, A.; Casado, C.; Etxeberría-Benavides, M.; Zuñiga, J.; Irabien, A. Comparison of flat and hollow-fiber mixed-matrix composite membranes for CO_2 separation with temperature. *Chem. Eng. Technol.* **2017**, *40*, 997–1007. [CrossRef]
33. Rezakazemi, M.; Amooghin, A.E.; Montazer-Rahmati, M.M.; Ismail, A.F.; Matsuura, T. State-of-the-art membrane based CO2 separation using mixed matrix membranes (MMMs): An overview on current status and future directions. *Prog. Polym. Sci.* **2014**, *39*, 817–861. [CrossRef]
34. Casado-Coterillo, C.; Fernández-Barquín, A.; Zornoza, B.; Téllez, C.; Coronas, J.; Irabien, Á. Synthesis and characterisation of MOF/ionic liquid/chitosan mixed matrix membranes for CO_2/N_2 separation. *RSC Adv.* **2015**, *5*, 102350–102361. [CrossRef]
35. Roslan, R.A.; Lau, W.J.; Sakthivel, D.B.; Khademi, S.; Zulhairun, A.K.; Goh, P.S.; Ismail, A.F.; Chong, K.C.; Lai, S.O. Separation of CO_2/CH_4 and O_2/N_2 by polysulfone hollow fiber membranes: Effects of membrane support properties and surface coating materials. *J. Polym. Eng.* **2018**, *38*, 871–880. [CrossRef]
36. Scholes, C.A.; Stevens, G.W.; Kentish, S. The effect of hydrogen sulfide, carbon monoxide and water on the performance of a PDMS membrane in carbon dioxide/nitrogen separation. *J. Membr. Sci.* **2010**, *350*, 189–199. [CrossRef]
37. Casado, C.; Fernández-Barquín, A.; Irabien, A. Effect of humidity on CO_2/N_2 and CO_2/CH_4 separation using novel robust mixed matrix composite hollow fiber membranes: Experimental and model evaluation. *Membranes* **2019**, *10*, 6. [CrossRef]
38. Galizia, M.; Chi, W.S.; Smith, Z.P.; Merkel, T.C.; Baker, R.W.; Freeman, B.D. 50th anniversary perspective: Polymers and mixed matrix membranes for gas and vapor separation: A review and prospective opportunities. *Macromolecules* **2017**, *50*, 7809–7843. [CrossRef]
39. Robeson, L.M.; Dose, M.E.; Freeman, B.D.; Paul, D.R. Analysis of the transport properties of thermally rearranged (TR) polymers and polymers of intrinsic microporosity (PIM) relative to upper bound performance. *J. Membr. Sci.* **2017**, *525*, 18–24. [CrossRef]
40. Molino, A.; Nanna, F.; Migliori, M.; Iovane, P.; Ding, Y.; Bikson, B. Experimental and simulation results for biomethane production using peek hollow fiber membrane. *Fuel* **2013**, *112*, 489–493. [CrossRef]
41. Lokhandwala, K.A.; Pinnau, I.; He, Z.; Amo, K.D.; Dacosta, A.R.; Wijmans, J.G.; Baker, R.W. Membrane separation of nitrogen from natural gas: A case study from membrane synthesis to commercial deployment. *J. Membr. Sci.* **2010**, *346*, 270–279. [CrossRef]
42. Casado, C.; Yokoo, T.; Yoshioka, T.; Tsuru, T.; Asaeda, M. Synthesis and characterization of microporous ZrO_2 membranes for gas permeation at 200 °C. *Sep. Sci. Technol.* **2011**, *46*, 1224–1230. [CrossRef]
43. Santos, E.; Rodríguez-Fernández, E.; Casado, C.; Irabien, Á. Hybrid ionic liquid-chitosan membranes for CO_2 separation: Mechanical and thermal behavior. *Int. J. Chem. React. Eng.* **2016**, *14*, 713–718. [CrossRef]
44. Jo, E.-S.; An, X.; Ingole, P.G.; Choi, W.-K.; Park, Y.-S.; Lee, H.K. CO_2/CH_4 separation using inside coated thin film composite hollow fiber membranes prepared by interfacial polymerization. *Chin. J. Chem. Eng.* **2017**, *25*, 278–287. [CrossRef]

45. Salih, A.A.; Yi, C.; Peng, H.; Yang, B.; Yin, L.; Wang, W. Interfacially polymerized polyetheramine thin film composite membranes with PDMS inter-layer for CO_2 separation. *J. Membr. Sci.* **2014**, *472*, 110–118. [CrossRef]
46. Casado, C.; Guerrero, M.L.; Irabien, A. Synthesis and characterisation of ETS-10/Acetate-based ionic liquid/chitosan mixed matrix membranes for CO_2/N_2 permeation. *Membranes* **2014**, *4*, 287–301. [CrossRef] [PubMed]
47. Scholes, C.A.; Bacus, J.; Chen, G.Q.; Tao, W.X.; Li, G.; Qader, A.; Stevens, G.W.; Kentish, S.E. Pilot plant performance of rubbery polymeric membranes for carbon dioxide separation from syngas. *J. Membr. Sci.* **2012**, *389*, 470–477. [CrossRef]

Publisher's Note: MDPI stays neutral with regard to jurisdictional claims in published maps and institutional affiliations.

© 2020 by the authors. Licensee MDPI, Basel, Switzerland. This article is an open access article distributed under the terms and conditions of the Creative Commons Attribution (CC BY) license (http://creativecommons.org/licenses/by/4.0/).

Article

The Impact of Various Natural Gas Contaminant Exposures on CO_2/CH_4 Separation by a Polyimide Membrane

Nándor Nemestóthy, Péter Bakonyi, Piroska Lajtai-Szabó and Katalin Bélafi-Bakó *

Research Group on Bioengineering, Membrane Technology and Energetics, University of Pannonia, 8200 Veszprém, Hungary; nemesn@almos.uni-pannon.hu (N.N.); bakonyip@almos.uni-pannon.hu (P.B.); tpiroska94@gmail.com (P.L.-S.)
* Correspondence: bako@almos.uni-pannon.hu

Received: 8 October 2020; Accepted: 29 October 2020; Published: 31 October 2020

Abstract: In this study, hollow fibers of commercial polyimide were arranged into membrane modules to test their capacity and performance towards natural gas processing. Particularly, the membranes were characterized for CO_2/CH_4 separation with and without exposure to some naturally occurring contaminants of natural gases, namely hydrogen sulfide, dodecane, and the mixture of aromatic hydrocarbons (benzene, toluene, xylene), referred to as BTX. Gas permeation experiments were conducted to assess the changes in the permeability of CO_2 and CH_4 and related separation selectivity. Compared to the properties determined for the pristine polyimide membranes, all the above pollutants (depending on their concentrations and the ensured contact time with the membrane) affected the permeability of gases, while the impact of various exposures on CO_2/CH_4 selectivity seemed to be complex and case-specific. Overall, it was found that the minor impurities in the natural gas could have a notable influence and should therefore be considered from an operational stability viewpoint of the membrane separation process.

Keywords: gas separation; polyimide membrane; natural gas separation; pollutant effects; stability measurements

1. Introduction

The applicability of membranes in the processing of natural gas has been shown widely [1]. As the composition of (raw) natural gas varies considerably in line with its source, there are numerous tasks for improving its quality (e.g., methane content) and to meet pipeline and utilization requirements [2]. In fact, gas separation membranes (alone or in combination with other systems) can contribute to major technological steps, such as the removal of CO_2, acidic components (particularly H_2S), longer-chain hydrocarbons, water (vapor), and N_2 [3]. To get the actual job done, gas separation membranes manufactured with the use of polymers gained broad recognition at laboratories, as well as on an industrial scale, and among the available materials, the glassy-polymer polyimide (commonly in hollow-fiber membrane modules) is one of the most well-known [4,5].

Polyimide is characterized by good CO_2 permeability and simultaneous retention of CH_4, resulting in sufficiently high CO_2/CH_4 selectivity [6]. However, a larger quantity of CO_2 can make the membrane materials, including polyimide, suffer from plasticization, especially under higher feed pressure conditions [7]. As this penetrant-induced plasticization phenomenon (occurring in the presence of more notably condensable, soluble molecules) undermines the sensitive balance between the productivity of the separation (reflected in the permeability) and the purity of the product (influenced by the selectivity) [8], actions are still needed to design and synthetize better derivatives of polymers (with enhanced resistance to plasticization) via approaches such as blending and chemical crosslinking, etc. [9]. Still, choices and decisions are frequently needed as to whether the component permeability or the separation selectivity is more important in the given situation [10]. The dilemma of this trade-off has been addressed and assessed in depth by different studies based on the upper-bound relationship [11,12]. Moreover, the polyimide, and in general the glassy polymers, can be prone to physical aging, which may appear as a drawback in the long term due to the decrease of achievable gas fluxes [13,14].

Besides the issues related to carbon dioxide, other accompanying impurities may also cause adverse effects and deteriorate the performance of the membrane unit. Among the aforementioned components, the aggressive compound, the hydrogen sulfide is also regarded as a plasticizing agent using glassy polymers. From mixed gas permeation measurement applying ternary $CH_4/CO_2/H_2S$, the relatively faster transportation of H_2S through polyimide was concluded and is beneficial for its removal [15]. Such a step, the removal of acidic substances from the natural gas, is also referred to as the "sweetening" [16]. Interestingly, a recent paper using polyimide membranes reported the unexpected advantage of plasticization in H_2S/CH_4 separation, thanks to enhanced sorption coefficient [17]. However, at the same time, in agreement with common literature observations, the plasticization depressed the separation efficiency for the CO_2/CH_4 gas pair. As a matter of fact, there might be a necessity to develop process configurations, where the separations of H_2S/CH_4 and CO_2/CH_4 are carried out in the cascade of different, appropriately selected membranes [18]. Furthermore, removal of hydrocarbons (mainly C_3+) from the raw natural gas should be taken into consideration [19]. The paraffin and olefin components have a higher commercial value and thus, their recovery is an economic interest. Additionally, the contact of aromatic hydrocarbons (containing the benzene-ring, e.g., toluene) and polyimide membranes was shown to affect membrane separation performance and the attainable CO_2/CH_4 separation selectivity [20].

In this work, we present the results of our study conducted on commercial (UBE Industries, LTD.) polyimide membrane fibers in a single-gas experimental permeation apparatus and comparatively evaluate the impacts linked to various exposures of H_2S, dodecane hydrocarbon, and a mixture of benzene, toluene, and xylene (BTX) on CO_2 and CH_4 permeability and CO_2/CH_4 selectivity. The aim of this work is to deliver some new insights to the behavior of polyimide gas separation membranes under conditions when impurities (that are typically contained by the natural gas) are present during the separation of methane from carbon dioxide.

2. Materials and Methods

In this work, the effect of pollutants on the permeability of carbon dioxide (99.5%) and methane (99.95%) was investigated. The examined pollutants were H_2S, a benzene-toluene-xylene mixture in a 1:1:1 ratio called BTX and n-dodecane. H_2S was generated, as already mentioned in our earlier paper [21], and diluted thereafter with nitrogen (99.995%) to adjust the required concentration (Table 1). Every gas (CO_2, CH_4, N_2) was used from a cylinder (Messer Hungarogáz Kft., Veszprém, Hungary). N-dodecane (98.0%) was provided by Sigma–Aldrich (Taufkirchen, Germany), benzene (99.5%) by Spektrum-3D Kft. (Debrecen, Hungary), toluene (99.8%) by Merck KGaA (Darmstadt, Germany), and xylene (98.5%) by Sigma–Aldrich (Taufkirchen, Germany).

Table 1. The experimental boundaries in this work.

Pollutant	C_{low} [ppm]	C_{cent} [ppm]	C_{high} [ppm]	t_1 [day]	t_2 [day]	t_3 [day]
H_2S	100,000	300,000	500,000	1	3.5	7
BTX	500	750	1000	1	3.5	7
dodecane	1000	5500	10,000	1	3.5	7

For the experiments, polyimide capillaries were taken from a hollow fiber gas separation membrane (synthesised by UBE). A module consisted of six capillaries, for which ends were closed to get a "sack" (dead-end) configuration (Figure 1). The scheme of the gas separation test system can be seen in Figure 2. The actual test gas was filled to the gas container (GC-1), the pressure of which was monitored by a digital (WIKA A-10 type) pressure transducer (PT-1). During the measurements, the feed pressure of the membrane module (MM-1) was regulated and fixed by valve PC-1.

Figure 1. The membrane module containing the polyimide capillaries.

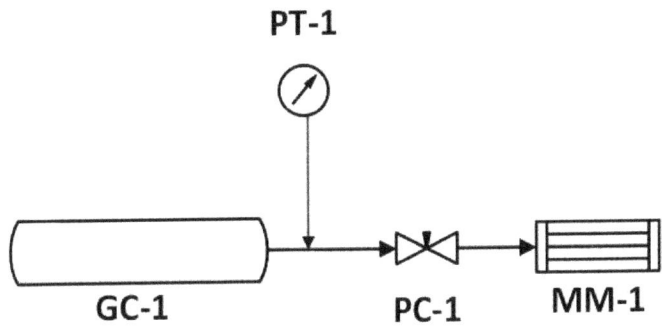

Figure 2. The layout of the experimental gas permeation apparatus.

Gas permeability measurements were carried out according to the constant pressure (CP) method [22] with a constant volume (CV) [23] pressure chamber. The amount of the permeated gas was calculated by the CV method [24] (Equation (1)):

$$n = \frac{P_{pc} \cdot V}{R \cdot T} \qquad (1)$$

where R is the gas universal constant, T is the temperature (K), and V is the volume of the gas chamber (m^3). P_{pc} is the pressure change in the chamber (Pa). The gas permeability coefficient, P, can be given by Equation (2):

$$P = \frac{n \cdot L}{A \cdot t \cdot P_d} \qquad (2)$$

where L (m) is the membrane thickness, A (m^2) is the area of the membrane for gas permeation, and P_d is the pressure difference (Pa) across the membrane. The P could be converted to the unit of Barrer (1 Barrer = 3.35×10^{-16} mol·m^{-1}·s^{-1}·Pa^{-1}).

To investigate the pollutant's effect, the membranes were put in a closed vessel for a given time (t_1, t_2, t_3), for which headspace contained a certain concentration of the given pollutant (C_{min}, C_{cent}, C_{max}), as displayed by Table 1, where it can be noticed that the concentration boundaries within a particular case were equally-spaced. The C_{cent} was repeated three times (to check the confidence of the measurements under fixed conditions), and a total number of seven data points with fairly balanced distributions could be considered in all cases, according to Figure 3. The vessels were incubated at constant temperature (27 °C). The permeability of every membrane module was measured before the experiments (pristine polyimide) and directly after the desired incubation, and then, a permeability change factor (Figures 4–6) was calculated as the ratio of respective gas permeabilities measured on the exposed and unexposed polyimide membranes. The parameter called exposure (the pollutant concentration multiplied by the time) was used as an independent variable to characterize the effects of pollutants on gas permeation (Figures 4–6) and separation selectivity behavior (Figures 7–9). For example, if the membrane is exposed to 1000 ppm of pollutant for 0.1 h, the exposure is equal to 100 ppm × h. The BTX and dodecane concentrations in Table 1 were estimated by the Antoine equation in CHEMCAD [25].

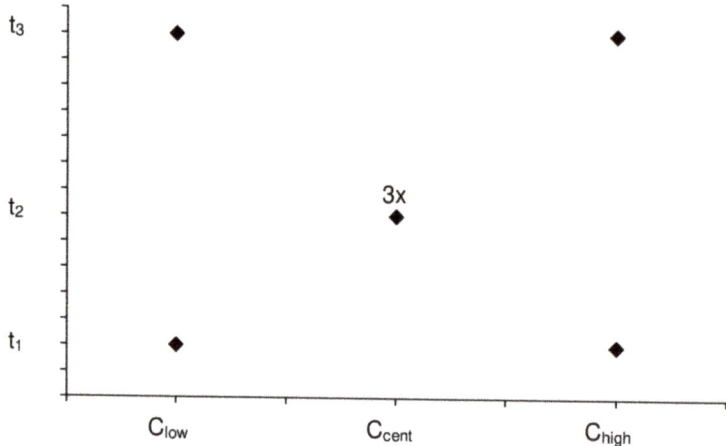

Figure 3. The layout of the experimental plan carried out in this study considering the conditions in Table 1.

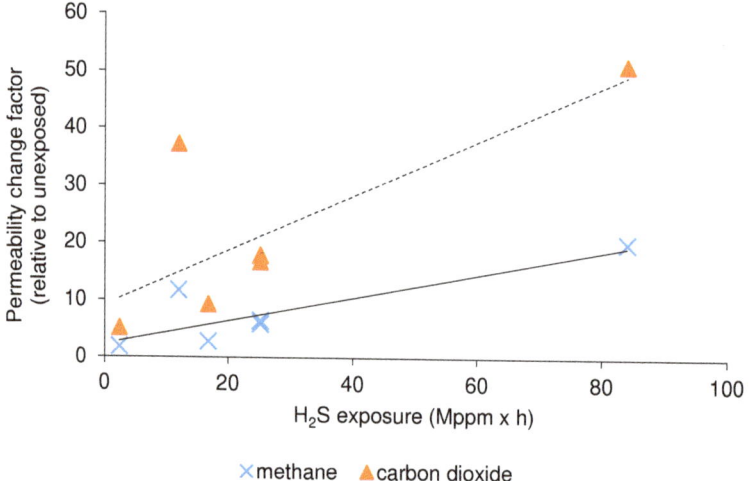

Figure 4. Alteration of gas permeabilities after exposures to H_2S (the dotted trend line belongs to CO_2).

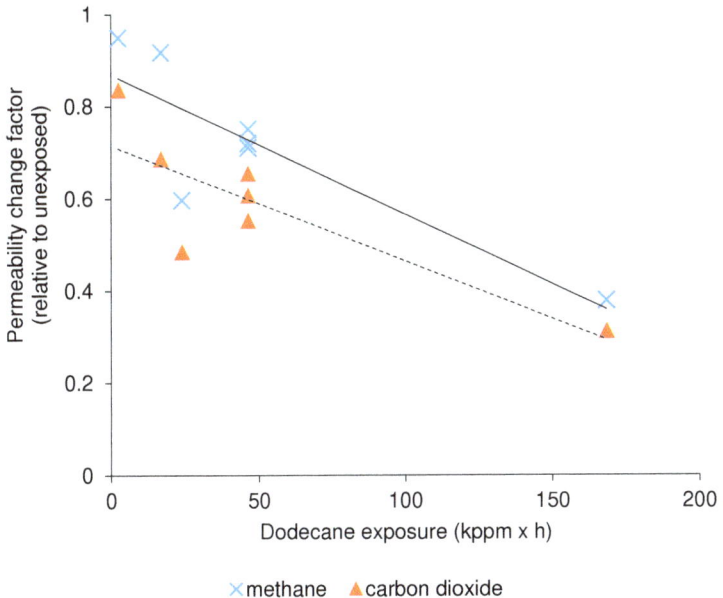

Figure 5. Alteration of gas permeabilities after exposures to dodecane (the dotted trend line belongs to CO_2).

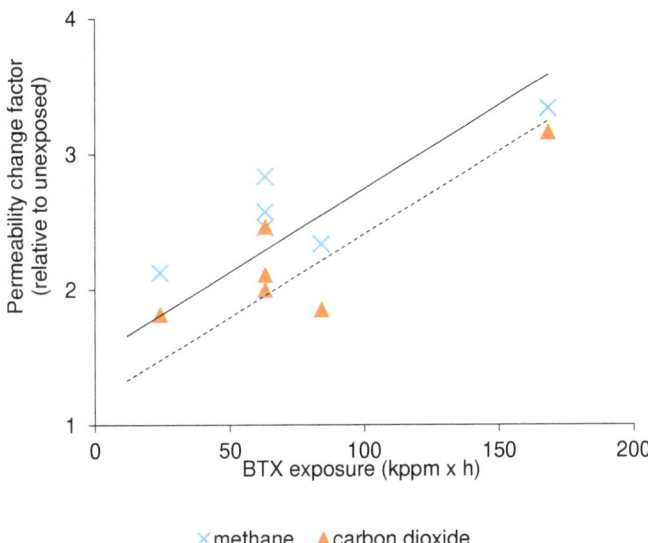

Figure 6. Alteration of gas permeabilities after exposures to benzene, toluene, and xylene (BTX) (the dotted trend line belongs to CO_2).

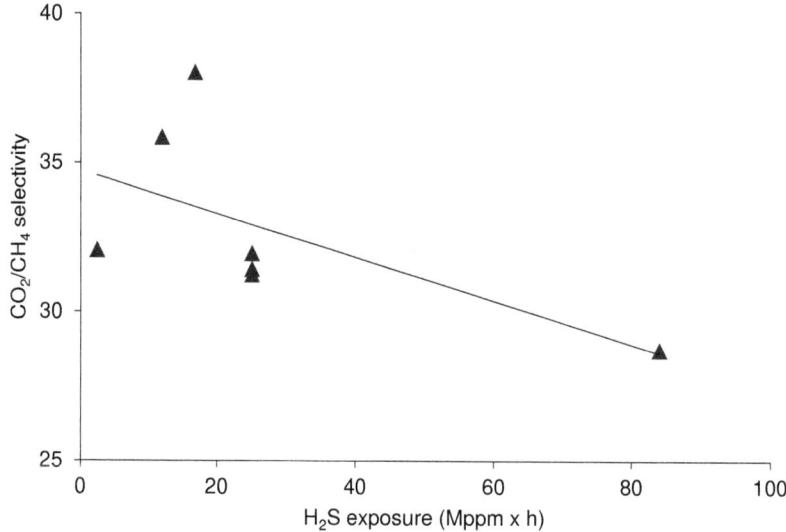

Figure 7. The effect of H_2S exposure on CO_2/CH_4 selectivity.

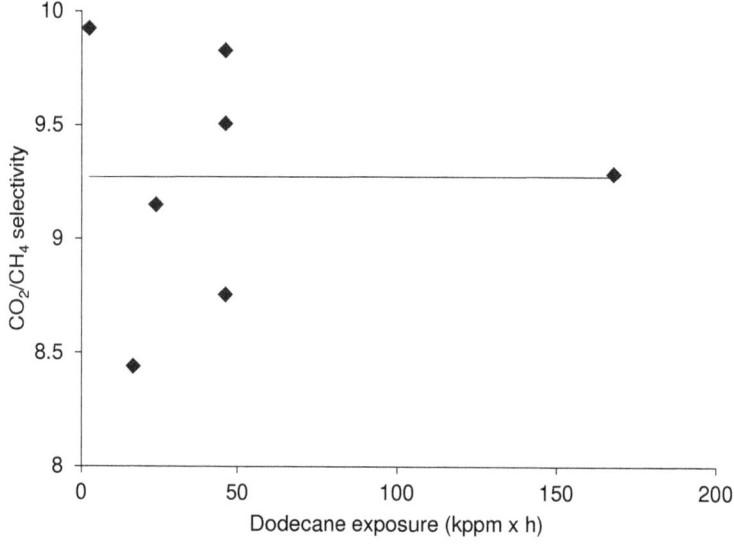

Figure 8. The effect of dodecane exposure on CO_2/CH_4 selectivity.

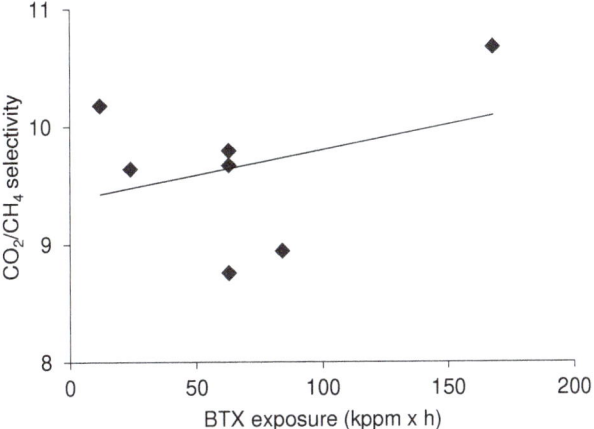

Figure 9. The effect of BTX exposure on CO_2/CH_4 selectivity.

3. Results and Discussion

First, the permeation of pure CO_2 and CH_4 gases was examined using the membrane prepared using the pristine polyimide hollow fibers. According to Figure 10, there was one order of magnitude difference in terms of the permeabilities: 0.156 Barrer and 1.76 Barrer for carbon dioxide and methane, respectively. Accordingly, the ideal CO_2/CH_4 selectivity (the ratio of the two permeabilities) was found as 11.28. This outcome coincided with the good mass of literature reporting the CH_4-rejective behavior of different polyimides. Typical CO_2/CH_4 selectivity data (obtained under mostly varying experimental conditions) were summarized in some articles for a wide range of polyimides, for instance: 16–64 [26] and 13.6–87 [27].

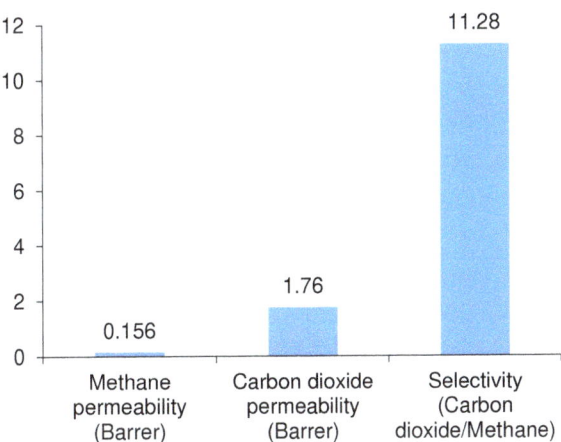

Figure 10. Permeation traits of CO_2 and CH_4 across fresh polyimide membrane.

In the next phase of the experiments, the change of gas permeation and separation performance were tested and assessed after various exposures (according to the experimental plan) to pollutants, such as H_2S, BTX, and dodecane. In all cases, a simple, linear-type association was assumed as a first approach to illustrate the trends in the change of permeability and selectivity using the polyimide membrane.

3.1. The Effect of H_2S Exposure on CO_2/CH_4 Separation

Shown in Figure 4, the impact of H_2S exposure on CO_2 and CH_4 permeability can be clearly drawn. From the experimental results, a linear-type correlation seems to be satisfactory to indicate that the contact of the polyimide membrane with higher concentrations of hydrogen sulfide for longer periods caused larger changes in the permeability of the two gases and vice versa. For both gases, it is illustrated in Figure 4 that the permeabilities were increased by the larger H_2S exposures.

However, considering the ideal selectivity values plotted in Figure 7, the tendency is the opposite compared to those experiences regarding the permeabilities. A significant decrease of CO_2/CH_4 separation performance was documented under greater H_2S exposures. This reverse influence of H_2S exposure on permeability and selectivity might be explained by the plasticization effect (resulting in the general faster permeation of components and the concurrent drop of selectivity) and/or the modification of the polymer structure. It follows the theory that when H_2S is present (together with water) in the membrane, it may induce, in some cases, the alteration of inherent material and gas permeation properties [19]. This will have to be further studied in addition to dissecting the reasons why the CO_2/CH_4 selectivity could have increased (by 3–4 times) relative to the polyimide unexposed to H_2S.

3.2. The Effect of Dodecane Exposure on CO_2/CH_4 Separation

The effect of dodecane exposure on CO_2 and CH_4 permeability is demonstrated in Figure 5. It can be inferred, based on the assumed linear relationships (fitted trendlines between the change of exposure and gas permeability), that the larger exposures led to more and more diminished gas permeations through the bunch of polyimide fibers. Concerning the selectivity shown in Figure 8, the tendency of the scattering experimental data reveals no obvious influence, and on average, the CO_2/CH_4 selectivity remained quite stable around 9.3 (in accordance with the fitted trendline in Figure 8). Nonetheless, compared to the pristine polyimide (Figure 10), some decline of the CO_2/CH_4 selectivity can be noted, and this means that the presence of dodecane had a real effect on the membrane performance.

To provide some plausible explanation, some findings of the relevant literature may be recalled here pertaining to the operational/testing experiences of gas separation membrane technology deployed for natural gas processing. The considerable swelling and, consequently, the change of the separation behavior over time could be concluded for silicone-based membranes upon exposure to heavier hydrocarbons [1,28]. In our opinion, one scenario could have been that the dodecane deposited on the membrane surface and formed a thin (microscopic-scale), fouling-layer like film. This may have automatically reduced the permeabilities of both gases, simply due to the increasing thickness of permeation pathway with greater exposure; however, in total, it did not really modify the separation selectivity.

3.3. The Effect of BTX Exposure on CO_2/CH_4 Separation

The permeability changes of CO_2 and CH_4 gases as a result of different BTX exposures are displayed in Figure 6. As a matter of fact, it can be concluded that the BTX exposure influenced the permeation of both gaseous compounds in a similar manner and the stronger BTX exposures were coupled with the more considerable increase of permeabilities. In terms of CO_2/CH_4 selectivity, to the naked eye, the various BTX exposures did not cause apparent changes, as respective values represented by the trend line in Figure 9 consistently spanned the narrow range of 9.4–10. In the literature, the presence of aromatic components, e.g., toluene, was found to impair the separation performance by altering the CO_2 and CH_4 permeabilities and depressing the CO_2/CH_4 separation selectivity [20]. Similar results were communicated more recently in other investigations [29,30]. Nevertheless, effects associated with toluene could be reversible [31], which is positive from the aspect of membrane stability.

4. Conclusions

In this work, the effect of some common natural gas pollutants (hydrogen sulfide, BTX, and dodecane) on the permeability of CO_2 and CH_4 gases was studied, applying polyimide hollow-fiber membrane. It was found that all of the investigated pollutants had an impact on the membrane's performance but in different ways and to different extents. The hydrogen sulfide increased the permeability of both CO_2 and CH_4 and the CO_2/CH_4 selectivity had a decreasing tendency as a function of increasing H_2S exposures. In the case of dodecane, permeability of CO_2 and CH_4 was decreased moderately by increasing the degree of exposure, while the CO_2/CH_4 selectivity, according to tendencies, was left unaffected. By contrast, larger exposures to BTX caused the increase of gas permeabilities; however, the corresponding trends indicated only marginal changes of CO_2/CH_4 selectivity. Though possible reasons to explain the dependency of permeability and selectivity on pollutant exposures using the polyimide membrane were implied, further exploration is intended to find out the underlying mechanisms taking place between the actual contaminant and the membrane and to get some insights into whether the observed influences are reversible or irreversible. In future studies, the scope might be expanded to other polymeric membranes, and when a good mass of data are collected, more generalized conclusions may be drawn.

Author Contributions: Conceptualization, N.N.; methodology, N.N.; writing, data curation, and editing, N.N., P.B., and P.L.-S.; review, N.N., and K.B.-B.; supervision, K.B.-B. All authors have read and agreed to the published version of the manuscript.

Funding: This research received no external funding.

Conflicts of Interest: The authors declare no conflict of interest.

References

1. Scholes, C.A.; Stevens, G.W.; Kentish, S.E. Membrane gas separation applications in natural gas processing. *Fuel* **2012**, *96*, 15–28. [CrossRef]
2. Adewole, J.; Ahmad, A.; Ismail, S.; Leo, C. Current challenges in membrane separation of CO_2 from natural gas: A review. *Int. J. Greenh. Gas Control.* **2013**, *17*, 46–65. [CrossRef]
3. Baker, R.W.; Lokhandwala, K. Natural gas processing with membranes: An overview. *Ind. Eng. Chem. Res.* **2008**, *47*, 2109–2121. [CrossRef]
4. Lokhandwala, K.A.; Pinnau, I.; He, Z.; Amo, K.D.; Dacosta, A.R.; Wijmans, J.G.; Baker, R.W. Membrane separation of nitrogen from natural gas: A case study from membrane synthesis to commercial deployment. *J. Membr. Sci.* **2010**, *346*, 270–279. [CrossRef]
5. Scholz, M.; Melin, T.; Wessling, M. Transforming biogas into biomethane using membrane technology. *Renew. Sustain. Energy Rev.* **2013**, *17*, 199–212. [CrossRef]
6. Sanders, D.F.; Smith, Z.P.; Guo, R.; Robeson, L.M.; McGrath, J.E.; Paul, D.R.; Freeman, B.D. Energy-efficient polymeric gas separation membranes for a sustainable future: A review. *Polymers* **2013**, *54*, 4729–4761. [CrossRef]
7. Wind, J.D.; Paul, D.R.; Koros, W.J. Natural gas permeation in polyimide membranes. *J. Membr. Sci.* **2004**, *228*, 227–236. [CrossRef]
8. Xiao, Y.; Low, B.T.; Hosseini, S.S.; Chung, T.S.; Paul, D.R. The strategies of molecular architecture and modification of polyimide-based membranes for CO2 removal from natural gas-A review. *Prog. Polym. Sci.* **2009**, *34*, 561–580. [CrossRef]
9. Sridhar, S.; Smitha, B.; Aminabhavi, T. Separation of carbon dioxide from natural gas mixtures through polymeric membranes-A Review. *Sep. Purif. Rev.* **2007**, *36*, 113–174. [CrossRef]
10. AlQaheem, Y.; Alomair, A.; Vinoba, M.; Pérez, A. Polymeric gas-separation membranes for petroleum refining. *Int. J. Polym. Sci.* **2017**, *2017*, 1–19. [CrossRef]
11. Robeson, L.M. The upper bound revisited. *J. Membr. Sci.* **2008**, *320*, 390–400. [CrossRef]
12. Freeman, B.D. Basis of permeability/selectivity tradeoff relations in polymeric gas separation membranes. *Macromolecules* **1999**, *32*, 375–380. [CrossRef]

13. Yong, W.F.; Kwek, K.H.A.; Liao, K.-S.; Chung, T.-S. Suppression of aging and plasticization in highly permeable polymers. *Polymers* **2015**, *77*, 377–386. [CrossRef]
14. Low, Z.-X.; Budd, P.M.; McKeown, N.B.; Patterson, D.A. Gas permeation properties, physical aging, and its mitigation in high free volume glassy polymers. *Chem. Rev.* **2018**, *118*, 5871–5911. [CrossRef] [PubMed]
15. Harasimowicz, M.; Orluk, P.; Zakrzewska-Trznadel, G.; Chmielewski, A. Application of polyimide membranes for biogas purification and enrichment. *J. Hazard. Mater.* **2007**, *144*, 698–702. [CrossRef]
16. Uddin, M.W.; Hägg, M.-B. Natural gas sweetening-The effect on CO_2-CH_4 separation after exposing a facilitated transport membrane to hydrogen sulfide and higher hydrocarbons. *J. Membr. Sci.* **2012**, 143–149. [CrossRef]
17. Liu, Y.; Liu, Z.; Liu, G.; Qiu, W.; Bhuwania, N.; Chinn, D.; Koros, W.J. Surprising plasticization benefits in natural gas upgrading using polyimide membranes. *J. Membr. Sci.* **2020**, *593*, 117430. [CrossRef]
18. Hao, J.; Rice, P.; Stern, S. Upgrading low-quality natural gas with H_2S- and CO_2-selective polymer membranes. *J. Membr. Sci.* **2008**, *320*, 108–122. [CrossRef]
19. Scholes, C.A.; Kentish, S.E.; Stevens, G.W. Effects of minor components in carbon dioxide capture using polymeric gas separation membranes. *Sep. Purif. Rev.* **2009**, *38*, 1–44. [CrossRef]
20. White, L.S.; Blinka, T.A.; Kloczewski, H.A.; Wang, I.-F. Properties of a polyimide gas separation membrane in natural gas streams. *J. Membr. Sci.* **1995**, *103*, 73–82. [CrossRef]
21. Toth, G.; Nemestóthy, N.; Bélafi-Bakó, K.; Vozik, D.; Bakonyi, P. Degradation of hydrogen sulfide by immobilized Thiobacillus thioparus in continuous biotrickling reactor fed with synthetic gas mixture. *Int. Biodeterior. Biodegrad.* **2015**, *105*, 185–191. [CrossRef]
22. Lashkari, S.; Tran, A.; Kruczek, B. Effect of back diffusion and back permeation of air on membrane characterization in constant pressure system. *J. Membr. Sci.* **2008**, *324*, 162–172. [CrossRef]
23. Mohammadi, A.T.; Matsuura, T.; Sourirajan, S. Design and construction of gas permeation system for the measurement of low permeation rates and permeate compositions. *J. Membr. Sci.* **1995**, *98*, 281–286. [CrossRef]
24. Ismail, A.F.; Khulbe, K.C.; Matsuura, T. Application of Gas Separation Membranes. In *Gas Separation Membranes*; Springer: Berlin, Germany, 2015; pp. 978–983.
25. Gaspar, D.J.; Phillips, S.D.; George, A.; Albrecht, K.O.; Jones, S.; Howe, D.T.; Landera, A.; Santosa, D.M.; Howe, D.T.; Baldwin, A.G.; et al. Measuring and predicting the vapor pressure of gasoline containing oxygenates. *Fuel* **2019**, *243*, 630–644. [CrossRef]
26. Cecopierigomez, M.; Palaciosalquisira, J.; Dominguez, J. On the limits of gas separation in CO_2/CH_4, N_2/CH_4 and CO_2/N_2 binary mixtures using polyimide membranes. *J. Membr. Sci.* **2007**, *293*, 53–65. [CrossRef]
27. Jeon, Y.-W.; Lee, D.-H. Gas Membranes for CO_2/CH_4 (Biogas) Separation: A Review. *Environ. Eng. Sci.* **2015**, *32*, 71–85. [CrossRef]
28. Ohlrogge, K.; Brinkmann, T. Natural gas cleanup by means of membranes. *Ann. New York Acad. Sci.* **2003**, *984*, 306–317. [CrossRef]
29. Lee, J.S.; Koros, W.J.; Koros, W.J. Antiplasticization and plasticization of Matrimid®asymmetric hollow fiber membranes-Part A. Experimental. *J. Membr. Sci.* **2010**, *350*, 232–241. [CrossRef]
30. Al-Juaied, M.; Koros, W.J. Performance of natural gas membranes in the presence of heavy hydrocarbons. *J. Membr. Sci.* **2006**, *274*, 227–243. [CrossRef]
31. Omole, I.C.; Bhandari, D.A.; Miller, S.J.; Koros, W.J. Toluene impurity effects on CO_2 separation using a hollow fiber membrane for natural gas. *J. Membr. Sci.* **2011**, *369*, 490–498. [CrossRef]

Publisher's Note: MDPI stays neutral with regard to jurisdictional claims in published maps and institutional affiliations.

© 2020 by the authors. Licensee MDPI, Basel, Switzerland. This article is an open access article distributed under the terms and conditions of the Creative Commons Attribution (CC BY) license (http://creativecommons.org/licenses/by/4.0/).

Article

Analysis of CO₂ Facilitation Transport Effect through a Hybrid Poly(Allyl Amine) Membrane: Pathways for Further Improvement

Bouchra Belaissaoui [1,*], **Elsa Lasseuguette** [2], **Saravanan Janakiram** [3], **Liyuan Deng** [3] **and Maria-Chiara Ferrari** [2]

1. LRGP-CNRS, University of Lorraine, ENSIC, 1 rue Grandville, 54001 Nancy, France
2. School of Engineering, University of Edinburgh, Robert Stevenson Road, Edinburgh EH9 3FB, UK; e.lasseuguette@ed.ac.uk (E.L.); M.Ferrari@ed.ac.uk (M.-C.F.)
3. Department of Chemical Engineering, Norwegian University of Science and Technology (NTNU), NO-7491 Trondheim, Norway; saravanan.janakiram@ntnu.no (S.J.); liyuan.deng@ntnu.no (L.D.)
* Correspondence: bouchra.belaissaoui@univ-lorraine.fr

Received: 21 October 2020; Accepted: 22 November 2020; Published: 25 November 2020

Abstract: Numerous studies have been reported on CO_2 facilitated transport membrane synthesis, but few works have dealt with the interaction between material synthesis and transport modelling aspects for optimization purposes. In this work, a hybrid fixed-site carrier membrane was prepared using polyallylamine with 10 wt% polyvinyl alcohol and 0.2 wt% graphene oxide. The membrane was tested using the feed gases with different relative humidity and at different CO_2 partial pressures. Selected facilitated transport models reported in the literature were used to fit the experimental data with good agreement. The key dimensionless facilitated transport parameters were obtained from the modelling and data fitting. Based on the values of these parameters, it was shown that the diffusion of the amine-CO_2 reaction product was the rate-controlling step of the overall CO_2 transport through the membrane. It was shown theoretically that by decreasing the membrane selective layer thickness below the actual value of 1 μm to a value of 0.1 μm, a CO_2 permeance as high as 2500 GPU can be attained while maintaining the selectivity at a value of about 19. Furthermore, improving the carrier concentration by a factor of two might shift the performances above the Robeson upper bound. These potential paths for membrane performance improvement have to be confirmed by targeted experimental work.

Keywords: facilitated transport; fixed site carrier membrane; polyallylamine-polyvinyl alcohol-graphene oxide membrane; modelling; carbon capture; gas permeation

1. Introduction

Post-combustion capture (PCC) is an efficient strategy to achieve greenhouse gas emission reductions, as it can be retrofitted to existing power stations or industrial plants and can be integrated into new ones. In the post-combustion framework, flue gases are treated at atmospheric pressure and carbon dioxide is diluted in nitrogen with a typical CO_2 volume fraction of 5% (natural gas turbine exhaust), 15% (coal combustion power plant), and 30% (steel plant or oxygen enriched air flue gas). As a result, CO_2 partial pressure in the flue gas is very low, creating a major engineering challenge, especially in terms of energy and membrane surface area requirements for the separation process [1,2]. Many studies have been dedicated to improving existing and already mature technologies (i.e., gas-liquid absorption in amine solvents, cryogenic separation, adsorption). Their success hinges on their ability to lower the cost of CO_2 capture while still attaining the targets for CO_2 purity and recovery ratio. Membrane separation processes represent an interesting alternative as a more energy-efficient process with no need for chemicals and no extra source of direct pollution. Commercial gas separation membranes are mainly based on dense polymers, such as polyimide, polysulfone, polycarbonate, polyphenyl oxide, cellulose derivatives, or poly(ethylene oxide) [3,4], and follow the solution-diffusion mechanism. The key shortcoming of solution-diffusion membranes is the trade-off between permeability and selectivity, governed by the Robeson's upper bound, as shown in Figure 1 [5,6]. The development of membrane materials with high permeability and selectivity is a key challenge for efficient CO_2 separation. High selectivity is essential to achieve the purity target at low energy cost, and high permeability is required to minimize the membrane surface area and related cost [7,8]. The challenge is then to push the material performances to the right upper side of the Robeson plot.

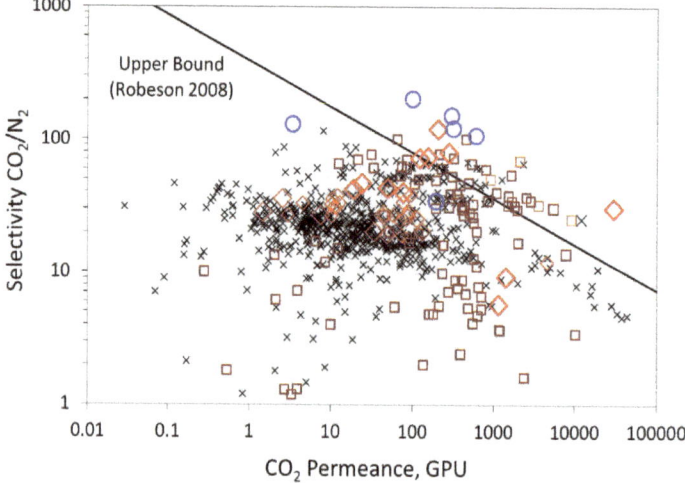

Figure 1. State of the art: trade-off curve showing the CO_2/N_2 selectivity (α) data for different membrane materials versus CO_2 permeance for a 1-μm-thick membrane. ×: polymeric membranes, □: inorganic membranes, ○: facilitated transport membranes, ◇: hybrid organic-inorganic membranes. The theoretical trade-off limit, calculated for a strict physical separation mechanism through a dense polymer (i.e., solution-diffusion), is also shown. A membrane thickness of one micron is considered for the upper bound to convert permeability in Barrer to permeance in GPU.

Among these membranes, facilitated transport membranes (FTMs) have gained much interest in recent years and have shown a promising performance beyond the Robeson upper bound region [3,4,9]. These membranes are based on a selective reversible reaction between the incorporated carrier agents and the target gas component. Facilitated transport membranes for CO_2 separation most commonly contain amino groups [10–12]. Polyvinylamine (PVAm) is one of the most intensively studied fixed-site carrier polymeric membrane materials [13,14].

In such membrane, the carrier is covalently bonded to the polymer backbone and a CO_2 molecule reacts with one carrier site in the presence of water, which induces the formation of bicarbonate (HCO_3^-) on the feed side interface of the membrane (Figure 2); the complex diffuses along its concentration gradient to the permeate side of the membrane, "hopping" through fixed carrier sites until it reaches the permeate side [15,16]. CO_2 is released on the permeate side, regenerating the carrier, which can react with another CO_2 molecule on the feed side. Therefore, a major part of CO_2 is transported by the carriers inside the membrane, in addition to the physical solution-diffusion mechanism, which other non-reactive gases also follow (Figure 2). A measure of the facilitation effect is the facilitation factor, defined as the ratio of total solute flux with the carrier present to the solute diffusional flux. It represents the contribution of the CO_2-carrier reaction to the overall transport. Thus, a high facilitation factor corresponds to high selectivity as the flux of the facilitated solute is enhanced in comparison to the diffusion flux of other components in the mixture to be separated.

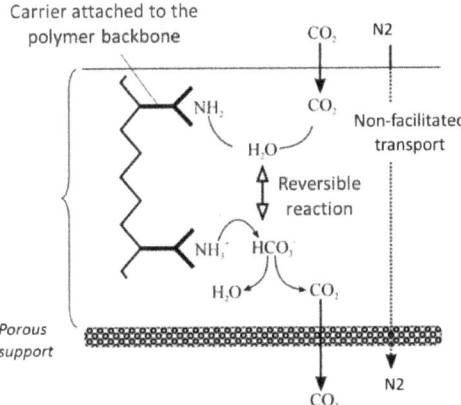

Figure 2. Illustration of the facilitated transport mechanism of fixed-site amine carrier membrane.

From the theoretical point of view, many studies have been carried out to calculate and predict the permeation rates and facilitation factors. The general formulation of the facilitated transport of a gas in a membrane can be described mathematically by means of non-linear differential equations expressing a diffusion reaction mass balance of the species involved. The treatment of the nonlinear diffusion reaction (NLDR) problem has been developed considering approximate analytical and/or numerical solutions. Despite the advance in the numerical techniques and computational power, the approximate analytical solutions are still useful because they are more flexible and reliable for extreme cases, such as the chemical equilibrium regime or/and excess carrier. As an example, assuming a large excess of carrier compared to solute is equivalent to assuming that the carrier concentration is constant across the membrane thickness. With this assumption, the NLDR problem describing this process becomes linear and can be solved easily. Several analytical solutions have been proposed to predict the facilitation factor of fixed-site carrier membranes [16–21]. In the conditions where these analytical models are applicable, they are accurate despite the simplicity of their mathematical description.

Generally, several studies were reported on membrane synthesis, but few works have dealt with the interaction between material synthesis and transport modelling aspects. In facilitated transport systems, the interaction between both chemistry and chemical engineering is key to understanding the relationship of facilitated transport system properties to separation performances. The performance of a facilitated transport membrane process is dependent upon a number of physical and chemical properties. These properties can be independently measured or retrieved thanks to appropriate analytical transport solutions at suitable operating conditions [22]. Indeed, the interaction of modelling and experimentation is very useful in improving the knowledge base and permits optimization of the facilitation transport mechanism at both the material and process scale [16,23].

For the same combination of carrier and key permeant (here CO_2), improvement of the membrane performance can possibly be achieved by the modification of the total carrier concentration and the selective layer thickness. Their effect on CO_2 permeability and selectivity is not trivial [23,24].

On the one hand, decreasing the membrane thickness will not systematically increase the permeance of CO_2. Indeed, even if the solution-diffusion permeance mechanism of both CO_2 and N_2 increases, the transport of CO_2 by the facilitation mechanism can be reduced significantly, due to reaction limitation (reduced inverse Damkoler number). The effects of the two mechanisms balance, and the facilitated effect (and thus selectivity) can be depleted. On the other hand, the possible positive effect of increasing the carrier concentration has to be weighted by the possible depletion of the reaction complex effective diffusivity. Indeed, the mobility of the complex is a function of the product of both an increasing carrier concentration and reaction complex effective diffusivity, as has been pointed out in the literature [25]. This has to be confirmed by targeted experimental work.

For these reasons, a combined modelling and experimental strategy was used in the present study, in order to show if improvement of the actual membrane performance through chemical and structural modifications is possible and how this could be achieved and evaluated. The results could be used to guide future experimental work. Such a combined analysis is infrequent in the literature.

In this work, first, a hybrid fixed-site carrier membrane using polyallylamine (PAA) with 10 wt% polyvinyl alcohol (PVA) and 0.2wt% graphene oxide was prepared. In order to increase membrane resistance to plasticization and compaction under relevant industrial conditions of CO_2 capture, hybrid membranes combining the aforementioned polymers and inorganic additives were recently investigated, with the inorganic fillers used as a reinforcing agent in the membrane structure [26,27]. Experimental measurements of CO_2 and N_2 fluxes through a hybrid fixed-site carrier membrane, based on poly(allyl amine) matrix, under different operation conditions of the feed relative humidity and CO_2 upstream partial pressure, are presented and discussed. Second, analytical solutions of the facilitation factor are presented and used to analyze the experimental results and estimate some key facilitation properties. A dedicated parametric analysis was performed to show how the facilitation effect of CO_2 transport through the studied membrane is affected by the key properties of the membrane, mainly total carrier concentration and membrane selective layer thickness. Indeed, their effect on CO_2 permeability and selectivity is not obvious. Moreover, the modelling results are compared to actual membrane performances to determine if significant improvements are possible in system performance and how this could be achieved and evaluated. Finally, pathways for membrane chemical and structural modifications are proposed in order to increase the CO_2 facilitated transport and improve the membrane separation performance.

2. Modelling Background

The most common and generalized reaction scheme for the facilitated transport mechanism of CO_2 across fixed-site amine carrier membranes is expressed as A + C = AC, in which three species co-exist: the key solute (A), here CO_2, being transported across the membrane; the active chemical carrier (C); and their reaction products or active carrier–solute complex (AC) [26,27].

The solute is transferred from one boundary to the other by two different mechanisms, pure diffusion in an unreacted state and diffusion as a complexed species. Considering a flat-plane geometry with one-dimensional transport (z-direction), Fick's law diffusion mechanism, and constant diffusion coefficient of the species, the total solute flux can be expressed as:

$$J_A = -D_A \frac{dC_A}{dz} - D_{Ac} \frac{dC_{Ac}}{dz}. \tag{1}$$

The facilitation factor F is an evaluation of the impact of the facilitation reaction compared to the pure diffusion mechanism. It is a measure of the increased selectivity resulting from the selective facilitation transport of the key component in a mixture. It is defined as the ratio of the total solute flux of A inside the membrane to the pure solution-diffusion (SD) flux, representing the contribution of the reaction to the overall transport:

$$F = \frac{J_A}{J_A^{SD}} = \frac{total\ solute\ flux}{solute\ solution\ diffusion\ flux} \tag{2}$$

The general formulation of the facilitated transport of a gas in a membrane can be described mathematically by means of differential equations expressing a steady-state nonlinear diffusion reaction (NLDR) problem. The species transport consists of simple diffusion coupled with a single chemical reversible reaction. The forward rate reaction is assumed to depend on both the carrier and solute concentration, while the backward reaction has a linear dependence on the reaction product concentration. Both kinetic constants, k_f and k_r, are considered concentration independent (with $K_{eq} = k_f/k_r$). The mathematical derivation of the mass balance for fixed-site carrier membranes is an analogue to the mobile carrier formulation while an excess of carrier is considered [27]. Indeed, assuming excess carrier, the concentration of the carrier can be considered constant and defined by the reaction equilibrium calculated at the membrane upstream side [18]. Accordingly, the differential equations that describe the steady-state solute transport for fixed-site carrier membranes are:

$$D_A \frac{\partial^2 C_A}{\partial z^2} = k_f \left(C_A C_C - \frac{1}{K_{eq}} C_{AC} \right) \qquad (3)$$

$$D_{AC} \frac{\partial^2 C_{AC}}{\partial z^2} = -k_f \left(C_A C_C - \frac{1}{K_{eq}} C_{AC} \right) \qquad (4)$$

With

$$C_C = \frac{C_T}{1 + K_{eq} C_{AO}}, \qquad (5)$$

Typical boundary conditions consist of the fact that the carrier and the product of the reaction are non-volatile and are constrained to stay confined inside the membrane. A constant source of solute A is considered at one boundary and A is removed continuously from the opposite boundary so that the concentration at that boundary is a constant. Only the solutes can cross the membrane boundaries:

$$\text{at } z = 0, \ C_A = C_{A0} \qquad \frac{\partial C_C}{\partial z} = \frac{\partial C_{AC}}{\partial z} = 0 \qquad (6)$$

$$\text{at } z = L, \ C_A = C_{AL} \qquad \frac{\partial C_C}{\partial z} = \frac{\partial C_{AC}}{\partial z} = 0 \qquad (7)$$

In a dimensionless form, the differential equation system becomes:

$$\frac{\partial^2 C_A^*}{\partial z^2} = \frac{\alpha_m K}{\varepsilon} \left(C_A^* C_C^* - \frac{1}{K} C_{AC}^* \right), \qquad (8)$$

$$\frac{\partial^2 C_{AC}^*}{\partial z^2} = -\frac{K}{\varepsilon} \left(C_A^* C_C^* - \frac{1}{K} C_{AC}^* \right), \qquad (9)$$

And:

$$C_C^* = \frac{1}{1 + K^*}. \qquad (10)$$

With the following boundary conditions:

$$\text{at } z^* = 0, \ C_A^* = 1 \qquad \frac{\partial C_{AC}^*}{\partial z^*} = 0 \qquad (11)$$

$$\text{at } z^* = 1, \ C_A^* = \frac{C_{AL}}{C_{A0}} \qquad \frac{\partial C_{AC}^*}{\partial z^*} = 0 \qquad (12)$$

The treatment of the nonlinear diffusion reaction (NLDR) problem was developed considering approximate analytical and/or numerical solutions [17,18,22,28]. Many analytical solutions have been proposed to predict the facilitation factor of fixed-site carrier membranes. The solution developed by Smith and Quinn [18] assumes an excess of carrier and zero downstream key permeant partial pressure (here CO_2). According to this model, the facilitation factor, F, is expressed in terms of the key dimensionless facilitated transport parameters, K, ε, α, and λ, as the following:

$$F = \frac{1 + \frac{\alpha_m K}{1+K}}{1 + \frac{\alpha_m K}{1+K}\left(\frac{\tanh\lambda}{\lambda}\right)} \quad (13)$$

In this model, the facilitated factor is expressed as a function of key physicochemical properties, such as the reaction rate constant, chemical equilibrium constant, diffusivities of the chemical species, and membrane thickness. They are combined in a number of dimensionless groupings, having physical significance, which are presented in Table 1.

The key dimensionless number appearing in the equations above are defined in Table 1. These properties of the facilitated transport may be combined in a number of dimensionless groupings, having physical significance. The mobility ratio, α_m, can be defined as the reactive versus the diffusive pathway. It is related to the ratio (D_{AC}/D_A) of the diffusion coefficient of the A-carrier reaction product and that of free solute A inside the membrane. The mobility ratio is also proportional to the initial carrier concentration. $k_{D,A}$ is the sorption coefficient of solute A in the membranes. C_{A0} is the feed molar concentration of solute A. K is a dimensionless equilibrium constant. ε is the inverse of a Damkohler number and is a measure of the characteristic reverse reaction time to the characteristic diffusion time; it serves the same function as a Thiele modulus in catalysis.

Table 1. Definition of the key dimensionless number in facilitation transport.

Dimensionless Numbers	Definition	Expression
ε	Inverse Damkohler number, ratio of the characteristic reverse reaction to diffusion time	$\varepsilon = \frac{D_{AC}}{k_r \ell^2}$
α_m	Mobility ratio of mobility of carrier to mobility of solute	$\alpha_m = \frac{D_{AC} C_T}{D_A C_{A0}}$, with $C_{A0} = k_{D,A}p'_A$
K	Dimensionless reaction equilibrium constant	$K = \frac{k_f C_{A0}}{k_r} = k_{eq} C_{A0}$
λ	Measure of the facilitation factor	$\lambda = \frac{1}{2}\sqrt{\frac{1+(\alpha+1)K}{\varepsilon(1+K)}}$

It can be seen from the definition of ε that the latter is thickness dependent. These dimensionless parameters allow for a simplified evaluation of the performance of particular carrier/solute combinations.

According to Equation (13), F is mainly determined by the value of $\tanh\lambda/\lambda$. The value of λ decreases with the K and α_m values, thus high K and α_m values are desired. The maximum F is obtained when λ tends to infinity (reaction equilibrium) and thus $\tanh\lambda/\lambda$ to zero while F decreases to one when λ tends to zero and thus $\tanh\lambda/\lambda$ to one (reaction kinetics limitations).

The value of $\tanh \lambda/\lambda$ is then a measure of the facilitation effect. Indeed, a simple and quick calculation of this one term can give the range of the facilitation effect factor [17]. Accordingly, one may determine the necessary property modifications to move toward reaction equilibrium by moving the above quantity toward zero and thus to rich high facilitation factors [17].

In the limit of the equilibrium reaction, Equation (13) becomes:

$$F = 1 + \frac{\alpha_m K}{1+K}. \quad (14)$$

Generally, the key permeant (here CO_2) flux through the membrane is expressed as a sum of two contributions: solution diffusion flux of free CO_2 and flux of the CO_2-amine complex. Assuming the reaction equilibrium approximation throughout the membrane, the total CO_2 flux through the membrane film can be calculated as follows [29]:

$$J_A = D_A \frac{k_{D,A}}{e}(p'_A - p''_A) + D_{AC}K_{eq}C_T \frac{k_{D,A}}{e}\left[\frac{p'_A}{1+K_{eq}k_{D,A}\,p'_A} - \frac{p''_A}{1+K_{eq}k_{D,A}\,p''_A}\right], \quad (15)$$

where p'_A and p''_A are the upstream and downstream partial pressure of A, respectively.

Accordingly, the facilitated factor can be expressed as the following:

$$F = 1 + \frac{D_{AC}K_{eq}C_T}{D_A(p'_A - p''_A)}\left[\frac{p'_A}{1+K_{eq}k_{D,A}\,p'_A} - \frac{p''_A}{1+K_{eq}k_{D,A}\,p''_A}\right] \quad (16)$$

In case of a downstream CO_2 pressure equal to 0, Equation (16) reduces to Equation (14).

Table 2 summarizes the hypothesis and the facilitation expression of each model, namely: (i) the simplified equilibrium model (Equation (14)); (ii) the equilibrium model expression, which considers a non-zero downstream concentration (Equation (16)); and (iii) the general model expression (Equation (13)).

Table 2. Summary of the hypothesis and the facilitation expression of each model.

Model Hypothesis		Model Name	Model Equation (F Expression)
Chemical equilibrium ($\tanh\lambda/\lambda \sim 0$)	Zero downstream CO_2 concentration ($p''_{CO_2} = 0$) (very dilute CO_2 gas)	Simplified equilibrium model	Equation (14)
	Non-zero downstream CO_2 concentration	Equilibrium model	Equation (16)
Non-chemical equilibrium (reaction limitation)	Zero downstream CO_2 concentration	General model	Equation (13)

3. Membrane Fabrication

PAA-PVA-GO membrane was prepared at the Norwegian University of Science and Technology (Group of Prof. L. Deng) in the framework of NanoMEMC^2 project. It is a composite membrane based on a selective layer composed of a mixture of 0.2 wt% graphene oxide (GO), 89.8 wt% poly(allyl amine) (PAA), and 10 wt% poly(vinyl alcohol) (PVA), coated onto a porous support made of polyvinylidene fluoride (PVDF). The choice of the polymer and nanofiller composition is attributed to the superior performance recorded from previous studies [29]. The use of PVA is to exploit its excellent film-forming capabilities while the use of GO as nanofiller enhances selective gas transport by reorienting polymer chain packing and increasing water distribution in the polymer matrix. Several tests have been conducted in these membranes recently, with studies explaining the role of the individual components present in such hybrid membranes [30]. These membranes have also been validated in industrial testing conditions [31].

Poly(allyl amine) was obtained by purification of poly(allyl amine hydrochloride) (Mw = 120,000 g/mol) (bought from Thermo Fisher Scientific, Stockholm, Sweden) using equivalent potassium hydroxide (KOH) in methanol at room temperature for 24 h. The purified poly(allyl amine) was recovered by centrifuging the supernatant comprising of PAA in methanol. The solution was then dried in a ventilated oven at 60 °C to result in dry polymer, ready to be re-dissolved in deionized water. About 6 wt% solution of PAA in water was obtained by stirring (at 500 rpm) for 24 h. In the case of

PVA, a 4 wt% polymer solution in water was prepared by dissolving PVA pellets in deionized water at 80 °C for 4 h under reflux conditions. GO dispersion was obtained as 2 mg mL^{-1} from Graphene-XT, Bologna, Italy.

In order to prepare the casting solution for membrane fabrication, the calculated amount of GO solution was first diluted using water. Drops of both 6 wt% PAA and 4 wt% PVA solutions were then carefully added to the diluted dispersion under stirring conditions to result in a casting solution, which contained 89.8 wt% PAA, 10 wt% PVA, and 0.2 wt% GO. The total solid contents (polymer + GO) in the casting solution were maintained at 2wt%.

The composite membrane was fabricated using the bar coating method mentioned elsewhere [32]. The porous support was washed with tap water at 45 °C for 1 h and DI water for 30 min to remove the pore protection agents. The support was then mounted on to a flat glass plate using aluminum tape. The polymer cast solution was then casted over the support and the membrane was dried at 60 °C in vacuum for 4–5 h. With the bar coating method, defect-free selective layer coating was obtained with the use of FC-72 electronic liquid (3M™ Fluorinert™) as the pore-filling agent. Scanning electron microscopy (Figure 3) revealed that the selective skin thickness was approximately 1 µm. Based on the physical properties of PAA (Mw = 12,000 g/mol, density = 1.02 g/cm^3), it was possible to determine a total carrier concentration in the membrane of 1.61×10^{-2} mol·cm^{-3}.

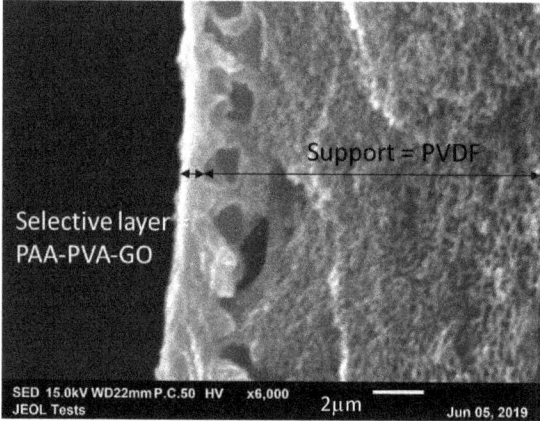

Figure 3. SEM images of a cross-section of polyallylamine (PAA)-poly(vinyl alcohol) (PVA)-graphene oxide (GO) membrane.

4. Experiments: Results and Discussion

Gas permeability measurements were performed with a mixed gas–continuous flow permeation cell designed at the University of Edinburgh; details on the permeation cell and the experimental procedure can be found in a previous article [33]. A flat circular membrane with an area of 2.835 cm^2 was used for the experiments.

The operating conditions used for our project are summarized in Table 3. The downstream pressure, feed and sweep flow rates, and sweep gas relative humidity were set constant, whereas the humidity content and the CO_2 partial pressure in the feed side were varied from 20% to 90% and from 0.15 to 3 bar, respectively.

The permeability $\wp_i$ and the permeance PM_i were calculated from the experimental data of the measured fluxes, J_i, as follows:

$$PM_{i,\, i=CO2,N2} = \frac{J_i}{S\left(p'_i - p''_i\right)} = \frac{\wp_{i,\, i=CO2,N2}}{e}, \tag{17}$$

where e is the thickness of the membrane (cm), S is the membrane surface area, J_i is the gas permeate flowrate of component j (cm^3 (STP)/s), A is the effective membrane area (cm^2), and p'_i and p''_i are the partial pressure of the gas in the feed and permeate stream, respectively (cmHg).

The CO_2/N_2 ideal selectivity $\alpha_{CO2/N2}$ between two gas species was calculated as the ratio of the two permeabilities:

$$\alpha_{CO2/N2} = \frac{\wp_{CO2}}{\wp_{N2}}. \tag{18}$$

The common unit of permeability is Barrer, which corresponds to 10^{-10} cm^3 (STP) cm cm^{-2} s^{-1} cmHg^{-1}. The permeance is defined as the ratio of permeability to the selective layer thickness. Its common unit is GPU (gas permeation unit), which corresponds to 10^{-6} cm^3 (STP) cm^{-2} s^{-1} cmHg^{-1}.

Table 3. Operating conditions.

Operating Conditions	Value
Temperature °C	40 and 50
Feed composition	10–100% CO_2/N_2
Sweep gas composition	Pure He
Feed pressure (bar)	1–3
CO_2 partial pressure, P_{CO2} (bar)	0.15–3
Sweep pressure (bar)	1
Feed Relative humidity, RH_f (%)	20–90
Sweep Relative humidity, RH_s (%)	50
Feed flow rate, Q_F (Ncm^3/min)	150
Sweep flow rate, Q_s (Ncm^3/min)	10

4.1. Effect of Feed Relative Humidity

A large relative humidity is required for the facilitated transport reaction mechanism to operate in the polymer [34]. Moreover, the impact of humidity on the solution diffusion mechanism can also be important. Indeed, a high water content is likely to swell rubbery membranes, increasing the polymeric chain mobility and as a result inducing increased gas permeabilities and possible loss in CO_2/N_2 selectivity. Moreover, a loss of CO_2 permeability can be observed in glassy polymer due to the competitive sorption between water and CO_2 and the free volume reduction effect [34–37].

Figure 4 shows the effect of variation of the feed relative humidity (RH_f) on the CO_2 and N_2 permeances and CO_2/N_2 membrane selectivity. The RH_f varied from 20% to 90% with a constant sweep gas relative humidity of 50% and a temperature of 50 °C. It can be seen that both CO_2 and N_2 permeances increase with RH_f, with CO_2 permeance increasing exponentially. Indeed, an increase in RH leads to a higher water content in the membrane matrix, which enhances the reaction between CO_2 and amine carriers and the mobilities of the reaction products but also unreacted CO_2 and N_2 [27,34]. Regarding the CO_2/N_2 selectivity, it shows a non-monotonic trend as it shows a slight maximum at a relative humidity of 70%. A selectivity drop at a high RH might be associated with enhanced penetration of non-reacting gas molecules relative to the contribution of facilitated transport due to the swelling of the membrane at high water content. The same trend was also observed by Sandru et al. [14] using PVAm/PPO for CO_2/N_2 separation at a feed pressure of 2.2 bar, 10% CO_2–90% N_2 feed at 25 °C. They showed that with increasing humidity content, the CO_2/N_2 selectivity increased to the maximum at 65% feed RH and then decreased. The same trend was also observed by Ansaloni et al. [27] using amino-functionalized multi-walled carbon nanotubes under high pressures (15–28 bar) and high temperatures (103–121 °C). They showed that the CO_2/CH_4 and CO_2/H_2 selectivity increased with feed RH until a maximum at 72% RH and then decreased. This trend has already been reported in the literature and can be explained by the occurrence of important water-induced swelling of the membrane at a high relative humidity content in the gas, which impaired the barrier property of the membrane toward N_2 [27].

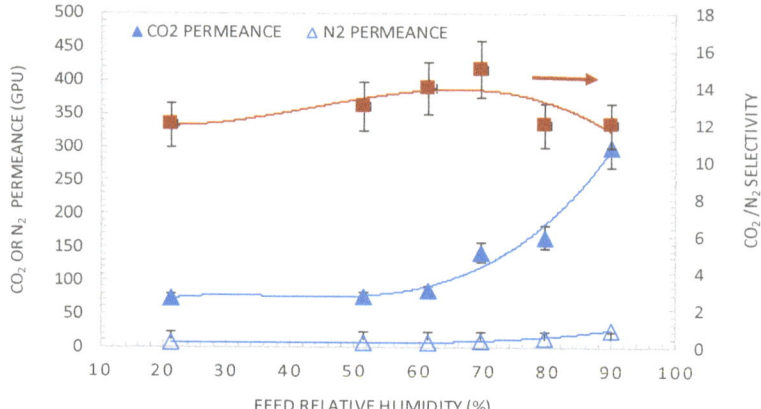

Figure 4. The effect of feed relative humidity on the CO_2 and N_2 permeances and CO_2/N_2 membrane selectivity, at a feed CO_2 partial pressure of 1.5 bar and temperature of 50 °C. Lines are to guide the eye.

4.2. Effect of CO_2 Partial Pressure

A general characteristic of facilitated transport membrane is that the CO_2 permeance is pressure dependent, as the carrier is saturated under high CO_2 partial pressure, resulting in decreasing CO_2 selectivity over other gases. Figure 5 shows the effect of varying the CO_2 upstream partial pressure on the CO_2 permeance and CO_2/N_2 membrane selectivity. The CO_2 upstream partial pressure varied from 0.15 to 3 bar with a feed relative humidity of about 85% and a temperature of 40 °C. The permeance of N_2 was found to be independent from the feed pressure and has a value of 13 GPU under the investigated operating conditions. As illustrated in this figure, CO_2 permeance and selectivity decrease with the feed pressure and then reach a constant value. This can be explained by the carrier saturation phenomenon [11,38]. For CO_2 partial pressure of 0.15 bar, the actual membrane CO_2 permeance is of 302 GPU and a CO_2/N_2 selectivity of 23.2.

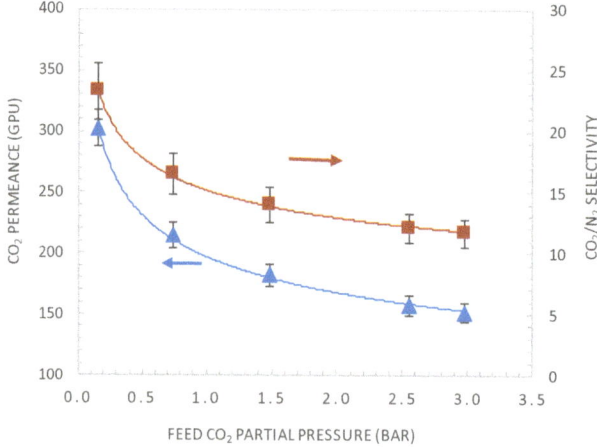

Figure 5. The effect of CO_2 upstream partial pressure on the CO_2 permeance and CO_2/N_2 selectivity, at a feed relative humidity of 85% and temperature of 40 °C.

From the above results, it can be seen that the best performances of the membrane correspond to a permeance of 300 GPU and CO_2/N_2 selectivity of 24. These performances are located below the upper-bound Robeson plot shown in Figure 2. The objective of the present work was to show through a dedicated combined modelling and experimental strategy if further improvement of the actual membrane performance is possible and how this could be achieved and evaluated. The results could be used to guide future experimental work.

5. Simulations: Results and Discussion

5.1. Retrieval of Key Facilitation Parameters

The physical and chemical properties of the facilitation factor can be independently measured or estimated. The analytical expression, being simple, offers the possibility to estimate some key system properties based on transport measurements and simple fitting techniques. Indeed, the analytical solution presented above can provide the basis for estimating the carrier gas complex diffusion coefficient and the reaction equilibrium constant. Assuming reaction equilibrium and a negligible CO_2 downstream concentration, the analytical solution of Smith and Quinn [18], Equation (14), can be rearranged as:

$$(F-1)^{-1} = E = \alpha_m^{-1} + (\alpha_m K)^{-1} = \left(\frac{D_A k_{D,A}}{D_{AC} C_T}\right) p'_A + \left(\frac{D_A}{D_{AC} C_T}\right) K_{eq} \qquad (19)$$

Thus, by varying the solute feed concentration and measuring the permeant flux, one can calculate the facilitation factor, F, and make a plot of $E = (F-1)^{-1}$ versus the solute upstream partial pressure. A straight-line relationship (as in Equation (19)) implies reaction equilibrium, i.e., diffusion-limited system (tanh $\lambda/\lambda \sim 0$). Deviation from a straight line indicates reaction limitations (tanh $\lambda/\lambda \neq 0$). Considering the straight-line part of the curve, at a low CO_2 upstream concentration, the slope and intercept can be used to determine the complex diffusion coefficient D_{AC} and the reaction equilibrium constant K_{eq}, respectively [19,37,38].

The total carrier concentration, C_T, for the tested membrane was calculated to be 1.61×10^{-2} mol·cm^{-3}. This value is in accordance with the literature and among the highest values [39,40]. The CO_2 permeance corresponding to the solution-diffusion mechanism, PM_{CO2}^{SD}, through the membrane was set at 145 GPU (+/−4), determined from experimental data under high permeant partial pressure corresponding to carrier saturation, conditions where CO_2 flux increase is only due to the pure solution–diffusion mechanism. Indeed, we suppose that saturation occurs at 3 bars, but this is an approximation, as saturation could occur at higher pressure; indeed, a starting pseudo-plateau can be observed for pressures above 2.5 bar. The CO_2 diffusion coefficient was set at a value of 1×10^{-6} cm^2·s^{-1}, which is in the range reported in the literature with respect to the order of magnitude of amine-functionalized carrier membranes [28,41–44].

Starting from the value of CO_2 permeance in the SD mechanism and the value of the CO_2 diffusion coefficient, the CO_2 sorption coefficient, k_d, can be calculated according to the following equation:

$$\wp_{CO2}^{SD} = k_{D,CO2} D_{CO2} = PM_{CO2}^{SD} * e. \qquad (20)$$

For our system, we found a value of $k_d = 4.84 \times 10^{-5}$ moL·cm^{-3}·bar^{-1}, which is comparable to values from the literature for amorphous polymers [45].

Thanks to the measurement of the CO_2 permeabilities, it is possible to calculate an experimental facilitation factor according to the following equation:

$$F_{exp} = \frac{PM_{CO2}}{PM_{CO2}^{SD}}. \qquad (21)$$

The plot of $E = (F-1)^{-1}$ versus upstream CO_2 partial pressure is shown in Figure 6.

Applying the analysis exposed above (Equation (7)), which is valid at very low CO_2 partial pressures, a straight-line fitting curve in this range (<1.5 bar) is plotted and added in Figure 6. Thus, the slope and the intercept of this line are used to estimate the complex diffusion coefficient and the reaction equilibrium constant, respectively.

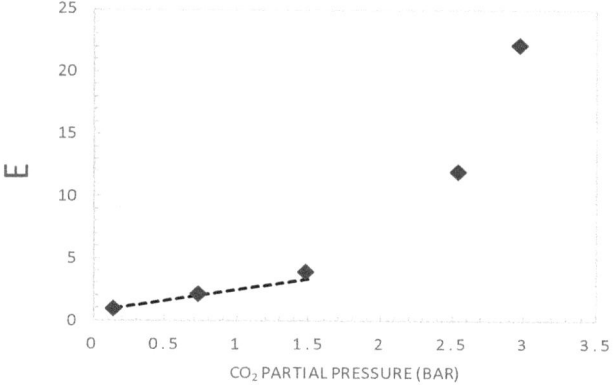

Figure 6. Plot of E as a function of the upstream CO_2 partial pressure. The dashed line corresponds to a straight-line fitting curve at very low CO_2 partial pressures (<1.5 bar), according to Equation (19).

The retrieved value of D_{AC} is found to be equal to 1.7×10^{-9} cm$^2 \cdot$s^{-1}, which is about one order of magnitude lower than that reported in the literature for CO_2-amine complex. In an ion exchange membrane with ethylene diamine as a carrier [41], a value of about 9.7×10^{-9} cm$^2 \cdot$s^{-1} was estimated. Cussler [15] showed that for fixed-site carrier membrane or "chained carrier" membrane, the order of magnitude of the apparent diffusion coefficient of the reaction complex is 10^{-8} cm$^2 \cdot$s^{-1}. This can explain the low permeance value even if the value of C_T is relatively high compared to typical values from the literature. The retrieved reaction equilibrium constant K_{eq} is 5.2×10^4 cm$^3 \cdot$moL^{-1}. It is found to be in accordance with typical values for the amine–CO_2 reaction from the literature [41,46]. The value of the reverse reaction constant, k_r, was set at 110 s^{-1}, similar to typical values from the literature for the CO_2–amine reaction [28,47,48].

5.2. Comparison of Experimental and Simulation Results

Table 4 summarizes the properties of the membrane used for the modelling analysis.

Table 4. Permeation properties used for the modelling analysis.

D_{CO2} (cm$^2 \cdot$s^{-1})	$k_{d,CO2}$ (mol·cm^{-3}·bar^{-1})	D_{AC} (cm$^2 \cdot$s^{-1})	K_{eq} (cm$^3 \cdot$mol^{-1})	K_r (s^{-1})	SD CO_2 Permeability (Barrer)	SD CO_2 Permeance (GPU), * e = 1 µm	N_2 Permeability (Barrer)	N_2 Permeance (GPU) * e = 1 µm
1×10^{-6}	4.84×10^{-5}	1.7×10^{-9}	5.20×10^4	110	145	145	13	13

* Relative to dimensionless.

From the predicted facilitation factor, the permeance of CO_2 and CO_2/N_2 selectivity is calculated, respectively, as follows:

$$PM_{CO2} = F \times PM_{CO2}^{SD} \qquad (22)$$

$$\alpha = \frac{PM_{CO2}}{PM_{N2}}. \qquad (23)$$

The experimental and modelling results are compared in terms of the facilitation factor, CO_2 permeance, and CO_2/N_2 selectivity as a function of the CO_2 upstream partial pressure in Figures 7 and 8, respectively.

The simplified equilibrium model, which assumes reaction equilibrium and negligible CO_2 permeate pressure (Equation (2)), shows good agreement with the experimental results at low feed CO_2 partial pressure, as expected. However, it diverges for higher values of the feed CO_2 partial pressure. Both the equilibrium model and general model show good agreement with the experimental results in the whole investigated CO_2 feed pressure range, with a maximum deviation between the experimental and predicted values below 5%. Figure 8 shows the comparison of CO_2 permeance and CO_2/N_2 selectivity between the general model prediction (Equation (13)) and experimental values. These results again demonstrate the good fitting ability of the general model.

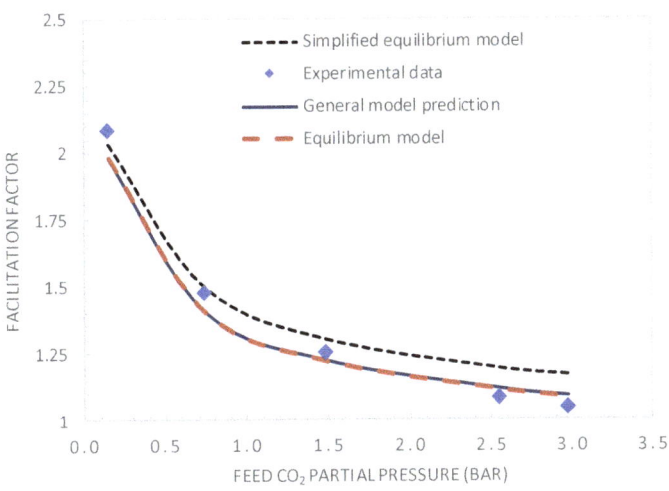

Figure 7. Comparison of the model predictions and experimental values of the facilitation factor.

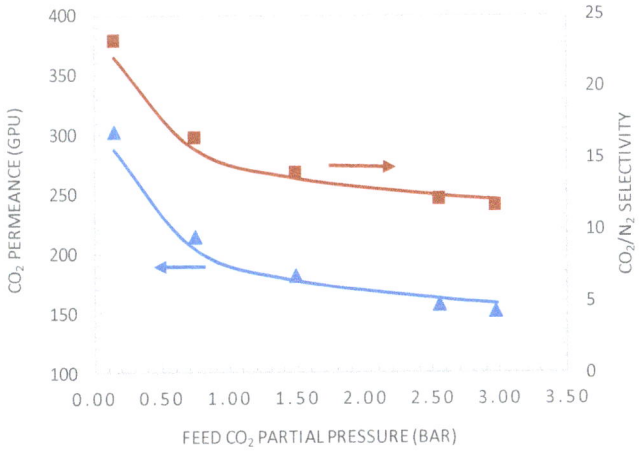

Figure 8. Comparison of CO_2 permeance and CO_2/N_2 selectivity between the general model simulation and experimental values. Simulations are shown in a continuous line.

For the sake of analysis, the dimensionless facilitated transport parameters for the actual membrane were calculated for two limits of the investigated range of upstream CO_2 partial pressure: P_{CO2} = 0.15 bar and 3 bar. The results are given in Table 5.

Table 5. Calculated dimensionless facilitated transport parameters, P_{CO2} = 0.15 and 3 bar.

P_{CO2} (bar)	α_m	ε	K	$\tanh\lambda/\lambda$
0.15	3.78	1.54×10^{-3}	0.377	1.52×10^{-3}
3	0.91	1.54×10^{-3}	7.49	2.64×10^{-3}

The value of $\tanh\lambda/\lambda$, in Equation (13), is a measure of solute facilitated transport. In the operating conditions of the experiments, the values of $\tanh\lambda/\lambda$ were all lower than 3×10^{-3}, indicating that the system is operating near the reaction equilibrium regime or diffusional limitation regime. This indicates that, under the actual operating conditions, the facilitation factor is already at its maximum for the actual membrane composition and selective layer thickness. For the actual membrane system, the values of K calculated for the two limits of the upstream CO_2 partial pressure lie globally within the optimal values range [1.5–10], predicted by the substantial numerical parametric analysis of Kemena et al [24,28,38].

The calculated ε was below 0.01, indicating again that compared to the reaction rate, the diffusion rate of the reaction complex was the rate-controlling step of the overall CO_2 transport across the membrane. This also means that there might be room for membrane separation performance improvement by modifying the value of ε and the mobility ratio α_m. These modifications can be examined by analyzing the effect of the membrane selective layer thickness and the initial carrier concentration in the membrane, C_T, on membrane performances.

5.3. Parametric Analysis

For the same combination of carrier and key permeant (here CO_2), improvement of the membrane performance can possibly be achieved by the modification of the total carrier concentration, C_T, the selective layer thickness, δ, and/or the operating parameters, such as temperature (T), CO_2 partial pressure (p_{CO2}), and relative humidity (RH). The membrane selective layer thickness and carrier concentration are two key characteristics that can be modified in order to improve performances, but their effects on CO_2 permeability and selectivity are not trivial [22,24]. In this section, a parametric analysis is achieved in order to show if improvement of the actual membrane performance is theoretically possible. The effect of the membrane selective layer thickness and total carrier concentration was evaluated through a parametric analysis, using the general model solution (Equation (13)), presented in the previous section. The membrane selective layer thickness and total carrier concentration were varied around their actual values of 1 μm and 1.61×10^{-2} mol·cm^{-3}, respectively.

The permeability of N_2 as well as the SD CO_2 permeability were kept constant at the experimentally measured values of 13 and 145 Barrer, respectively, regardless of the total carrier concentration inside the membrane and CO_2 upstream partial pressure. The SD permeance for a given membrane thickness was then calculated according to Equation (20). From the predicted values of the facilitated factor, the permeance of CO_2 for a given membrane thickness was calculated according to Equation (22).

5.3.1. Effect of Membrane Thickness

The analytical expression of the facilitation factor clearly indicates, from the definition of ε, the thickness dependence of the facilitation factor. Figure 9 shows the variation of the CO_2 permeance and CO_2/N_2 selectivity with the membrane thickness for a CO_2 partial pressure of 0.75 bar. Indeed, considering that the CO_2 concentration in the feed gas car varies depending on the emission sources, from 5% (natural gas turbine exhaust) to 30% (steel plant or oxygen enriched air flue gas), and that the flue gas can be pressurized to about 4 bar, a CO_2 partial pressure of 0.75 bar was taken as an average value.

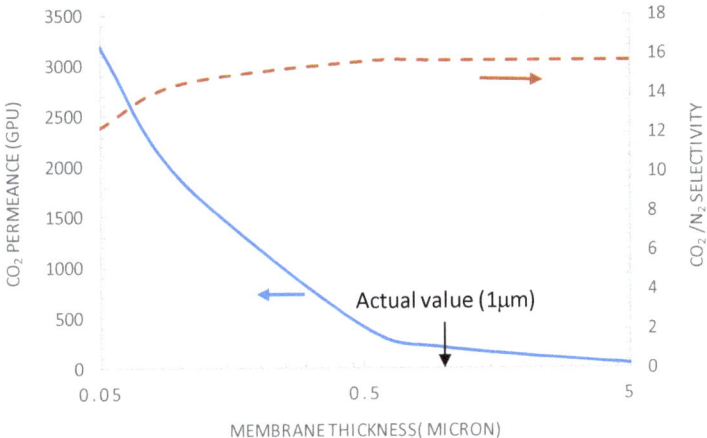

Figure 9. CO_2 permeance and CO_2/N_2 selectivity as a function of the selective layer membrane thickness (general model simulation results). $p'_{CO2} = 0.75$ bar, $CT = CT_0 = 1.61 \times 10^{-2}$ mol·cm^{-3}.

The selective layer thickness varied in the range 0.05–5 µm. The results in Figure 9 show that the permeance decreases monotonically with increasing membrane thickness. This trend can be explained by the fact that the mass flux through a membrane is inversely proportional to the selective layer thickness. On the contrary, the selectivity shows an increasing trend with an increasing selective layer thickness. Actually, the species diffusion becomes more limiting relative to the reaction, translating into decreased ε values, as the latter is proportional to the inverse of the square of the membrane thickness. Consequently, the facilitation effect and CO_2/N_2 selectivity increase with an increased selective layer thickness.

Moreover, from Figure 9, it is clear that there is no benefit from increasing the membrane thickness above the actual value of 1 µm, as CO_2 permeance will deplete without any substantial positive effect on membrane selectivity. Decreasing the membrane selective layer thickness from the actual value to a value around 0.1 µm permits a significant increase of the CO_2 permeance without a significant effect on membrane selectivity. As an example, for a membrane thickness of 0.1 µm and CO_2 upstream partial pressure of 0.75 bar, a CO_2 permeance of 1880 GPU can be reached while maintaining the selectivity at a value of about 14.47. These performances are already located on the upper bound of the Robeson plot. It is found that the actual value of ε is of about 10^{-3}, indicating that the system is under a diffusional limitation regime and experiences the maximum facilitation factor and membrane selectivity [22]. This explains why an increase in membrane thickness above the actual value has no benefit on membrane selectivity as shown in Figure 9.

Figures 10 and 11 show the variation of CO_2/N_2 selectivity and CO_2 permeance as a function of the CO_2 partial pressure, respectively, for different values of membrane thickness. First, it can be seen that the increase of the membrane thickness above the actual value of 1 µm results in an insignificant improvement of the membrane selectivity (see Figure 10). Furthermore, decreasing the membrane thickness below the actual value to 0.1 µm has a negligible effect on the membrane selectivity and a positive effect on the CO_2 permeance (see Figure 11). Moreover, the calculated values of tanh λ/λ are below 0.1, indicating that the system is near reaction equilibrium and thus near the maximum facilitation factor, explaining the negligible effect of increasing the selective layer thickness on membrane selectivity. As an example, for a membrane thickness of 0.1 µm and CO_2 upstream partial pressure of 0.15 bar, a CO_2 permeance as high as 2500 GPU can be attained while maintaining the selectivity at a value of about 19. These performances are interestingly above the Robeson upper bound for CO_2/N_2 separation. These results highlight the thickness dependence characteristic of the facilitation factor, which has rarely been reported in the literature [10,27]. Generally, the facilitation factor does not directly correspond to absolute permeate flux or capacity. Increasing the membrane thickness (i.e., decreasing ε) will continuously increase F toward its maximum value. However, the permeate solution diffusion flux (the denominator in F) decreases as the membrane thickness increases. Moreover, the maximal facilitation factor will give the reader a measure of the best selectivity obtainable for a given set of operating conditions, and this has to be weighed along with the permeate flux obtained when designing such a system [24]. Consequently, in order to increase the selectivity of the investigated membrane system, the modifications could be achieved by increasing the mobility ratio through increased carrier solubility or/and amine-carrier complex diffusivity. The effect of the total carrier concentration modification on the membrane performances is analyzed and discussed in the next section.

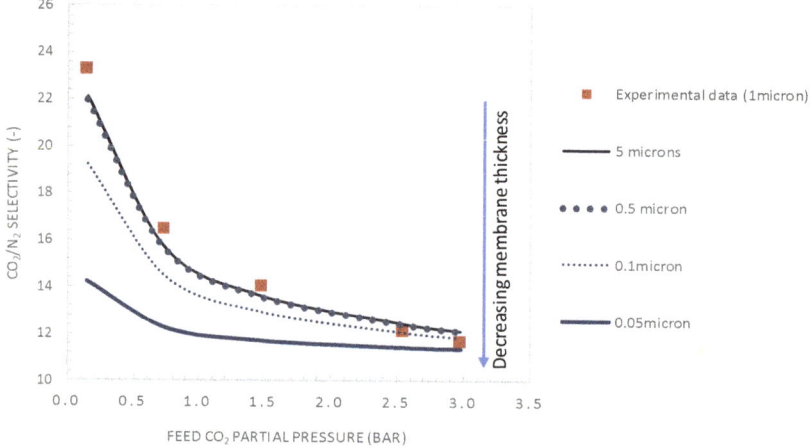

Figure 10. CO_2/N_2 selectivity as a function of the feed CO_2 partial pressure. Results are given for different values of selective layer membrane thickness (general model simulation results). $CT = CT_0 = 1.61 \times 10^{-2}$ mol·cm^{-3}.

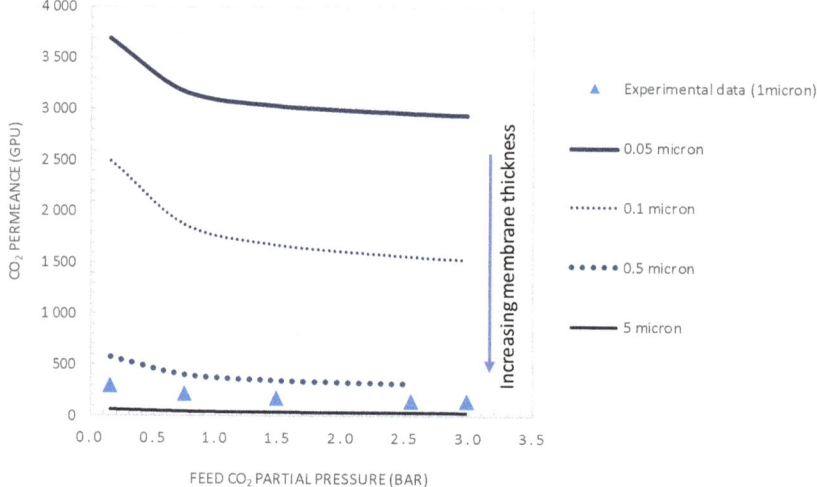

Figure 11. CO_2 permeance as a function of the feed CO_2 partial pressure. Results are given for different values of selective layer membrane thickness (general model simulation results). $CT = CT_0 = 1.61 \times 10^{-2}$ mol·cm^{-3}.

5.3.2. Effect of Carrier Concentration

The facilitation factor is a function of the mobility ratio α_m, which depends on the ratio (D_{AC}/D_A) of the diffusion coefficient of the CO_2-amine reaction product to that of CO_2 inside the membrane and on the initial carrier concentration, C_T, for a given CO_2 upstream partial pressure. Figure 12 shows the variation of CO_2 permeance and CO_2/N_2 selectivity with the total carrier concentration for a CO_2 partial pressure of $p_{CO2} = 0.75$ bar. The carrier concentration was varied around the actual value of 1.61×10^{-2} mol·cm^{-3}, noted as C_{T0} in Figure 12. The permeance and selectivity were found to increase linearly with an increasing total carrier concentration. This trend can be explained by the fact that an increase in the mobility ratio, α_m, with the total carrier concentration induces an increase in CO_2 facilitation transport. Many authors have shown experimentally that the facilitated flux increases linearly as a function of the carrier concentration through different FTMs [41,49,50]. The concentration of carriers in the membrane matrix can be increased by the addition of mobile carriers [30].

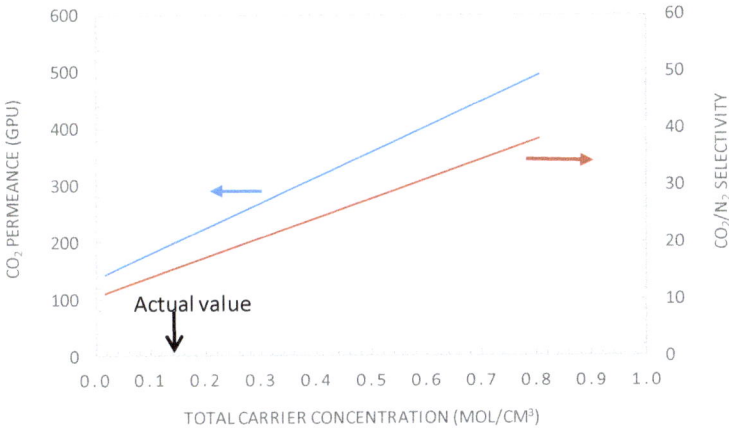

Figure 12. CO_2 permeance and CO_2/N_2 selectivity as the total carrier concentration (general model simulation results). $p'_{CO2} = 0.74$ bar, e = 1 µm.

Figures 13 and 14 show the evolution of the CO_2/N_2 selectivity and CO_2 permeance as a function of the upstream CO_2 partial pressure, respectively, for different values of the initial carrier concentration. As expected, the results indicate that increasing the C_T significantly increases the CO_2 permeance and membrane selectivity. As an example, for a CO_2 partial pressure of 0.15 bar, increasing the carrier concentration by a factor of 2 or 5 will significantly increase the membrane performance from its actual performance ($PM_{CO2} = 302$ GPU and $\alpha_{CO2/N2} = 23.2$), to $PM_{CO2} = 429$GPU and $\alpha_{CO2/N2} = 33$ or $PM_{CO2} = 891$ GPU and $\alpha_{CO2/N2} = 69$, respectively. These performances are above the Robeson upper bound of the CO_2/N_2 pair.

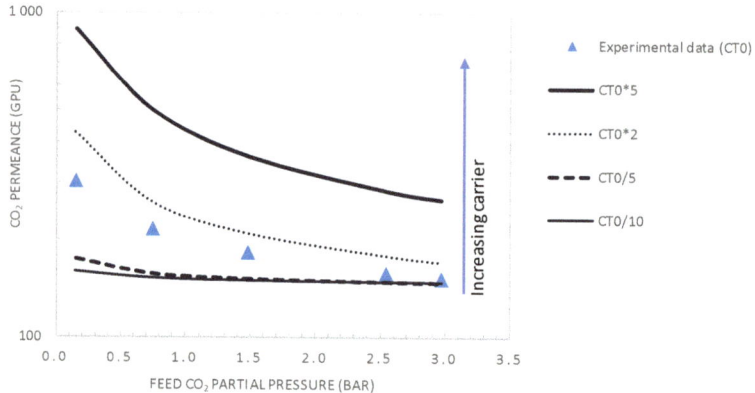

Figure 13. CO_2 permeance as a function of the feed CO_2 partial pressure (general model simulation results). Results are given for different values of the total carrier concentration around the actual value of $C_{T0} = 1.61 \times 10^{-2}$ mol·cm^{-3}.

It is expected that decreasing the selective layer thickness to 0.1 µm together with doubling of the total carrier concentration will shift the membrane performance far above the Robeson upper bound for the CO_2/N_2 pair. However, it is worth noticing that the product C_T*D_{AC} affects the mobility ratio, α_m, and thus the value of the facilitation effect. An increase in C_T could be offset by a decrease in D_{AC}, subsequent to the change in membrane morphology, as pointed out in the theoretical analysis of Noble [25]. Such a decrease in D_{AC} with carrier loading has been reported in the literature for O_2 transport by Tsuchida and co-workers [25]. Moreover, on the one hand, an increase of the carrier concentration increases CO_2-amine complex formation. On the other hand, the salting-out effect might occur at a high carrier concentration. Under these conditions, the formed ionic species (e.g., carbamate, bicarbonate or zwitterion) tend to surround the carrier and the polymeric chain, making it difficult for CO_2 molecules to access the carrier, depleting CO_2 solubility and reaction complex diffusion through the membrane [10]. The result of the two opposite effects indicates that an optimal carrier concentration exists. The occurrence of an optimal amine concentration has also been observed experimentally for CO_2 facilitated transport [51–53].

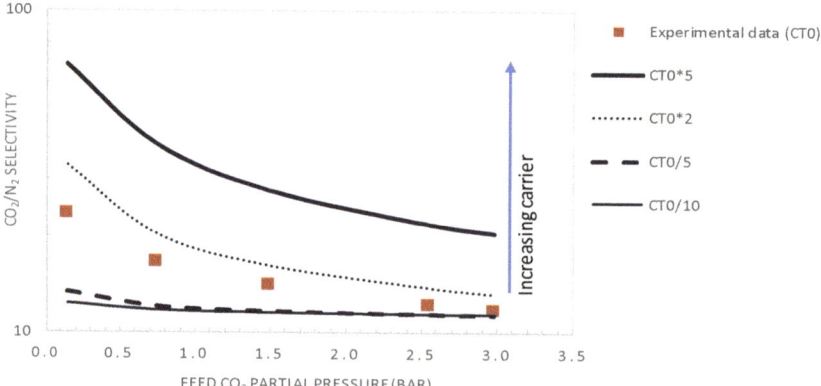

Figure 14. CO_2/N_2 selectivity as a function of the feed CO_2 partial pressure (general model simulation results). Results are given for different values of the total carrier concentration around the actual value of $C_{T0} = 1.61 \times 10^{-2}$ mol·cm^{-3}.

6. Conclusions

Based on combined transport measurement and modelling methods, the key membrane system properties were retrieved, and pathways for membrane performance enhancement through chemical and structural modifications (i.e., membrane selective layer thickness and total carrier concentration) were proposed. Experimental measurement of CO_2 and N_2 fluxes through a PAA-PVA-GO hybrid fixed-site carrier membrane under different operation conditions was performed. The effects of the humidity content and the CO_2 partial pressure were investigated. The values of the CO_2-amine reaction product diffusivity and the reaction equilibrium constant were found to be equal to 1.7×10^{-9} cm$^2 \cdot$s^{-1} and 5.2×10^4 cm$^3 \cdot$mol^{-1}, respectively. For a CO_2 partial pressure of 0.15 bar, the actual membrane CO_2 permeance was 302 GPU and CO_2/N_2 selectivity was 23.2, performances that were below the well-known Robeson upper bound. The comparison of experimental results and the analytical model predictions showed very good agreement. The major conclusions can be summarized as follows:

1. It was demonstrated that the current system operated near the reaction equilibrium regime (i.e., diffusion limitation), maximizing the facilitated transport of CO_2. This indicated that, under the investigated operating conditions, the membrane thickness was already at its optimal value maximizing the facilitation factor.
2. Increasing the membrane selectivity of the actual membrane by increasing the mobility ratio through increasing the carrier concentration and amine-CO_2 complex diffusivity is key to improving the membrane performances.

Furthermore, a parametric analysis regarding the membrane thickness and total carrier concentration was also performed. The main results were:

1. It was shown that after decreasing the membrane selective layer thickness below the actual value of 1 µm to a value of 0.1 µm and CO_2 upstream partial pressure of 0.15 bar, a CO_2 permeance as high as 2500 GPU can be attained while maintaining the selectivity at a value of about 19.
2. Moreover, increasing the carrier solubility by a factor of two permitted the attainment of a CO_2 permeance of 429 GPU and CO_2/N_2 selectivity of 33, performances that are above the Robeson upper bound of the CO_2/N_2 pair.

Moreover, it is expected that decreasing the selective layer thickness to 0.1 µm together with doubling of the total carrier concentration will theoretically shift the membrane performance far above the Robeson upper bound for the CO_2/N_2 pair. However, this potential path for membrane performance improvement has to be weighted by the possible depletion in the reaction complex effective diffusivity, as pointed out in the literature. Finally, it is important to emphasize that the analysis set forth in this paper provides some guidance for membrane performance enhancement through chemical and structural modifications. These potential improvement pathways have to be confirmed by targeted experimental work.

Author Contributions: Conceptualization, B.B., E.L. and M.-C.F.; Investigation, B.B. and E.L.; data curation, E.L. and S.J.; Writing—original draft preparation, B.B., E.L. and M.-C.F.; writing—review and editing, B.B., E.L., S.J., M.-C.F. and L.D. All authors have read and agreed to the published version of the manuscript.

Funding: This research was funded by European Union's Horizon 2020 Research and Innovation program under Grant Agreement No. 72773, and by Widen Horizon mobility program of the University of Lorraine under grant agreement: No. 200335.

Acknowledgments: The corresponding author gratefully acknowledges the financial support from the Widen Horizon mobility program of the University of Lorraine under grant agreement: No. 200335. She also thanks the University of Edinburgh and the School of Engineering for their invitation. This work acknowledges the financial support from the European Union's Horizon 2020 Research and Innovation program under Grant Agreement No. 727734.

Conflicts of Interest: The authors declare no conflict of interest.

Nomenclature

D	Diffusion coefficient ($cm^{-2} \cdot s^{-1}$)
e	Membrane selective layer thickness (cm)
F	Facilitation factor (dimensionless)
RH	Relative humidity (dimensionless)
K_{eq}	Reaction equilibrium constant ($cm^3 \cdot mol^{-1}$)
$\wp_i$	Permeability of gas i (Barrer)
p	Partial pressure (bar)
PM	Permeance (GPU)
K	Dimensionless equilibrium constant
k	Reaction rate constant (s^{-1})
J_i	Permeate flowrate of gas i (cm^3 (STP)$\cdot s^{-1}$)
D_{AC}	Diffusion coefficient of the carrier-permeant complex ($cm^{-2} \cdot s^{-1}$)
p'	upstream or feed partial pressure (bar)
p''	downstream partial pressure (bar)
S	Effective membrane surface area (cm^2)
C_T	Total carrier concentration ($mol \cdot cm^{-3}$)
$k_{d.,CO2}$	Sorption coefficient of CO_2 in the membranes ($mol \cdot cm^{-3} \cdot bar^{-1}$)
C	Molar concentration ($mol \cdot cm^{-3}$)
z	distance from the upstream side of the membrane (m)

Greek symbols

α_m	Mobility ratio (dimensionless)
α	Membrane selectivity (dimensionless)
ε	Inverse Damkhöler number (dimensionless)
K	Reaction equilibrium number (dimensionless)
$\alpha_{i/j}$	CO_2/N_2 ideal selectivity between tow gas species i and j (dimensionless)
λ	A measure of the facilitation factor (dimensionless)
$\wp_i$	Permeability of gas i (Barrer)

Subscripts

f	Feed
A	Solute
AC	Carrier-solute complex
r	Reverse
f	Forward
i	Compound
0	Upstream side ($z = 0$)

Superscripts

SD	Relative to solution-diffusion
*	Relative to dimensionless

References

1. Davidson, O.; Metz, B. *Special Report on Carbon Dioxide Capture and Storage*; International Panel on Climate Change: Geneva, Switzerland, 2005.
2. Belaissaoui, B.; Favre, E. Membrane Separation Processes for Post-Combustion Carbon Dioxide Capture: State of the Art and Critical Overview. *Oil Gas Sci. Technol. Rev.* **2013**, *69*, 1005–1020. [CrossRef]
3. Ebner, A.D.; Ritter, J.A. State-of-the-art Adsorption and Membrane Separation Processes for Carbon Dioxide Production from Carbon Dioxide Emitting Industries. *Sep. Sci. Technol.* **2009**, *44*, 1273–1421. [CrossRef]
4. Luis, P.; Van Gerven, T.; Van Der Bruggen, B. Recent developments in membrane-based technologies for CO_2 capture. *Prog. Energy Combust. Sci.* **2012**, *38*, 419–448. [CrossRef]
5. Robeson, L.M. The upper bound revisited. *J. Membr. Sci.* **2008**, *320*, 390–400. [CrossRef]
6. Freeman, B.D. Basis of Permeability/Selectivity Tradeoff Relations in Polymeric Gas Separation Membranes. *Macromolecules* **1999**, *32*, 375–380. [CrossRef]

7. Steeneveldt, R.; Berger, B.; Torp, T. CO_2 Capture and Storage. *Chem. Eng. Res. Des.* **2006**, *84*, 739–763. [CrossRef]
8. Belaissaoui, B.; Willson, D.; Favre, E. Membrane gas separations and post-combustion carbon dioxide capture: Parametric sensitivity and process integration strategies. *Chem. Eng. J.* **2012**, 122–132. [CrossRef]
9. Ramasubramanian, K.; Ho, W.S.W. Recent developments on membranes for post-combustion carbon capture. *Curr. Opin. Chem. Eng.* **2011**, *1*, 47–54. [CrossRef]
10. Matsuyama, H.; Terada, A.; Nakagawara, T.; Kitamura, Y.; Teramoto, M. Facilitated transport of CO_2 through polyethylenimine/poly(vinyl alcohol) blend membrane. *J. Membr. Sci.* **1999**, *163*, 221–227. [CrossRef]
11. Zou, J.; Ho, W.W. CO_2-selective polymeric membranes containing amines in crosslinked poly(vinyl alcohol). *J. Membr. Sci.* **2006**, *286*, 310–321. [CrossRef]
12. Huang, J.; Zou, J.; Ho, W.S.W. Carbon Dioxide Capture Using a CO_2-Selective Facilitated Transport Membrane. *Ind. Eng. Chem. Res.* **2008**, *47*, 1261–1267. [CrossRef]
13. Kim, T.-J.; Li, B.; Hägg, M.-B. Novel fixed-site-carrier polyvinylamine membrane for carbon dioxide capture. *J. Polym. Sci. Part B Polym. Phys.* **2004**, *42*, 4326–4336. [CrossRef]
14. Sandru, M.; Haukebø, S.H.; Hägg, M.-B. Composite hollow fiber membranes for CO_2 capture. *J. Membr. Sci.* **2010**, *346*, 172–186. [CrossRef]
15. Cussler, E.; Aris, R.; Bhown, A. On the limits of facilitated diffusion. *J. Membr. Sci.* **1989**, *43*, 149–164. [CrossRef]
16. Noble, R.D. Generalized microscopic mechanism of facilitated transport in fixed site carrier membranes. *J. Membr. Sci.* **1992**, *75*, 121–129. [CrossRef]
17. Rea, R.; De Angelis, M.G.; Baschetti, M.G. Models for Facilitated Transport Membranes: A Review. *Membranes* **2019**, *9*, 26. [CrossRef] [PubMed]
18. Smith, U.R.; Quinn, J.A. The prediction of facilitation factors for reaction augmented membrane transport. *AIChE J.* **1979**, *25*, 197–200. [CrossRef]
19. Noble, R.D.; Way, J.D.; Powers, L.A. Effect of external mass-transfer resistance on facilitated transport. *Ind. Eng. Chem. Fundam.* **1986**, *25*, 450–452. [CrossRef]
20. Jemaa, N.; Noble, R.; Koval, C. Combined mass and energy balance analysis of an electrochemically modulated equilibrium stage process. *Chem. Eng. Sci.* **1992**, *47*, 1469–1479. [CrossRef]
21. Paul, D.R.; Koros, W.J. Effect of partially immobilizing sorption on permeability and the diffusion time lag. *J. Polym. Sci. Polym. Phys. Ed.* **1976**, *14*, 675–685. [CrossRef]
22. Noble, R.D. Relationship of system properties to performance in facilitated transport systems. *Gas Sep. Purif.* **1988**, *2*, 16–19. [CrossRef]
23. Koval, C.A.; Reyes, Z.E. Chemical aspects of facilitated transport through liquid membranes. In *Liquid Membranes: Theory and Applications*; ACS Symposium Series No. 347; American Chemical Society: Washington, DC, USA, 1987; pp. 27–38.
24. Kemena, L.; Noble, R.; Kemp, N. Optimal regimes of facilitated transport. *J. Membr. Sci.* **1983**, *15*, 259–274. [CrossRef]
25. Noble, R.D. Analysis of facilitated transport with fixed site carrier membranes. *J. Membr. Sci.* **1990**, *50*, 207–214. [CrossRef]
26. Rafiq, S.; Deng, L.; Hägg, M.-B. Role of Facilitated Transport Membranes and Composite Membranes for Efficient CO_2 Capture—A Review. *ChemBioEng Rev.* **2016**, *3*, 68–85. [CrossRef]
27. Ansaloni, L.; Zhao, Y.; Jung, B.T.; Ramasubramanian, K.; Baschetti, M.G.; Ho, W.W. Facilitated transport membranes containing amino-functionalized multi-walled carbon nanotubes for high-pressure CO_2 separations. *J. Membr. Sci.* **2015**, *490*, 18–28. [CrossRef]
28. Gottschlich, D.; Roberts, D.; Way, J. A theoretical comparison of facilitated transport and solution-diffusion membrane modules for gas separation. *Gas Sep. Purif.* **1988**, *2*, 65–71. [CrossRef]
29. Janakiram, S.; Espejo, J.L.M.; Yu, X.; Ansaloni, L.; Deng, L. Facilitated transport membranes containing graphene oxide-based nanoplatelets for CO_2 separation: Effect of 2D filler properties. *J. Membr. Sci.* **2020**, *616*, 118626. [CrossRef]
30. Janakiram, S.; Espejo, J.L.M.; Høisæter, K.K.; Lindbråthen, A.; Ansaloni, L.; Deng, L. Three-phase hybrid facilitated transport hollow fiber membranes for enhanced CO_2 separation. *Appl. Mater. Today* **2020**, *21*, 100801. [CrossRef]

31. Janakiram, S.; Santinelli, F.; Costi, R.; Lindbråthen, A.; Nardelli, G.M.; Milkowski, K.; Ansaloni, L.; Deng, L. Field trial of hollow fiber modules of hybrid facilitated transport membranes for flue gas CO_2 capture in cement industry. *Chem. Eng. J.* **2020**, *2020*, 127405. [CrossRef]
32. Janakiram, S.; Yu, X.; Ansaloni, L.; Dai, Z.; Deng, L. Manipulation of Fibril Surfaces in Nanocellulose-Based Facilitated Transport Membranes for Enhanced CO_2 Capture. *ACS Appl. Mater. Interfaces* **2019**, *11*, 33302–33313. [CrossRef]
33. Dhuiège, B.; Lasseuguette, E.; Brochier-Salon, M.-C.; Ferrari, M.-C.; Missoum, K. Crosslinked Facilitated Transport Membranes Based on Carboxymethylated NFC and Amine-Based Fixed Carriers for Carbon Capture, Utilization, and Storage Applications. *Appl. Sci.* **2020**, *10*, 414. [CrossRef]
34. Deng, L.; Hägg, M.-B. Swelling Behavior and Gas Permeation Performance of PVAm/PVA Blend FSC Membrane. *J. Membr. Sci.* **2010**, *363*, 295–301. [CrossRef]
35. Chen, G.Q.; Scholes, C.A.; Doherty, C.M.; Hill, A.J.; Qiao, G.G.; Kentish, S.E. The thickness dependence of Matrimid films in water vapor permeation. *Chem. Eng. J.* **2012**, *209*, 301–312. [CrossRef]
36. Pfister, M.; Belaissaoui, B.; Favre, E. Membrane Gas Separation Processes from Wet Postcombustion Flue Gases for Carbon Capture and Use: A Critical Reassessment. *Ind. Eng. Chem. Res.* **2017**, *56*, 591–602. [CrossRef]
37. Lasseuguette, E.; Carta, M.; Brandani, S.; Ferrari, M.-C. Effect of humidity and flue gas impurities on CO_2 permeation of a polymer of intrinsic microporosity for post-combustion capture. *Int. J. Greenh. Gas Control.* **2016**, *50*, 93–99. [CrossRef]
38. Schultz, J.S.; Goddard, J.D.; Suchdeo, S.R. Facilitated Transport via Carrier-Mediated Diffusion in Membranes Part 1. Mechanistic Aspects, Regimes. *AIChE J.* **1974**, *20*, 417–445. [CrossRef]
39. Han, Y.; Wu, D.; Ho, W.S.W. Simultaneous effects of temperature and vacuum and feed pressures on facilitated transport membrane for CO_2/N_2 separation. *J. Membr. Sci.* **2019**, *573*, 476–484. [CrossRef]
40. Zhao, Y.; Ho, W.W. Steric hindrance effect on amine demonstrated in solid polymer membranes for CO_2 transport. *J. Membr. Sci.* **2012**, 132–138. [CrossRef]
41. Way, J.D.; Noble, R.D.; Reed, D.L.; Ginley, G.M.; Jarr, L.A. Facilitated transport of CO_2 in ion exchange membranes. *AIChE J.* **1987**, *33*, 480–487. [CrossRef]
42. Way, J.; Noble, R.D. Competitive facilitated transport of acid gases in perfluorosulfonic acid membranes. *J. Membr. Sci.* **1989**, *46*, 309–324. [CrossRef]
43. Yamaguchi, T.; Boetje, L.M.; Koval, C.A.; Noble, R.D.; Bowman, C.N. Transport Properties of Carbon Dioxide through Amine Functionalized Carrier Membranes. *Ind. Eng. Chem. Res.* **1995**, *34*, 4071–4077. [CrossRef]
44. Sharifzadeh, M.M.M.; Amooghin, A.E.; Pedram, M.Z.; Omidkhah, M. Time-dependent mathematical modeling of binary gas mixture in facilitated transport membranes (FTMs): A real condition for single-reaction mechanism. *J. Ind. Eng. Chem.* **2016**, *39*, 48–65. [CrossRef]
45. Lin, H.; Freeman, B.D. Gas solubility, diffusivity and permeability in poly(ethylene oxide). *J. Membr. Sci.* **2004**, *239*, 105–117. [CrossRef]
46. Jensen, A.; Christensen, R.; Bonnichsen, R.; Virtanen, A.I. Studies on Carbamates. XI. The Carbamate of Ethylenediamine. *Acta Chem. Scand.* **1955**, *9*, 486–492. [CrossRef]
47. Wang, X.; Conway, W.; Fernandes, D.; Lawrance, G.; Burns, R.; Puxty, G.; Maeder, M. Kinetics of the Reversible Reaction of CO_2(aq) with Ammonia in Aqueous Solution. *J. Phys. Chem. A* **2011**, *115*, 6405–6412. [CrossRef]
48. Versteeg, G.; Swaaij, W. On the kinetics between CO_2 and alkanolamines both in aqueous and non-aqueous solutions—I. Primary and secondary amines. *Chem. Eng. Sci.* **1988**, *43*, 573–585. [CrossRef]
49. Yoshikawa, M.; Shudo, S.; Sanui, K.; Ogata, N. Active transport of organic acids through poly(4-vinylpyridine-co-acrylonitrile) membranes. *J. Membr. Sci.* **1986**, *26*, 51–61. [CrossRef]
50. Ogata, N.; Sanui, K.; Fujimura, H. Active transport membrane for chlorine ion. *J. Appl. Polym. Sci.* **1980**, *25*, 1419–1425. [CrossRef]
51. Tsuchida, E.; Nishide, H.; Ohyanagi, M.; Kawakami, H. Facilitated transport of molecular oxygen in the membranes of polymer-coordinated cobalt Schiff base complexes. *Macromolecules* **1987**, *20*, 1907–1912. [CrossRef]

52. Francisco, G.J.; Chakma, A.; Feng, X. Membranes comprising of alkanolamines incorporated into poly(vinyl alcohol) matrix for CO_2/N_2 separation. *J. Membr. Sci.* **2007**, *303*, 54–63. [CrossRef]
53. Cai, Y.; Wang, Z.; Yi, C.; Bai, Y.; Wang, J.; Wang, S. Gas transport property of polyallylamine–poly(vinyl alcohol)/polysulfone composite membranes. *J. Membr. Sci.* **2008**, *310*, 184–196. [CrossRef]

Publisher's Note: MDPI stays neutral with regard to jurisdictional claims in published maps and institutional affiliations.

© 2020 by the authors. Licensee MDPI, Basel, Switzerland. This article is an open access article distributed under the terms and conditions of the Creative Commons Attribution (CC BY) license (http://creativecommons.org/licenses/by/4.0/).

Article

Enhancing the Separation Performance of Glassy PPO with the Addition of a Molecular Sieve (ZIF-8): Gas Transport at Various Temperatures

Francesco M. Benedetti [1,2,*], Maria Grazia De Angelis [1,*], Micaela Degli Esposti [1], Paola Fabbri [1], Alice Masili [3], Alessandro Orsini [3] and Alberto Pettinau [3]

1. Department of Civil, Chemical, Environmental and Materials Engineering, University of Bologna, 40131 Bologna, Italy; micaela.degliesposti@unibo.it (M.D.E.); p.fabbri@unibo.it (P.F.)
2. Department of Chemical Engineering, Massachusetts Institute of Technology, Cambridge, MA 02139, USA
3. Sotacarbo S.p.A., Grande Miniera di Serbariu, 09013 Carbonia, Italy; alice.masili@sotacarbo.it (A.M.); alessandro.orsini@sotacarbo.it (A.O.); alberto.pettinau@sotacarbo.it (A.P.)
* Correspondence: fmben@mit.edu (F.M.B.); grazia.deangelis@unibo.it (M.G.D.A.); Tel.: +39-051-209-0410 (M.G.D.A.)

Received: 9 March 2020; Accepted: 24 March 2020; Published: 27 March 2020

Abstract: In this study, we prepared and characterized composite films formed by amorphous poly(2,6-dimethyl-1,4-phenylene oxide) (PPO) and particles of the size-selective Zeolitic Imidazolate Framework 8 (ZIF-8). The aim was to increase the permselectivity properties of pure PPO using readily available materials to enable the possibility to scale-up the technology developed in this work. The preparation protocol established allowed robust membranes with filler loadings as high as 45 wt% to be obtained. The thermal, morphological, and structural properties of the membranes were analyzed via DSC, SEM, TGA, and densitometry. The gas permeability and diffusivity of He, CO_2, CH_4, and N_2 were measured at 35, 50, and 65 °C. The inclusion of ZIF-8 led to a remarkable increase of the gas permeability for all gases, and to a significant decrease of the activation energy of diffusion and permeation. The permeability increased up to +800% at 45 wt% of filler, reaching values of 621 Barrer for He and 449 for CO_2 at 35 °C. The ideal size selectivity of the PPO membrane also increased, albeit to a lower extent, and the maximum was reached at a filler loading of 35 wt% (1.5 for He/CO_2, 18 for CO_2/N_2, 17 for CO_2/CH_4, 27 for He/N_2, and 24 for He/CH_4). The density of the composite materials followed an additive behavior based on the pure values of PPO and ZIF-8, which indicates good adhesion between the two phases. The permeability and He/CO_2 selectivity increased with temperature, which indicates that applications at higher temperatures than those inspected should be encouraged.

Keywords: gas separation; CO_2 capture; mixed-matrix membranes

1. Introduction

Hydrogen purification was among the first commercial applications that provided potential for large-scale membrane gas separation technologies [1–5]. Polymers entered that market in the 70s due to their low cost, processability, and mechanical properties. However, for polymeric membranes, there is a tradeoff between the permeability, which measures the productivity of the process, and the selectivity, which determines the process efficiency [6,7]. Consequently, there is an upper bound to the performance of polymeric membranes, which makes it difficult to simultaneously enhance the permeability and selectivity. Research on membranes is constantly seeking new materials to improve the membrane performance [8–12]. One way to circumvent the intrinsic limit of the polymers is to combine polymeric materials with selective nanoporous particles. Such fillers can improve the polymer permeability and/or selectivity given their intrinsic superior properties, without compromising those features that make polymeric systems the best choice for industrial applications. The composite membranes thus obtained are usually called mixed-matrix membranes (MMMs). With this aim, many different materials have been dispersed in organic polymers, such as silica particles, zeolites, graphene sheets, carbon molecular sieves (CMS), carbon nanotubes, metal organic frameworks (MOFs), and more recently, covalent organic frameworks (COFs) [13–22]. In view of all these alternatives, the choice of polymers and fillers that can synergistically combine in MMMs with enhanced properties is of great importance. The use of commercially available materials can address the urgent request to apply membrane technologies on an industrially relevant scale [5,23,24].

Since the discovery of MOFs by Yaghi and co-workers about 15 years ago [25], such materials have attracted the attention of the scientific community because of their exceptional properties and structural tunability, which mean that they can be employed in a virtually infinite range of design. In particular, they have been revealed to be great materials for gas storage and separation applications [24,26]. The first mixed-matrix membrane containing an MOF (i.e., Cu BPDC-TED/PAET) was reported in 2004, and it was tested via single gas permeation measurements [27]. A crucial aspect of the fabrication of MMMs is to ensure good adhesion between the two phases, in order to prevent the formation of non-selective voids at the interface, which can cause an undesired loss in selectivity [28,29]. On the other hand, it is essential to avoid interpenetration between the two phases and keep filler porosities available for gas diffusion, in order to fully exploit their separation ability. MOFs have been proven to have a higher affinity with organic polymer matrices with respect to zeolites, given the organic nature of the linkers that connect the metal clusters to one another [30]. Nevertheless, different strategies have been developed to further increase the interfacial compatibility between the components of MOF-based MMMs [31–33].

Zeolitic Imidazole Frameworks (ZIFs) belong to a particular class of MOFs that presents an isomorphism with zeolites [34]. However, the completely inorganic aluminum-silicate structure is replaced by imidazole organic linkers coordinated with metal ions to form ordered frameworks. The crystallographic structure of ZIFs provides them with a monomodal pore size distribution, which is a remarkable feature for separating small gas molecules in the Angstrom scale [35,36]. However, pure MOF membranes cannot reach the expected ideal selectivity towards molecules smaller than the pore diameter because the presence of imperfections such as pinholes and cracks is hard to completely avoid [30]. Such a drawback can be prevented by dispersing ZIFs in a polymer matrix, but in this case, the theoretical separation, obtained based on pore dimension considerations, is hard to reach experimentally due to the flexibility of the metal-organic cage, explained by the so-called "breathing" phenomenon [37,38]. Like all MOFs, ZIFs provide a wide range of designs that can be obtained by changing the imidazolate/imidazolate-like linkers and the coordination metal (e.g., zinc(II) or cobalt(II)) [34]. This leads to different topologies (e.g. *sod*, *rho*, *lta*, *gme*, *gis* etc.) and different dimensions of the pores, which range from 0.7 Å in the case of ZIF-61 to 13.1 Å in the case of ZIF-70. ZIF-8, in particular, features a diameter of the pore (i.e., diameter of the largest sphere that can pass through the entrance of the framework) of 3.4 Å, which lies exactly in between the effective diameter of H_2 (i.e., 2.90 Å) and that of gases like CO_2, N_2, and CH_4 (i.e. 3.63, 3.66, and 3.81 Å, respectively) [39]. As a consequence, ZIF-8 turns out to be H_2-selective in terms of its permeability in comparison to those gases [40–44]. Furthermore, ZIF-8 is commercially available, which means that it is potentially readily available for real applications. The combination of these features led us to choose ZIF-8 as a filler in the development of H_2-selective mixed-matrix membranes. The polymer chosen as the matrix—poly(2,6-dimethyl-1,4-phenylene oxide) (PPO)—is also commercial and is already used in some industrial gas separation membrane modules. It is a glassy amorphous polymer with a good thermal resistance and high permeability in comparison to the standard materials industrially used for gas separation, such as cellulose acetate or polysulfone [45].

The aim of the study was to develop thermally-resistant, H_2-selective materials to be used in the purification of syngas in processes such as the Integrated Gasification Combined Cycle (IGCC).

A membrane-based pre-combustion separation step can reduce the energy consumption associated with the compression of CO_2 and provide a hydrogen-enriched stream ready to be used as fuel for power generation [46,47].

The membrane preparation was optimized to allow the formation of films with up to 45 wt% of filler, which is a remarkably high amount considering that embrittlement and agglomeration formation become harder to prevent at high loadings [16,48]. The permeation and diffusion of He (used as a model for H_2), N_2, CH_4, and CO_2 were investigated at 35, 50, and 65 °C to allow for the calculation of the activation energy. The maximum temperature used in this work was not dictated by the resistance of the membranes, which could work at higher temperatures, but by the operative limit of the permeation apparatus. The optimal, maximum operating temperature for an H2-selective membrane in an IGCC process is around 150 °C [46,47]. Ideally, a temperature of 200 °C would allow the value of the low-temperature shift reactor exhaust gas temperature to be matched and optimize the thermal efficiency. The activation energy results obtained in this work are thus of great importance, because they allow for the extrapolation of transport properties at temperatures higher than experimental ones. Furthermore, pure and composite materials were characterized from a morphological point of view by means of SEM analysis. Microscope images allowed the dispersion and adhesion of the MOF to PPO in the composite films to be assessed. The absence of voids was also verified by performing density measurements of the composite films at different loadings, by comparing the extrapolated value of pure ZIF-8 with that of the theoretical crystal [49]. Thermal properties were investigated by means of TGA and DSC techniques.

2. Materials and Methods

2.1. Materials

Poly(2,6-dimethyl-1,4-phenylene oxide) was purchased in powder form from *Sigma Aldrich* (*St. Louis, MO, USA*), and used as received. It is commonly indicated as poly(phenylene oxide) or PPO, and it is a commercial aromatic glassy polymer. In the literature, it has also been referred to by other acronyms, such as PDMPO and PMPO [50,51]. Some of the relevant physical properties of this material are summarized in Table 1.

Table 1. Bulk physical properties of amorphous poly(2,6-dimethyl-1,4-phenylene oxide) (PPO).

Polymer	ρ (25 °C)	T_g	%FFV [52]	Refractive Index [52]	Average Molecular Weight [53]
PPO	g/cm^3	°C			g/mol
(structure)	1.06	213	19	1.573	59,000

The commercial sieve selected to produce the MMMs was ZIF-8 (*Basolite® Z1200, Cat. 691348* produced by *BASF*). In Figure 1, the structure of ZIF-8 is shown. Table 2 summarizes the physical properties, structure, and composition of ZIF-8. The company provided particle size information for ZIF-8, which had a D50 of 4.90 μm. No further grinding was performed on the fillers used to produce MMMs, because the results showed that only small decrements of the particle size could be obtained after the milling process, which is quite invasive and might break the crystalline structure of the materials. Figure 1 highlights the six-membered ring in red, in which the six ZnN_4 tetrahedra, represented in blue, are connected to one another through organic linkers of 2-methylimidazolate. Carbon atoms are shown in grey, nitrogen atoms are shown in green, and hydrogen atoms are not represented.

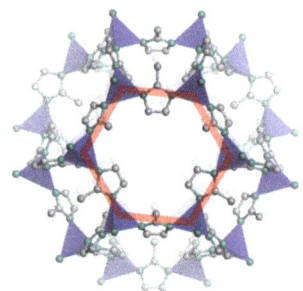

Figure 1. Structure of the Zeolitic Imidazolate Framework 8 (ZIF-8) framework.

ZIF-8 presents a regular zeolite-like sodalite (*sod*) structure. Despite their similar topology, zeolites and ZIFs have a very different chemistry. Zeolites are aluminum-silicate, and thus fully inorganic materials [54]. Conversely, ZIF-8 is a hybrid organic-inorganic material in which metal cations (i.e., Zn^+) are linked by organic molecules (i.e., 2-methylimidazolate) to form a crystalline and regular structure [30,34,40,49]. ZIF-8 is easier embed in organic polymers, compared with purely inorganic materials, due to the organic part. The vast and recent studies regarding MMMs conducted using ZIF-8 as a filler present evidence for the previous statement [48,55–63]. Applications at an industrial scale require materials capable of resisting harsh operative conditions, which can preserve their initial properties over time.

Table 2. Physical properties, composition, and reticular structure of ZIF-8.

Filler	Composition [49]	Net	d_a [34]	d_p [34]	Surface Area (BET) [49]	Theoretical Density	Thermal Stability [49]	Hydrophilicity [49,64]
			Å	Å	m²/g	g/cm³	°C	
ZIF-8	Zn(MeIM)$_2$	sod	3.4	11.6	1630	0.95 [55] 0.93 [61]	550	Hydrophobic

Park et al. [49] investigated the thermal and chemical stability of zeolitic imidazole frameworks, focusing on ZIF-8 due to its exceptional properties. ZIF-8 was demonstrated to possess a high hydrothermal stability, maintaining its architecture, as shown by the PXRD analysis, and its porosity (i.e., sorption capacity) after being exposed to 550 °C in N_2 atmosphere and after being boiled in water for 7 days. This exceptional result was attributed to the hydrophobicity of ZIFs, which can repel water molecules and avoid attacks of ZnN_4 units, which can otherwise jeopardize the framework integrity [49]. Küsgens at al. [64] also reported the water sorption isotherm of ZIF-8, which resulted in a negligible amount of water being adsorbed up to $p_w/p_w^0 = 0.6$, where p_w is the actual partial pressure and p_w^0 is the saturation partial pressure of water vapor. The behavior was also successfully modeled by using a molecular simulation [65]. Finally, the high Brunauer, Emmett, and Taller (BET) surface area of ZIF-8 of 1630 m²/g allows the MOF to have a high sorption capacity [49].

2.2. Experimental Methods

2.2.1. Mixed-Matrix Membrane Preparation

The production of homogeneous and stable membranes with a sufficient mechanical resistance, and the development of a simple and reproducible protocol, were crucial steps in this study. It was necessary to optimize several factors, such as the size of the filler particles; the solvent; the solution casting temperature, which directly affects the evaporation rate during casting and the state of the polymer; and finally, the thermal annealing treatment conditions.

Membrane Casting

Self-standing PPO and MMM films were obtained through the solution casting technique. A target thickness ranging from 80 to 120 µm was selected. Chloroform ($CHCl_3$), 1,1,2-trichloroethylene (TCE), and toluene (TOL) were widely tested in this work under different conditions for both polymeric membranes and mixed matrices, since it has been shown that the choice of the solvent affects the performance of the membranes in terms of gas separation [66,67]. Studies on PPO membrane formation have shown that a decrease in the boiling point (BP) of the solvent leads to a decrease of the permeability and an increase of the selectivity [66]. In addition, PPO may crystallize if the evaporation occurs too slowly, so a more volatile solvent ensures the formation of fully amorphous membranes [66]. Finally, chloroform (*Sigma Aldrich*, purity ≥ 99.8%, *St. Louis, MO, USA*) was selected as the optimal solvent, according to the criteria previously mentioned. Therefore, a similar methodology to the one developed by Aguilar-Vega and Paul [50] was applied to PPO and extended to the MMMs.

The preparation of each solution in this work began by dissolving a 5% by weight of PPO in $CHCl_3$. The use of a concentrated solution significantly reduced the sedimentation and the agglomeration of ZIF-8 during the casting step. This was because the high viscosity of the suspension sufficiently reduced the particle mobility, as suggested by Das et al. [68]. Complete dissolution of the polymer in the solvent was reached through magnetic stirring at room temperature for at least 2 h. This solution may be used as such to prepare pure PPO membranes, and as a precursor, in which different quantities of ZIF-8 can be added to obtain MMMs in various percentages by weight of filler in PPO. Prior to use, ZIF-8 was activated at 200 °C under vacuum overnight. At the end of the thermal treatment, the powder was promptly mixed with the polymeric solution under stirring. When the suspension reached a homogeneous condition, it was further sonicated for 4 h (*Lavo, Ultrasonic Vibrator ST-3*). The suspension was poured onto a Petri dish with a diameter of 10 cm, in order to avoid any edge effect on the center of the membrane. The dish was heated at 50 °C and kept on a hot plate to induce quick evaporation of the solvent, which was necessary for obtaining a defect-free material with a good filler dispersion and ensuring the formation of fully amorphous films.

Thermal Annealing

Thermal treatment at 150, 200, and 250 °C (with the latter being above T_g, i.e., 213 °C) was applied to the films to remove the residual solvent and stabilize the gas transport properties over time. A high temperature accelerates the kinetics of the aging process, which would otherwise play a more significant role, although the membranes produced were rather thick (i.e., 80–120 µm), and thus subject to a slower physical aging process.

2.2.2. Density Measurement

Determination of the density of the membranes, ρ_{MEM}, was performed by means of the buoyancy method, based on the Archimedes' principle, using a density kit (*MS-DNY-54*) on a high precision balance (*Mettler Toledo, NewClassic MF MS105DU, Columbus, OH, USA*). Deionized water (*Culligan, M1 Series Commercial Reverse Osmosis Water System*) was used to determine the hydrostatic weight of the sample. A wetting agent (*Pervitro 75% 72409*, included in the density kit) was used to avoid the formation of air bubbles on the submerged film, which might have affected the measurements, introducing a negligible change in the water density. The temperature of the fluid was monitored with a thermometer (± 0.1 °C) to determine the proper density, ρ_{H_2O}, taken from Perry's tables [69], in order to calculate the sample density as follows:

$$\rho_{MEM} = \frac{m^{Air}_{MEM}}{\left(m^{Air}_{MEM} - m^{H_2O}_{MEM}\right)} \rho_{H_2O}(T) \qquad (1)$$

where m_{MEM}^{Air} is the weight of the sample measured in air, while $m_{MEM}^{H_2O}$ is the weight measured when the sample was soaked in water. Accurate density values are essential for investigating the volumetric behavior of polymer-filler mixtures and estimating their mixing volume. In particular, the presence of voids inside MMMs would cause a lower-than-additive value of density.

2.2.3. Morphological Characterization

Gas transport properties in composite membranes significantly depend on the filler distribution within the matrix and the adhesion between the particles and polymer. To evaluate these morphological features, field-emission gun-scanning electron microscopy (FEG–SEM) was used (*Fei Company – Bruker Corporation, Nova NanoSEM 450, Hillsboro, OR, USA*). All the film samples were fractured in liquid nitrogen, in order to generate a brittle fracture on the cross section to be analyzed. The samples were gold sputtered (*Emitech K550*).

2.2.4. Thermal and Calorimetric Properties

Differential scanning calorimetry (DSC) (*Q20, TA Instrument, New Castle, DE, USA*) was used to optimize the annealing procedure; assess the state of PPO; and evaluate the presence of residual solvent and humidity in PPO, ZIF-8 powder, and MMMs. Thermal transitions were recorded by heating samples under N_2 flux at a 10 °C/min rate, from 25 to 300 °C. Two heating scans were recorded for each sample.

Thermal gravimetric analysis (*Q50, TA Instrument, New Castle, DE, USA*) experiments were performed to investigate the thermal stability of pristine materials and MMMs. Samples of about 20 mg were held in a platinum pan and were heated to 200 °C in N_2 atmosphere and left for 1 h in these conditions, in order to make sure that all residual water and gases were removed. After that, the temperature was increased to 800 °C in the same atmosphere at a constant rate of 10 °C/min.

2.2.5. Gas Permeability Experiments

The permeability of He, N_2, CH_4, and CO_2 was evaluated at different temperatures (i.e., 35, 50, and 65 °C) for pure PPO and MMMs at different loadings of the filler up to 45 wt% of ZIF-8. All gases were purchased from *S.I.A.D. Spa* (Bergamo, Italy) with a purity of or above 99.99% and used as received. The order in which gases were tested was as follows: He, N_2, CH_4, and CO_2. This was pursued to prevent any conditioning effect of the sample, as PPO undergoes plasticization when exposed to high pressures of CO_2 [70–72]. Each permeability experiment was performed at an absolute upstream pressure of 1.3 bar; thus, the pressure was low enough to prevent the plasticization effect. However, to make sure that the membrane permeability was unchanged after tests with CO_2, the helium permeability was measured again at the same conditions. To investigate the effect of the temperature at different loadings of the sieve, the permeability of the four gases was evaluated at 35 °C, 50 °C, and eventually 65 °C for the same sample, avoiding any difference due to the change of the membrane. The fixed-volume variable-pressure manometric technique previously described elsewhere [73] was implemented to perform the experiments. The essential layout of the equipment is shown in Figure 2.

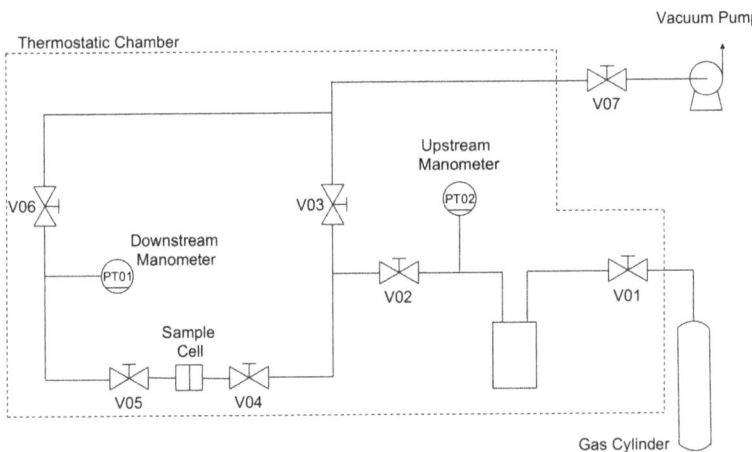

Figure 2. Layout of the permeation equipment. The outer black dashed line indicates the volume in which temperature is controlled.

A circular self-standing film was placed in the sample holder, which was a stainless-steel cell, and sealed by means of an O-ring made of Viton® to ensure that the system was leak-tight. A forced ventilation thermostatic chamber (*Type M 150-TBR, MPM Instruments S.r.l., Bernareggio, Italy*) was used to control the air temperature with an accuracy of ±0.1 °C. The specimen was conditioned under dynamic vacuum overnight to remove any possible species from the matrix, such as gases and humidity coming from the brief exposition to air. Once equilibrium conditions were achieved, V04 was opened to start the experiment. The increase of the downstream pressure in the calibrated closed volume, V_d, was monitored by a capacitance manometer (PT01 - *Barocel® Edwards, Burgess Hill, UK*) with a sensitivity of 10^{-2} mbar and an accuracy of 0.15% of the reading. Since the initial downstream pressure, p_d, was a vacuum, the permeability could be evaluated at the steady state by means of the following equation:

$$\mathcal{P} = \frac{V_d}{RT}\frac{l}{A}\frac{1}{(p_u - \bar{p}_d)}\left[\left(\frac{dp_d}{dt}\right)_{SS} - \left(\frac{dp_d}{dt}\right)_{leak}\right] \qquad (2)$$

in which R is the gas constant, T is the operative temperature, l is the membrane thickness, and $\bar{p}_d$ is the average downstream pressure of the considered gas. $\left(\frac{dp_d}{dt}\right)_{SS}$ and $\left(\frac{dp_d}{dt}\right)_{leak}$ are the changes in pressure at a steady state (SS) and when the equipment was sealed under static vacuum (leak). The uncertainty of the permeability values was calculated by considering the experimental error made to measure l, $\bar{p}_d$, and V_d, by means of the propagation of error approach [74]. The ideal selectivity between gas A and B, $\alpha_{A/B}$, could be calculated for each gas pair as follows:

$$\alpha_{A/B} = \frac{y_{A,d}/y_{B,d}}{y_{A,u}/y_{B,u}} \cong \frac{\mathcal{P}_A}{\mathcal{P}_B} = \frac{\mathcal{D}_A}{\mathcal{D}_B}\frac{\mathcal{S}_A}{\mathcal{S}_B} = \alpha^{\mathcal{D}}_{A/B}\alpha^{\mathcal{S}}_{A/B} \qquad (3)$$

where $y_{A,d}$ and $y_{B,d}$ are the molar fraction on the downstream side of the membrane of gas A and B, respectively, while $y_{A,u}$ and $y_{B,u}$ are those on the upstream side of the film. $\mathcal{P}_A$ is the permeability of the more permeable gas of the pair and $\mathcal{P}_B$ is that of the less permeable one. Furthermore, the ideal selectivity could be split into two contributions: $\alpha^{\mathcal{D}}_{A/B}$, which is the diffusivity selectivity, and $\alpha^{\mathcal{S}}_{A/B}$, which is the solubility selectivity.

The time-lag, θ_L, was evaluated for all the gases by using the time-lag method. The time-lag is a measure of the characteristic time required for the gas molecule to dissolve in the polymer matrix and diffuse through the film. Considering the operative conditions under which the experiments were performed, i.e., a zero initial concentration of gas across the membrane, θ_L can be related to the diffusivity, $\mathcal{D}$, through the following equation [75,76]:

$$\mathcal{D} = \frac{l^2}{6\theta_L} \quad (4)$$

The diffusivity and permeability can be described by an Arrhenius-like equation [58,77], and analogous formulations can be provided as follows:

$$\mathcal{P} = \mathcal{P}_\infty exp\left(-\frac{E_\mathcal{P}}{RT}\right) \quad (5)$$

$$\mathcal{D} = \mathcal{D}_\infty exp\left(-\frac{E_\mathcal{D}}{RT}\right) \quad (6)$$

where $E_\mathcal{D}$ is the activation energy of the diffusion process, which is the barrier that needs to be overcome by a gas molecule to make a diffusive jump from one cavity to another, and $\mathcal{D}_\infty$ is the temperature-independent pre-exponential term, which represents the diffusion coefficient at an infinite temperature. Similar considerations hold for permeability, with the essential difference that permeation is not a thermally-activated process, since permeability is a combination of a kinetic and thermodynamic factor, and both increasing and decreasing trends can be experienced as a function of temperature. However, the energetic constant of permeation, $E_\mathcal{P}$, can be calculated by means of (5), and the pre-exponential factor, $\mathcal{P}_\infty$, as for diffusion, is temperature-independent and represents the permeation coefficient at an infinite temperature. Proper fitting provides the possibility to extrapolate experimental permeability and diffusivity values at higher temperatures than those investigated. This is very useful information, since these MMMs might be required to work at a high temperature, with H_2/CO_2 separation favored under those conditions and materials compatible with such a high temperature. Eventually, the heat of sorption, ΔH_S, of each gas in the MMMs can be calculated by simply subtracting the two contributions as follows:

$$\Delta H_S = E_\mathcal{P} - E_\mathcal{D}. \quad (7)$$

3. Results and Discussion

3.1. Aspect of the MMMs

The PPO-based MMMs at different loadings of ZIF-8 were prepared following the optimized protocol described above. In Figure 3, it is possible to appreciate the transparency of pure PPO membranes, while MMMs with ZIF-8 developed a slight haze due to the presence of fillers. Composite membranes were found to be macroscopically homogeneous, testifying to the overall good dispersion of the particles inside the polymer matrix.

Figure 3. Membrane samples used for permeation tests. (**a**) PPO and (**b**) 25 wt% ZIF-8/PPO.

Membranes made of polymer have a higher flexibility. MMMs with a filler content up to 15 wt% are still robust and preserve this feature. At intermediate loads (e.g., 25 wt%), membranes are less robust. Films begin to become more brittle when the particle content increases up to 45 wt%. These materials revealed a lower resistance to bending, which was expected since, for instance, a ZIF-8/PPO membrane at 45 wt% contains about a 48% volume of the MOF. A detailed and quantitative characterization of the effect of filler on the mechanical properties of the membranes was beyond the scope of this work, which was aimed at preliminarily assessing the effect of filler on the transport properties. For this purpose, it is a requirement that the composite membranes resist the pressure difference applied across the membrane and the stress imposed by leak-proof tightening of the permeation cell, without inducing pinholes or cracks. This requirement was fulfilled by all membranes with up to 45 wt% of filler.

3.2. Density

The density values of pure materials and composite membranes versus filler content are reported in Figure 4. The density of the MMMs decreased when increasing ZIF-8 loading. By plotting the same data in terms of the specific volume of each film as a function of weight filler loading, the expected linear correlation ($\hat{v}_{MMM} = w_{PPO}\hat{v}_{PPO} + w_{ZIF}\hat{v}_{ZIF}$) was obtained with a correlation coefficient of $R^2 = 0.98$ (represented with a dashed line in Figure 4). Hence, it was possible to estimate the density of ZIF-8 by extrapolating the linear function. Surprisingly, the ZIF-8 density was 0.96 g/cm^3, which is a value very close to that of the theoretical density of the regular ZIF-8 crystal, i.e., 0.93–0.95 g/cm^3 [49,55,61]. This may indicate that the presence of voids inside the MMMs is negligible, since interfacial adhesion defects would be detected by the lower-than-ideal values of density of the membranes. The additive rule for composite materials is shown in (8):

$$\rho_{MMM} = \frac{\rho_{PPO}\rho_{ZIF}}{w_{PPO}\rho_{ZIF} + w_{ZIF}\rho_{PPO}} \tag{8}$$

and it is represented as a solid red line in Figure 4. The additive rule was implemented to compare the experimental values with the ones predicted by the ideal combination of the two phases. The consistency between the additive rule and the actual density of the composites indicates that the polymer and filler phase display good adhesion. This does not necessarily indicate that the filler phase is evenly distributed in the matrix, but rather that there is no void formation at the polymer/filler interface, which may compromise the intrinsic selectivity of the composite materials.

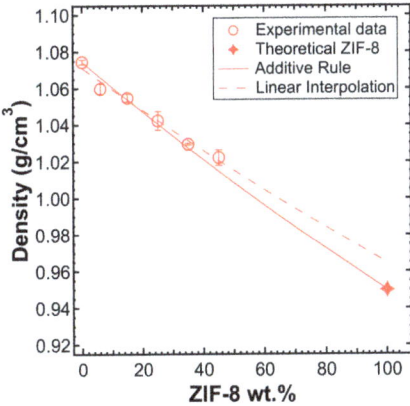

Figure 4. Density of the mixed-matrix membranes (MMMs) versus ZIF-8 weight fraction in the film (empty circles), measured with the buoyancy technique in water. Solid line represents the additive rule and was evaluated by the experimental density of PPO and the theoretical density of ZIF-8. Dashed line represents the linear interpolation of the experimental values measured, extrapolated to pure ZIF-8.

3.3. SEM Analysis

The morphology of ZIF-8/PPO MMMs was investigated at different loadings of the MOF by means of SEM analysis. This enabled us to determine the quality of adhesion between the particles and polymer matrix, as well as evaluate the dispersion of the filler. The SEM images are reported in Figure 5, which generally shows that ZIF particles and the polymer are compatible and had good adhesion. This is consistent with the partially organic nature of the filler, which improves the affinity with the polymer matrix, as described above. Additional SEM pictures are reported in the Supplementary Information in Figure S3.

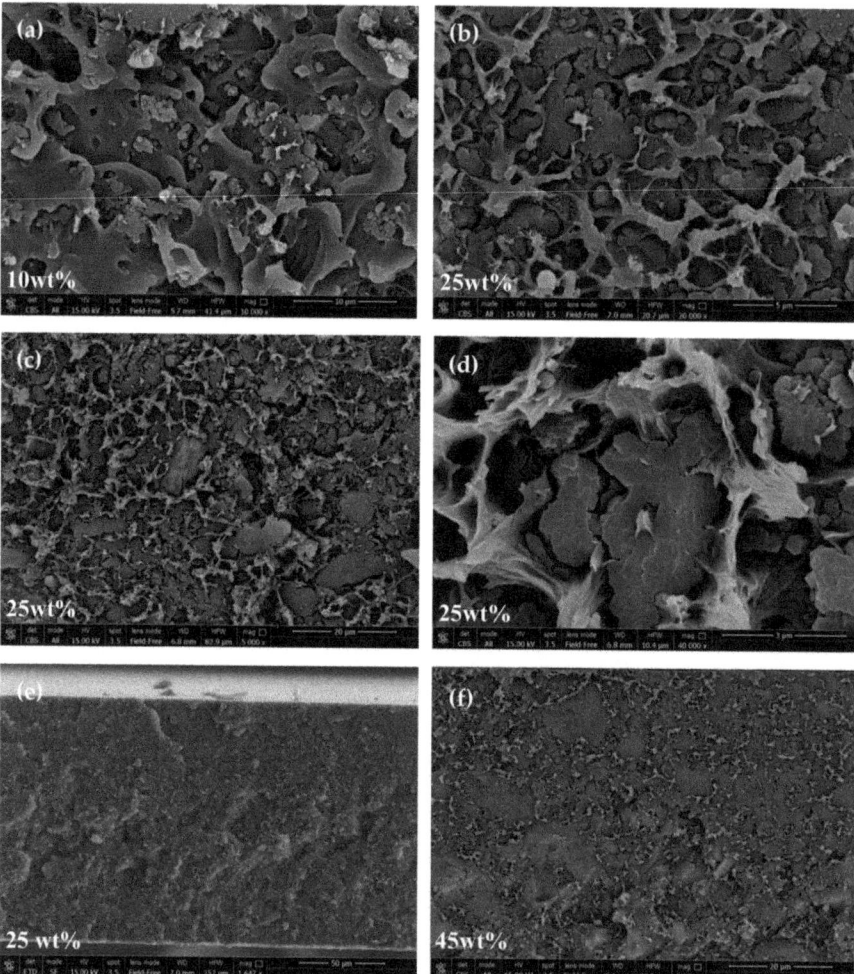

Figure 5. SEM images of the cross-section of ZIF-8/PPO mixed-matrix membranes at different loadings and magnitudes: (**a**) 10 wt%, (**b–e**) 25 wt%, and (**f**) 45 wt%. Other images can be found in the Supplementary Information.

However, detachment at the interphase appears to happen in some cases, and this can be explained by the following. It is common knowledge that glassy polymers with a rigid backbone and a high glass transition temperature, such as PPO, vitrify when the solvent evaporates [18]. The evaporation-induced transition from a rubbery state to glassy state can cause significant stress to the system. As also pointed out by Koros et al. [28], this phenomenon can happen during composite membrane formation before all of the solvent has left the film, making further evaporation beyond this point crucial for detachment of the polymer chain from the filler. This could lead to the formation of non-selective voids, which might prevent the membranes from being as selective as expected, but, on the other hand, make them more permeable. In this work, slight delamination between the two phases can be seen, especially at high loadings (i.e., ≥25 wt%). However, the presence of non-selective voids can be excluded, since no anomalous selectivity loss was observed with increasing filler loading, as will be discussed in the following sections, and the composite density closely followed the volume additivity, as shown above. As pointed out by Ordonez et al. [48], delamination can also be induced by fracturing the membranes with liquid nitrogen prior to SEM analysis. The latter contribution would not affect the transport properties, being caused artificially during the preparation of the sample for the analysis. However, we believe that the mechanical stress imposed to break the films would be responsible for the detachment to some extent. Chung et al. [18] reported the use of a higher-than-ambient temperature to promote fast evaporation during film formation, and in the case reported here, a net heat flux and temperature gradient between the bottom and top of the nascent membrane were generated. This way of heating was found to promote convective fluxes inside the fluid suspension, leading to inhomogeneous thicknesses and irregular distributions of the filler in the resulting MMMs. However, these kinds of consequences were not experienced by the ZIF-8/PPO membranes, probably because of the high viscosity of the casting solution. Overall, the distribution of filler within the matrix was ubiquitous, although in a small number of cases, larger aggregates were visible. The quick solution casting (i.e., 15 min) and the high viscosity of the solution, were not sufficient to completely eradicate this phenomenon. Nevertheless, the majority of particles were smaller (e.g., 1–4 µm) and many ZIF nanocrystal cubes with a side of ≈200 nm could be observed (Figure S3).

3.4. DSC Tests

DSC analysis was carried out in two subsequent runs on films of PPO, ZIF-8 powder as received, and mixed-matrix films, between 25 and 300 °C. Figure 6 shows that the pure PPO films obtained via solvent casting in chloroform at 50 °C were fully amorphous, as they showed the typical glassy transition peak at about 213 °C, consistent with literature values, in both scans [50,53]. It must be noted, however, that slower casting at room temperature (i.e., complete evaporation in 3 days) results in the formation of semi-crystalline PPO samples, as verified via DSC analysis (Figure S2a), and rupture of the membrane during film formation (Figure S2b). A specific analysis was carried out in this work on the effect of the casting temperature on the final properties, although it is not reported here because it is beyond the scope of the paper. The value of 50 °C appears to be the optimal one for producing robust and amorphous PPO films. There is almost no difference between the first and second scan in the amorphous PPO, because the sample was previously treated at 200 °C under vacuum, proving that the treatment can remove any residual solvent.

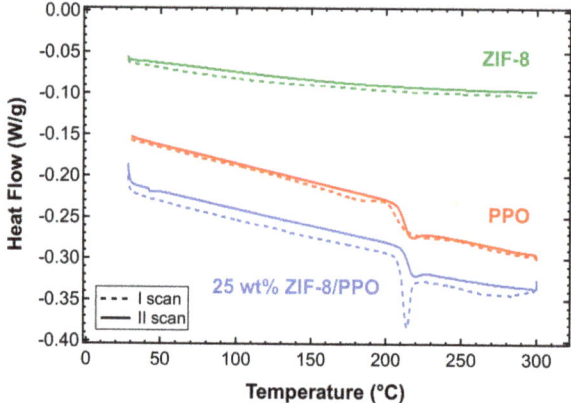

Figure 6. Differential scanning calorimetry (DSC) tests on pure amorphous PPO film after a thermal annealing treatment at 200 °C under vacuum (red), ZIF-8 powder as received (green), and 25 wt% ZIF-8/PPO mixed-matrix film after undergoing a thermal annealing treatment at 200 °C under vacuum (blue).

Almost no difference was observed between the two scans carried out on ZIF-8, which indicates that the material does not retain a significant amount of water due to its hydrophobicity, as indicated in the literature (Figure 6) [49,64,65]. The observation comes from the fact that hydrophilic materials, such as zeolites, show broad endothermic peaks in DSC scans performed in the same range of temperatures, as well as when specimens are stored in environmental conditions prior to the test [78].

The same analysis was carried out on MMMs containing PPO and different amounts of ZIF-8. Figure 6 shows the results relative to the sample containing 25 wt% of ZIF-8. One can notice a sharper peak upon transition in the first scan, which indicates enthalpic relaxation at T_g. Such a sharp peak is absent in the second scan.

The addition of an increasing amount of ZIF-8 increases, albeit slightly, the glass transition temperature of the PPO, as shown in Figure 7. Previous works that tried to correlate the variation of permeability and selectivity with the variation of T_g of MMMs did not provide clear and univocal conclusions. In particular, Moaddeb et al. [77] observed that the reduction in the mobility of 6FDA-IPDA chains conferred an enhanced O_2/N_2 selectivity, while the permeability was barely altered or negligibly reduced. The opposite was observed by Song et al. [55]. In their work, ZIF-8/Matrimid® membranes showed an enhanced permeability up to about 3–4 times that of pure Matrimid®, while the H_2/CO_2 selectivity remained constant with the filler loading (i.e., with increasing T_g). These results are in contrast to what was reported by Díaz et al. for ZIF-8/PPEES systems, where no changes of T_g were observed [60]. In the case of the composite materials studied in the present work, we believe that the slight increase of T_g should not affect the gas transport properties in the polymer phase, because at temperatures far below T_g and pressures far below the plasticization point, the different rigidity of the various matrices should not play a role.

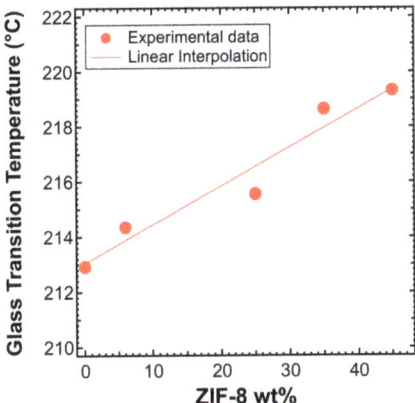

Figure 7. Trend of the glass transition temperature in ZIF-8/PPO MMMs as a function of the filler loading.

3.5. Permeability and Permselectivity

3.5.1. Gas Transport in PPO and ZIF-8

PPO has been widely studied in the framework of gas separation because of its excellent sorption and transport properties [50,52,79,80]. As mentioned by Toi et al. [80], PPO shows a higher permeability and sorption than other commercial glassy polymers with a rigid chain backbone. The high extent of sorption can be ascribed to the high glass transition temperature ($T_g \approx 213$ °C) [50,52,80], which indicates a high amount of non-equilibrium excess-free volume. The relatively high permeability is related to the high diffusion coefficients of low-weight penetrants, which stem from the high Fractional Free Volume (FFV) of the polymer, which was measured to be about 19% by Huang and Paul [52]. Along with these properties, other features that make PPO suitable for industrial applications are its relatively low cost, compared to other common techno-polymers, and the possibility to work with it at high temperature [53]. This is an important aspect when the separation process is controlled by the diffusivity of the gas species in the membrane, as in this case. Despite gas permeability values showing discrepancies in the literature, different studies have revealed that PPO is highly permeable to H_2 (i.e., 86.9–112.8 Barrer), instead demonstrating a moderate selectivity for the H_2/CO_2 couple at room temperature (i.e., ideal perm-selectivity range between 1.49 and 1.54) [51,81,82]. PPO, in particular, behaves as a molecular sieve, and permeability values can be arranged in the following order: $\mathcal{P}_{H_2} > \mathcal{P}_{He} > \mathcal{P}_{CO_2} > \mathcal{P}_{N_2} \cong \mathcal{P}_{CH_4}$, which is almost the opposite trend of the kinetic diameter, for which the order is $d_{He} < d_{H_2} < d_{CO_2} < d_{N_2} < d_{CH_4}$ [50,51]. It is worth stressing that hydrogen is the most permeable gas in PPO, which makes the use of helium as a model for hydrogen, in this work, a conservative choice for estimating the performance of H_2/CO_2 separation.

ZIFs, such as ZIF-7 and ZIF-8, and zeolites, such as Zeolite 3A, have pore sizes that approach the size of gas molecules, which is a feature that makes them theoretically capable of performing gas separation with a very high selectivity towards smaller gases. In particular, for ZIF-8, the diameter of the apertures is estimated to be 3.4 Å based on crystallographic data, and is thus larger than H_2's effective diameter (2.90 Å), but smaller than that of CO_2, N_2, and CH_4 (3.63 Å, 3.66 Å, and 3.81 Å, respectively) [39]. Bux et al. [40,42], as well as McCarthy et al. [41], reported that ZIF-8 is an H_2-selective material, as far as H_2/CO_2 separation is concerned, having performed experiments with both pure and mixed gases. The measurements showed that there is no sharp cut-off between molecules smaller and bigger than the pores. The fact that molecules with a kinetic diameter larger than the pores can permeate through metal-organic networks was studied in detail by Caro [30]. He found that MOFs often exhibit a pronounced structural flexibility, which makes the framework of these materials less rigid than that of zeolites. Furthermore, Bux and coworkers [40] noticed that the H_2 flux in ZIF-8 membranes in the presence of co-permeating CH_4 is only slightly affected by the presence of the larger molecule in the mixture, and the permeation results are comparable to those of single-gas permeability. This behavior was ascribed to the fact that, even though the pore size of ZIF-8 is small, the space inside the largest cage of the system is far larger, thus accommodating a sphere with a diameter as big as 11.6 Å. The values of d_a (diameter of the aperture by which molecules can enter the framework) and d_p (diameter of the largest sphere that can fit into the largest cavity of the crystalline structure) are reported in Table 2 [34]. Therefore, once CH_4 enters the cage and frees the 3.4 Å-wide pore of ZIF-8, H_2 can diffuse through the network. The organic nature of the filler makes the framework flexible, causing values of selectivity that are lower than expected. However, its organic nature is essential for having a good compatibility with the polymer [37,38]. The permeability across a pure ZIF-8 membrane has been measured by various authors and the results are reported in Table 3. The H_2 permeability ranges between 4916 and 10,333 Barrer. However, the selectivity values are moderate, possibly due to the flexible morphology of this material. Therefore, a dramatic increase of the membrane selectivity, at least at low and moderate temperatures, is not to be expected upon the addition of ZIF-8.

Table 3. The single-gas permeability and ideal selectivity of pure ZIF-8 membranes [55].

Thickness (μm)	~30	~20	~20
Ref.	[40]	[41]	[42]
Permeance (10^{-8} mol m^{-2} s^{-1} Pa^{-1})			
H_2	6.04	17.3	8.23
N_2	0.52	1.49	0.69
CH_4	0.48	1.33	0.63
CO_2	1.33	4.45	/
Permeability (Barrer)			
H_2	5411	10,333	4916
N_2	466	890	412
CH_4	430	794	376
CO_2	1192	2658	/
Ideal Selectivity			
H_2/CO_2	4.54	3.89	/
CO_2/N_2	2.56	2.99	/
CO_2/CH_4	2.77	3.35	/
H_2/CH_4	12.6	13.0	13.1
H_2/N_2	11.6	11.6	11.9

3.5.2. Thermal Annealing

Physical aging is a phenomenon that occurs in all amorphous materials in a glassy state which evolve towards an equilibrium point and is accelerated by high temperatures. Indeed, it was shown by Ansaloni et al. [83], who performed studies on another glassy polymer, Matrimid® polyimide, that the increase of the thermal treatment temperature led to a larger reduction of the FFV and a stabilization of transport properties over time. Savoca et al. [84] observed a considerable decrease of permeability in poly(1-trimethylsilyl-1-propyne) (PTMSP) films when increasing the temperature of the thermal treatment, testifying that the sample returned to its original permeability after dissolving and recasting the membrane. Hung and Paul's studies have demonstrated that the aging rate of PPO is faster the higher the aging temperature, and slower the thicker the membrane, by monitoring key parameters, such as the FFV, gas permeability, and refractive index of the polymer [52,85,86]. At 35 °C, a PPO film with a thickness of ≈400 nm, experienced a loss of permeability of ≈65% over 4000 h (≈6 months); conversely, a ≈25 μm film revealed a decrease of ≈20% over 10000 h (≈14 months) [85].

For these reasons, whenever starting a comprehensive experimental campaign on a glassy membrane system, it is necessary to stabilize the membrane properties. We carried out a specific study to locate the minimum annealing temperature required to reach a stable permeability and solvent-free films. The study was carried out by measuring the permeability of He and CO_2, at 35 °C, after thermal treatment carried out under vacuum at various temperatures, from 150 to 250 °C, overnight. The results of gas permeability plotted against the pretreatment temperature are reported in Figure 8. It shows that the permeability decreased when increasing the treatment temperature, due to the accelerated aging induced by such high temperatures, and reached a plateau at a temperature of 200 °C, which is also below the T_g of PPO. Therefore, such tests allowed us to identify the optimal treatment temperature as being equal to 200 °C.

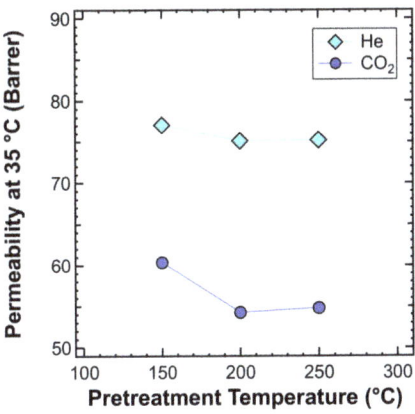

Figure 8. Effect of the thermal annealing temperature on the pure PPO permeability at 35 °C.

It must also be noted that high-temperature treatment can replace a stability test on the membrane; indeed it induces accelerated ageing, which decreases the permeability of the membranes by quickly compacting and densifying the polymer chains. Such treatment thus reduces the need to perform a time-consuming lifetime test that will require the membrane to be naturally ageing across periods of months or even years. Filled membranes have similar responses to thermal treatment than unfilled ones, so the addition of filler does not seem to impact the expected durability of the material.

3.5.3. Effect of the Filler Loading

Pure gas permeability tests with He, N_2, CH_4, and CO_2 were performed for several membranes at 35 °C, covering the whole range of filler loadings investigated, namely 0%, 3%, 6%, 10%, 15%, 25%, 35%, and 45% by weight. Figure 9 shows that the permeability increases monotonously with the filler loading. The permeability enhancement is extremely high and, in particular, adding 45 wt% of ZIF-8 to PPO enhances the He permeability by a factor of about 8. Table 4 presents the ideal selectivity for relevant gas pairs (i.e., He/CO_2, CO_2/N_2, CO_2/CH_4, He/CH_4, and He/N_2), evaluated by means of Equation (3). The uncertainty of the calculated permeability primarily originated from the variation of membrane thickness. However, the overall variability was always kept within ±3% for MMMs up to 25 wt% and within ±8% for higher loadings.

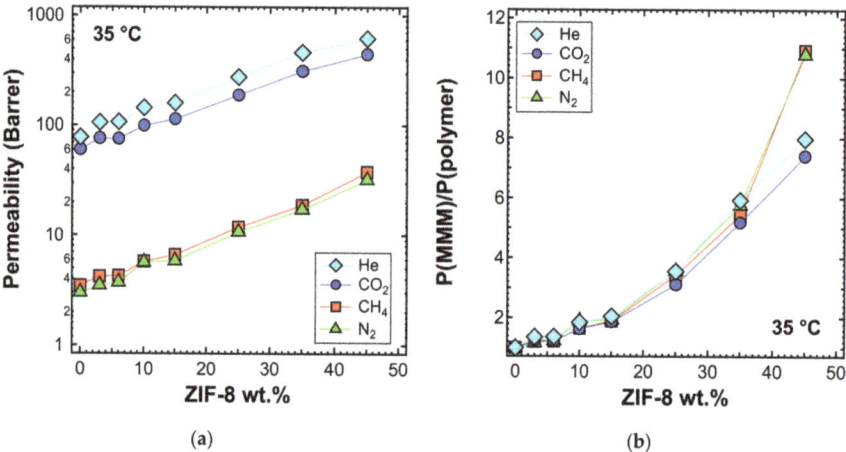

Figure 9. (a) Permeability and (b) relative permeability increase of various gases at 35 °C with an upstream pressure of 1.3 bar in ZIF-8/PPO MMMs.

The significantly enhanced permeability was accompanied by a modest increase in selectivity for the He/CO_2 gas pair (i.e., up to 15% more than that of pure PPO). Similar results were achieved by other authors [55,56,60,87]. The MMM permeability results are in line with what was expected for the transport properties of pure PPO measured in this work, and ZIF-8 permeability and ideal selectivity data from the published literature summarized in Table 3. In fact, the remarkable enhancement of permeability can be attributed to the very high permeability of ZIF-8, which is about two orders of magnitude higher than that of pure PPO. This result demonstrates that the filler actively contributes to the transport of gas molecules and that its pores are not blocked by the polymer phase. Furthermore, ZIF-8 shows a higher H_2/CO_2 ideal selectivity than PPO, and this also led to a small improvement of He/CO_2 selectivity in the ZIF-8/PPO composite membranes up to 35 wt% (Figure 10a). Conversely, CO_2/N_2, CO_2/CH_4, He/CH_4, and He/ZIF-8 ideal selectivity for ZIF-8 was lower than that of PPO, so a slight decreasing selectivity was expected by MMMs. At loadings up to 35 wt%, the results matched the expectations, while when the 45 wt% loading was reached, a more pronounced decrease was measured (Figure 10b), likely due to the formation of a small amount of non-selective voids. This was further confirmed by the anomalous permeability increase for N_2 and CH_4—the bigger gas molecules—presented in Figure 9b.

Table 4. Pure gas permeability and ideal selectivity in PPO and ZIF-8/PPO MMMs. Tests were performed at 35 °C and an upstream pressure of 1.3 bar.

ZIF-8 Loading (wt%)	Pure Gas Permeability (Barrer[a])				Ideal Selectivity				
	He	N_2	CH_4	CO_2	He/CO_2	CO_2/N_2	CO_2/CH_4	He/CH_4	He/N_2
0 (PPO) [88]	77.9 ± 2.3	2.99 ± 0.07	3.47 ± 0.09	60.6 ± 1.5	1.29	20.2	17.4	22.3	26.0
3	105.8 ± 2.5	3.49 ± 0.08	4.20 ± 0.10	76.1 ± 1.9	1.39	21.8	18.1	25.2	30.3
6	106.7 ± 2.4	3.71 ± 0.08	4.25 ± 0.10	75.3 ± 1.7	1.42	20.3	17.7	25.4	28.8
10	144.3 ± 3.6	5.67 ± 0.13	5.76 ± 0.14	99.5 ± 2.4	1.45	17.5	17.3	24.9	25.4
15	159.7 ± 1.5	5.83 ± 0.06	6.61 ± 0.07	114.1 ± 1.1	1.40	19.6	17.3	24.2	27.4
25	276.4 ± 10.4	10.7 ± 0.4	11.9 ± 0.5	189.0 ± 7.2	1.46	17.7	15.9	23.2	25.9
35	462.0 ± 31.0	17.2 ± 1.1	18.9 ± 1.3	314.2 ± 21.0	1.47	18.2	16.6	24.4	26.8
45	620.9 ± 54.0	32.3 ± 2.8	37.9 ± 3.3	448.7 ± 38.9	1.38	13.9	11.8	16.4	19.2

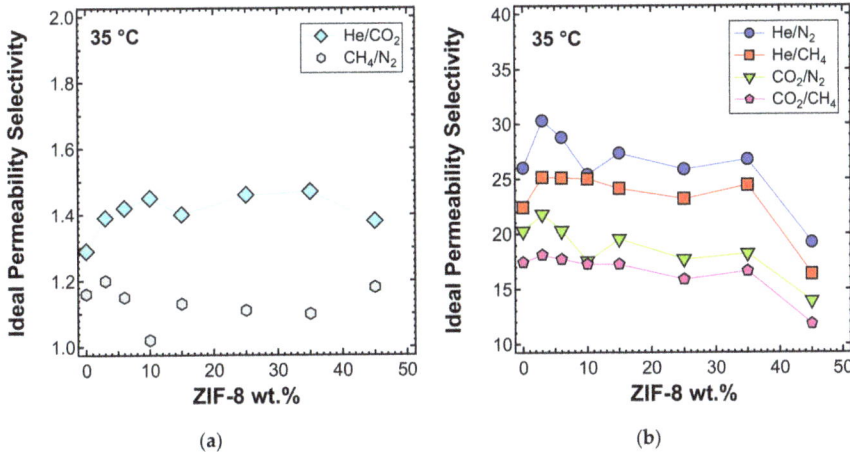

Figure 10. Ideal selectivity for (**a**) He/CO$_2$ and (**b**) other gas couples at 35 °C with an upstream pressure of 1.3 bar in ZIF-8/PPO MMMs.

Permeability tests were also performed at 50 and 65 °C for a selected list of MMM samples. When viewing the behavior at different temperatures, one can notice that the qualitative trends were similar, albeit with generally higher values of permeability, which are shown in Figure 11a (50 °C) and Figure 12a (65 °C). The relative increase of permeability, which is displayed in Figure 11b (50 °C) and Figure 12b (65 °C), was slightly lower than what was observed at 35 °C in Figure 9a. In particular, we noticed that at higher temperatures, the effect of adding ZIF-8 on the permeability is around the same for He and CO$_2$, while less marked for N$_2$ and CH$_4$.

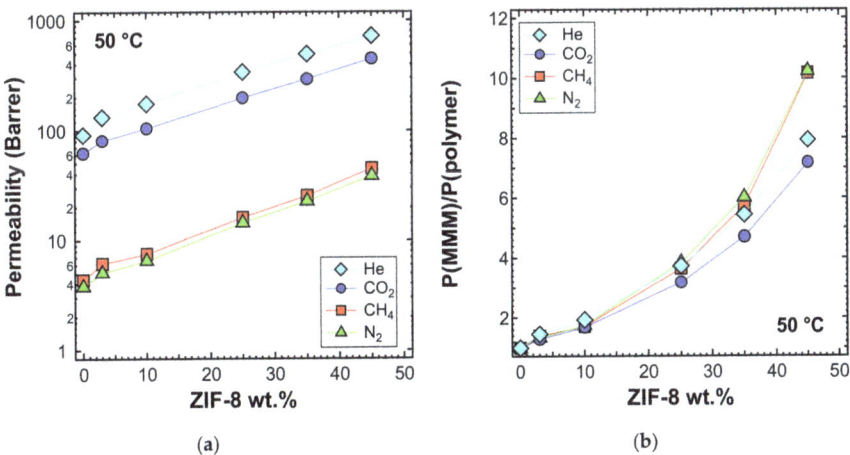

Figure 11. (**a**) Permeability and (**b**) relative permeability increase of various gases at 50 °C with an upstream pressure of 1.3 bar in ZIF-8/PPO MMMs.

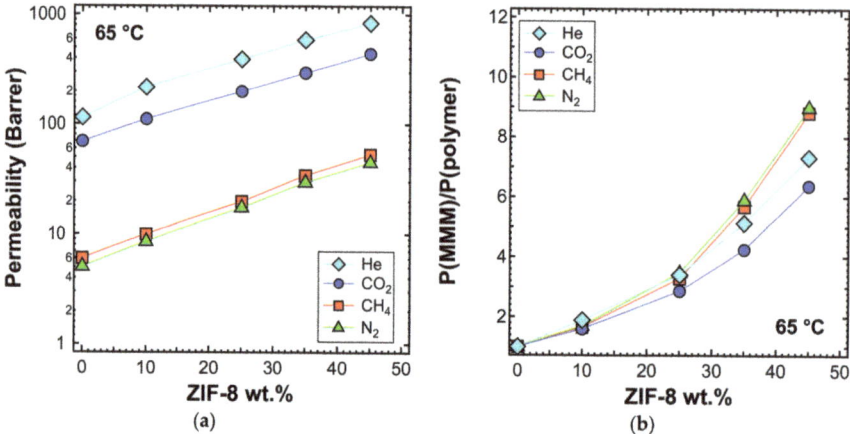

Figure 12. (a) Permeability and (b) relative permeability increase of various gases at 65 °C with an upstream pressure of 1.3 bar in ZIF-8/PPO MMMs.

The ideal selectivity was also estimated at 50 and 65 °C. In Figure 13a, the He/CO$_2$ selectivity versus filler loading at the three different temperatures inspected is presented. A consistent increase of selectivity with temperature was observed. The shape of the curve remains similar at all temperatures, with an initial higher increase of selectivity for loadings below 10 wt%, followed by a stable trend and then a slight decrease at a filler loading of 45 wt%, for the reasons mentioned above. He/CO$_2$ is the only gas pair, along with H$_2$/CO$_2$, for which the selectivity is enhanced by temperature [89]. This is because of the different nature of the two gases. Helium is a very small and non-condensable gas (T$_c$ = 5.2 K), for which the permeability is controlled by diffusivity. CO$_2$ is bigger and much more condensable (T$_c$ = 304.2 K) and its permeability has a higher solubility contribution. Temperature enhances diffusivity, which is a kinetic property, but compromises solubility, which is a thermodynamic contribution. Overall, this leads to an increased He/CO$_2$ selectivity. Given that a high temperature promotes the He/CO$_2$ separation performance, these MMMs have potential applications at high temperatures. To test the thermal stability, TGA experiments were performed on the MMMs and the results are reported in Figure S1 of the Supplementary Information.

The selectivity of the MMMs inspected with respect to other gas couples is reported in Figure 13b and c for the temperatures of 50 and 65 °C, respectively. The optimal selectivity is obtained for a filler loading of 10% at both temperatures, for the gas He/CH$_4$ and He/N$_2$, respectively. On the other hand, the CO$_2$/N$_2$ and CO$_2$/CH$_4$ selectivity decreases with the filler content at both temperatures.

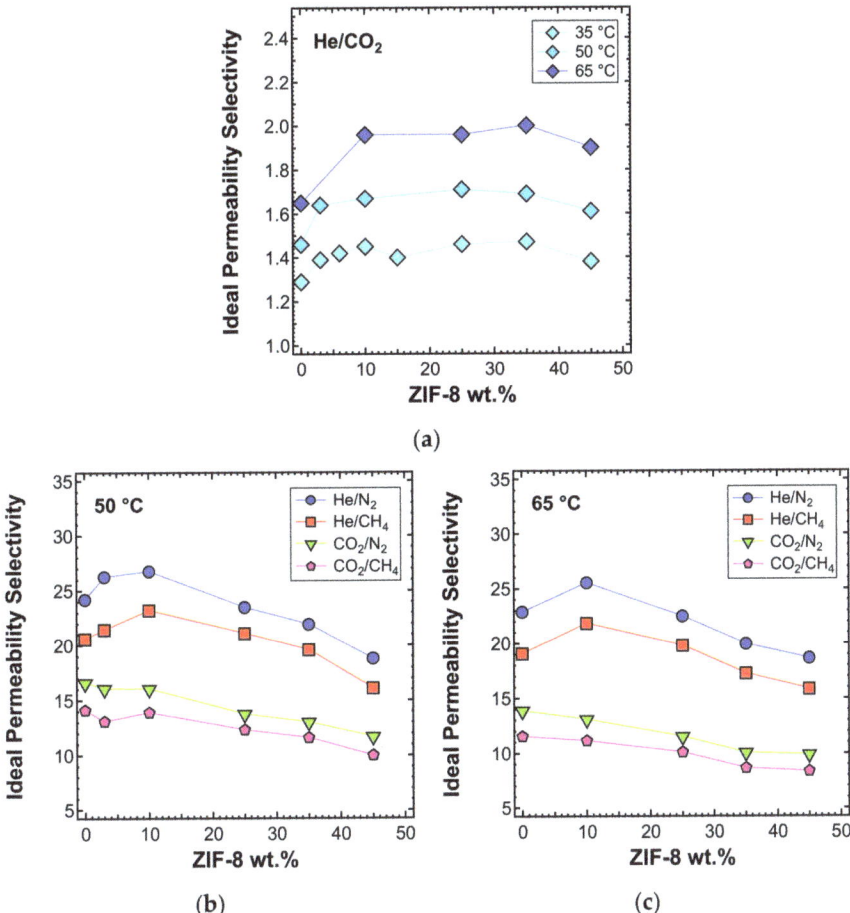

Figure 13. Ideal selectivity for (**a**) He/CO$_2$ at 35, 50, and 65 °C and other gas couples at (**b**) 50 and (**c**) 65 °C in ZIF-8/PPO MMMs, with an upstream pressure of 1.3 bar.

3.5.4. Effect of Temperature

The effect of temperature on the transport properties was also investigated by plotting permeability data as a function of temperature for various filler loadings. This is reported in Figure 14 for each gas. These data can be further elaborated to obtain a more quantitative indication of the effect of temperature on permeability, namely, the energy contributions associated with the permeation process. The values calculated based on the data measured at three different temperatures are reported in Figure 15 and Table 5.

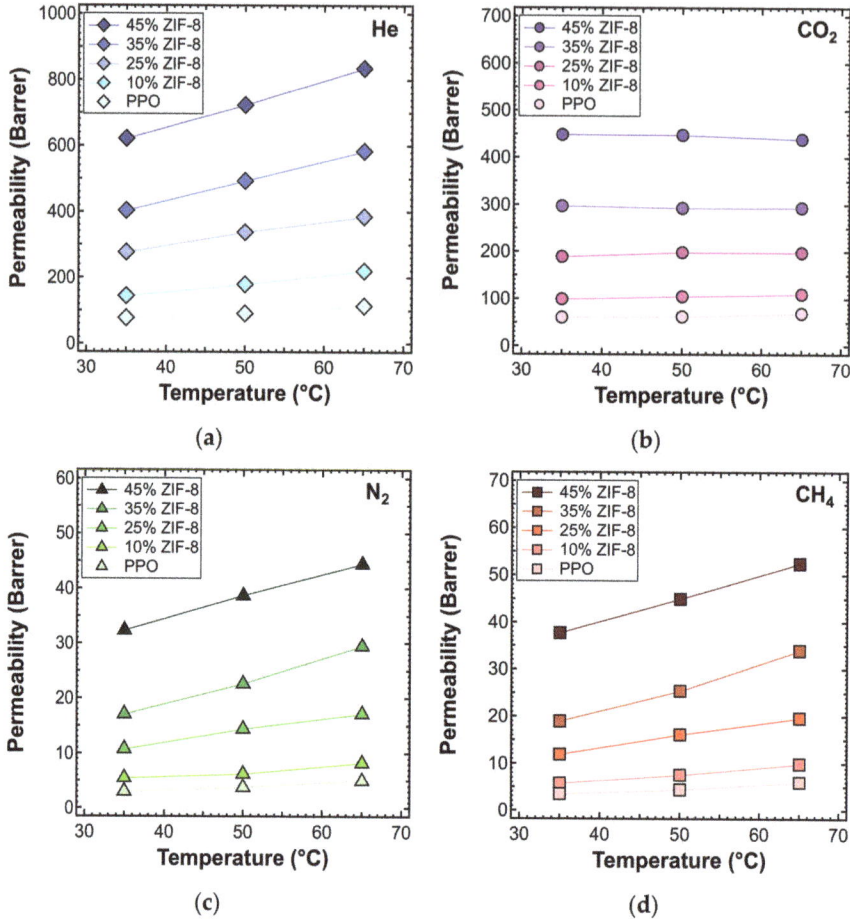

Figure 14. Permeability of (**a**) He, (**b**) CO_2, (**c**) N_2, and (**d**) CH_4 at different temperatures, in the ZIF-8/PPO MMMs of different weight fractions of ZIF-8, from 0% to 45%.

Table 5. Activation energy of permeability in the range of 35–65 °C for ZIF-8/PPO MMMs at different filler loadings.

ZIF-8 Loading (wt%)	E_P (kJ/mol)			
	He	N_2	CH_4	CO_2
0 [50]	9.7	9.8	12.1	1.5
0	11.07	14.81	15.78	3.84
10	12.11	11.74	15.59	2.99
25	9.64	13.68	14.46	1.34
35	10.66	15.76	16.85	0.33
45	8.53	9.15	9.52	0.59

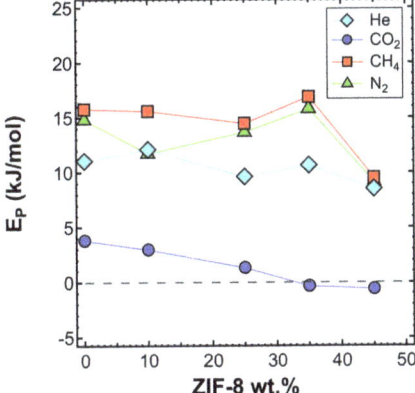

Figure 15. Activation energy of permeability as measured in the interval 35–65 °C for four gases in the various MMMs inspected, as a function of filler loading.

It can be noticed that the addition of ZIF-8 particles made the permeability a weaker function of the temperature. The plot clearly indicates that CO_2 has a slight dependence on temperature (i.e., smaller values of E_P), which becomes negligible at high filler loadings. At 35 wt% and 45 wt%, E_P had negative values, which means that the permeability decreases with an increasing temperature. This further enhances the He/CO_2 selectivity, since the helium permeability has a larger dependence on temperature and remarkably increases with it. The main reason why CO_2 permeability behaves in this way is because ZIF-8 provides a large sorption contribution at high loadings, and the loss of sorption outweighs the increase of diffusion while the temperature increases. N_2 and CH_4 show higher values for E_P, with CH_4 consistently being the highest. All values drop in the case of the membrane containing 45 wt% of ZIF-8, consistent with the possible presence of voids, as previously discussed.

3.6. Diffusivity

3.6.1. Effect of Filler Loading

The diffusivity of the different gases in the various MMMs was estimated from the permeation output through the time-lag method represented by Equation (4). Figure 16a shows the diffusivity at 35 °C as a function of ZIF-8 loading. Diffusivity values were in agreement with the following order: $D(He) > D(CO_2) > D(N_2) > D(CH_4)$, which is consistent with the values of the effective diameter of all the gas molecules investigated. In Figure 16b, the diffusivity ratio between each mixed-matrix membrane and the polymeric phase is reported. The diffusivity ratio was generally higher than unity for all gases and filler loadings. This indicates that the addition of filler enhanced the diffusion coefficient of gases, which possibly takes advantage of the fastest filler diffusive paths offered by the ZIF-8. Helium showed an inconsistent trend for MMMs at 3%, 6%, and 10% ZIF-8 loading, but we attribute this to the fact that the time-lag was often shorter than 2–3 s, which made it difficult to conduct accurate estimations in some cases. Overall, the trend remains clear. The enhancement of diffusivity induced by the filler addition is not as high as the one recorded for permeability: such a phenomenon indicates that the filler also promotes the gas solubility (Figure S5). A detailed analysis of the sorption properties is currently being conducted and will be the object of a forthcoming study.

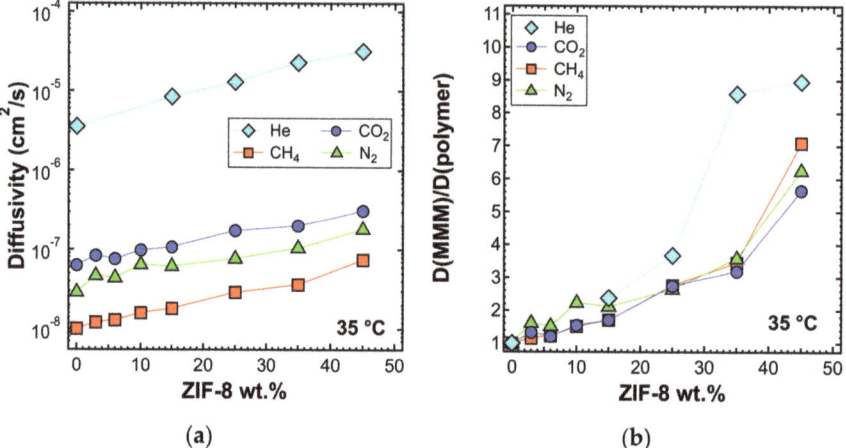

Figure 16. (a) Diffusivity and (b) relative diffusivity increase of various gases at 35 °C in ZIF-8/PPO mixed-matrix membranes (MMMs), estimated with the time-lag method.

In Figure 17, the estimated values of diffusivity-selectivity for the couples CO_2/N_2 and CO_2/CH_4 are reported. In both cases, it can be seen that the diffusivity-selectivity is lower than the overall selectivity (Figure 10b), due to the fact that solubility plays a synergic role in separations involving CO_2. The slightly decreasing trend of diffusivity-selectivity with the filler loading for both CO_2/CH_4 and CO_2/N_2 separations is consistent with the more preferable diffusion path offered by ZIF-8 in the MMMs.

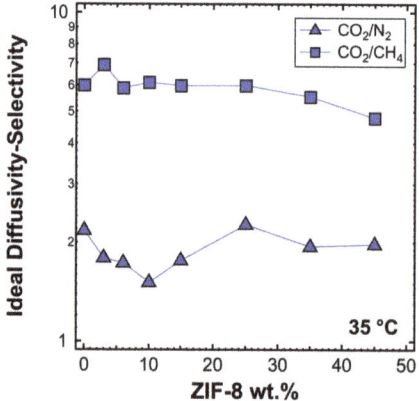

Figure 17. Ideal diffusivity-selectivity for CO_2/N_2 and CO_2/CH_4 at 35 °C in ZIF-8/PPO MMMs.

The diffusivity was also estimated with the time-lag method at higher temperatures, namely 50 and 65 °C, and is reported in Figures 18a and 19a, respectively. In Figures 18b and 19b, the values of the diffusivity ratio between the MMMs and the pure polymer at 50 and 65 °C, respectively, are reported. It can be noticed, from a qualitative point of view, that the diffusivity followed a similar behavior at all temperatures. However, the enhancement of diffusivity induced by the filler seems less remarkable at the higher temperatures. This could be explained by the lower activation energy of diffusion required for the MMMs than for PPO, which is indeed an aspect that will be analyzed quantitatively in the next section.

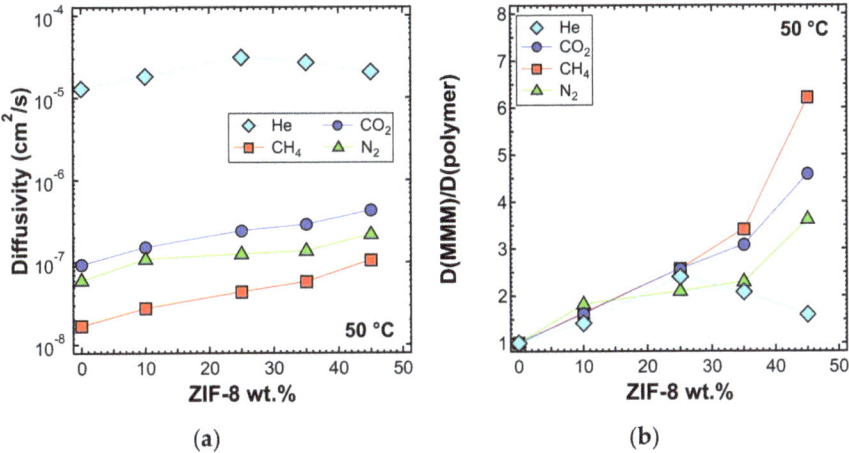

Figure 18. (**a**) Diffusivity and (**b**) relative diffusivity increase of various gases at 50 °C in ZIF-8/PPO MMMs, estimated with the time-lag method.

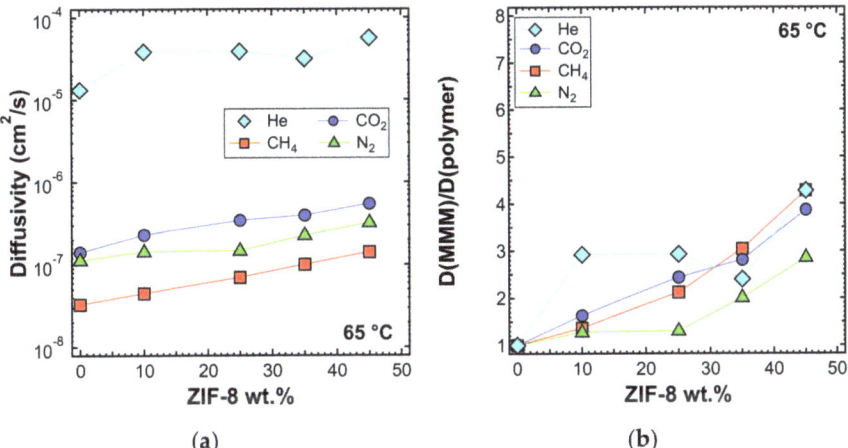

Figure 19. (**a**) Diffusivity and (**b**) relative diffusivity increase of various gases at 65 °C in ZIF-8/PPO MMMs, estimated with the time-lag method.

3.6.2. Effect of Temperature

Figure 20 shows the diffusivity as a function of temperature, for the MMMs inspected and all gases, except helium. This is because higher temperatures result in diffusion occurring more quickly than at 35 °C, making the time-lag even shorter and less accurate. On the other hand, the diffusivity of the other gases, with a few exceptions, followed Arrhenius law in all the mixed matrices considered. Therefore, we were able to calculate the activation energy of diffusion, which is reported in Figure 21 and Table 6. The values of activation energy for diffusion were higher than the respective values of $E_\mathcal{P}$, since the sorption process is always exothermic and involves negative values of sorption enthalpy (ΔH_S). Furthermore, as in the case of $E_\mathcal{P}$, the values decreased with an increasing ZIF content. This aspect indicates that the addition of filler to the polymer lowered the energetic barrier of the diffusion process, possibly due to the presence of filler pores available for diffusion.

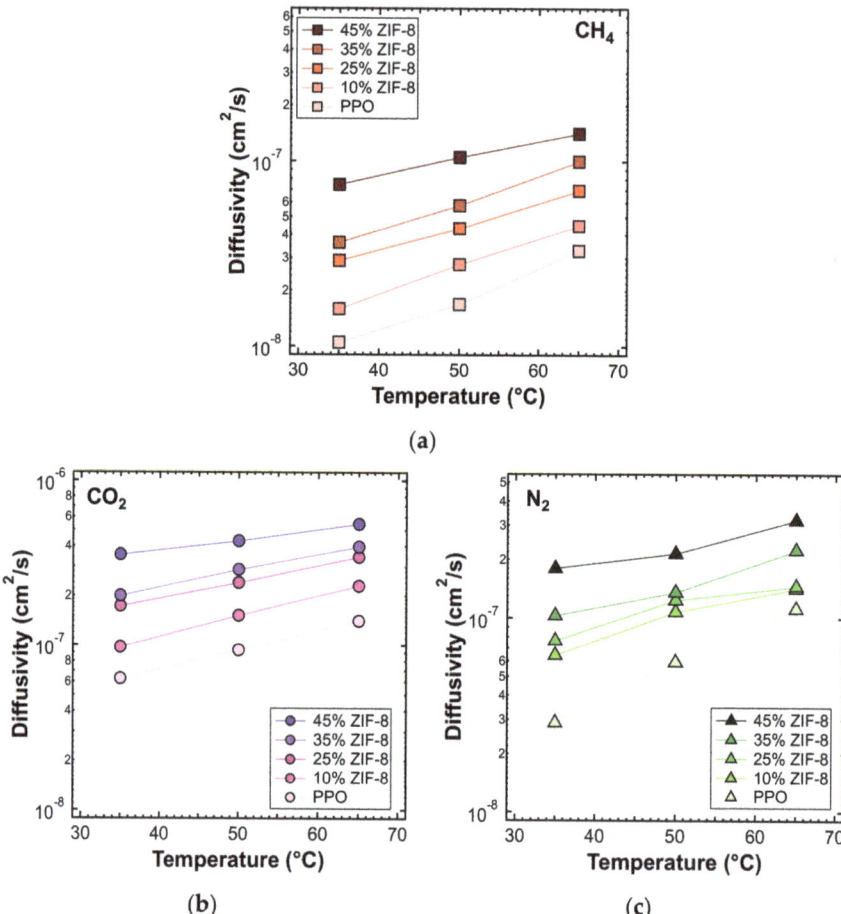

Figure 20. Diffusivity of (**a**) CH_4, (**b**) CO_2, and (**c**) N_2 at different temperatures, in the ZIF-8/PPO MMMs of different weight fractions of ZIF-8, from 0% to 45%.

Table 6. Activation energy of diffusion in the range of 35–65 °C for ZIF-8/PPO MMMs at different filler loadings.

ZIF-8 Loading (wt%)	E_D (kJ/mol)		
	N_2	CH_4	CO_2
0 [50]	22.4	29.4	23.4
0	39.10	32.87	23.33
10	28.62 *	29.99	24.78
25	26.73 *	25.37	19.82
35	22.32	29.28	19.73
45	16.42	18.47	12.35

* Activation energy of diffusion was calculated using diffusivity data at 35 and 50 °C.

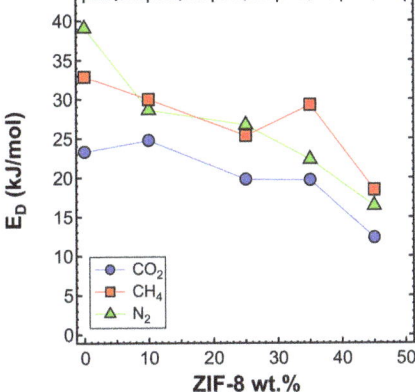

Figure 21. Activation energy of diffusion measured in the interval 35–65 °C for CO_2, CH_4, and N_2 in the MMMs as a function of filler loading.

3.7. Estimated Solubility

The solubility coefficient was calculated as the ratio between permeability and diffusivity, assuming the validity of the solution-diffusion model. The values are shown in Figure 22a for the experiment performed at 35 °C. A slight increase of the solubility coefficient at high filler loadings was recorded. Figure 22b shows that CO_2, N_2, and CH_4 experienced a similar enhancement of solubility, while for He, there was no such effect. The data indicate that the filler enhanced the gas permeability of the polymer by mainly acting on the diffusivity but also, to a non-negligible extent, on the ability of the membrane to absorb gas.

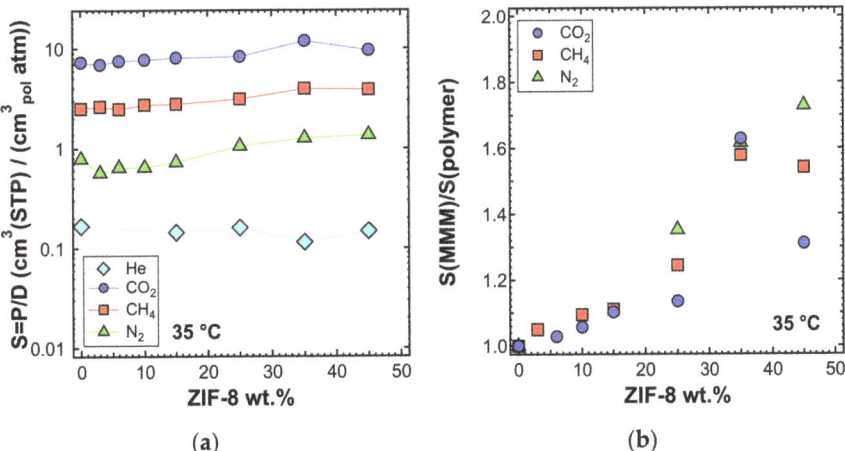

Figure 22. Estimated solubility (**a**) and solubility ratio (**b**) at 35 °C in the mixed-matrix membranes inspected. Values evaluated as P/D.

Effect of Temperature: Sorption Enthalpy (ΔH_S)

The sorption enthalpy can be estimated by subtracting the activation energy of diffusion from that of permeation according to Equation (7). Figure 23 shows the trend of ΔH_S with the ZIF-8 content, and Table 7 reports its values. CO_2 showed, on average, the most negative values among all gases, which means that CO_2 sorption is more favorable than that of CH_4 and N_2. This is related to the higher condensability of CO_2, as expected due to being the most soluble gas, as observed in Figure 22a. The absolute value of the sorption enthalpy decreases with an increasing filler content.

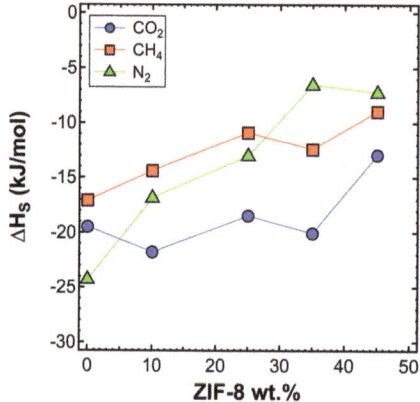

Figure 23. Heat of sorption estimated in the interval 35–65 °C for three gases in the various mixed-matrix membranes inspected, as a function of filler loading.

Table 7. Heat of sorption ΔH_S in the range of 35–65 °C for ZIF-8/PPO MMMs at different filler loadings.

ZIF-8 Loading (wt%)	ΔH_S (kJ/mol)		
	N2	CH4	CO_2
0 [50]	−12.6	−17.3	−21.9
0	−24.32	−17.10	−19.49
10	−16.88 *	−14.40	−21.79
25	−13.05 *	−10.91	−18.48
35	−6.56	−12.43	−20.06
45	−7.27	−8.95	−12.94

* Activation energy of diffusion was calculated using the activation energy of diffusion calculated using data at 35 and 50 °C.

3.8. Performance Evaluation Using a Robeson Upper Bound Plot

The performances of the composite membranes developed in this work were plotted on a Robeson plot for the He/CO_2 pair, featuring both the 1991 [6] and 2008 [7] upper bounds. The results are reported in Figure 24. The ZIF-8/PPO MMMs at 35 wt% and 45 wt% overcame the 1991 upper bound for the test at room temperature, while at 65 °C, the film at 25 wt% also surpassed the limit. These results were achieved because MMMs were simultaneously more permeable and more selective when compared to PPO. It must be noted, however, that the Robeson plots were obtained based on room temperature data for this gas pair. Robeson plots for He/N_2, He/CH_4, CO_2/N_2, and CO_2/CH_4 gas pairs are reported in Figure S4 of the Supplementary Information.

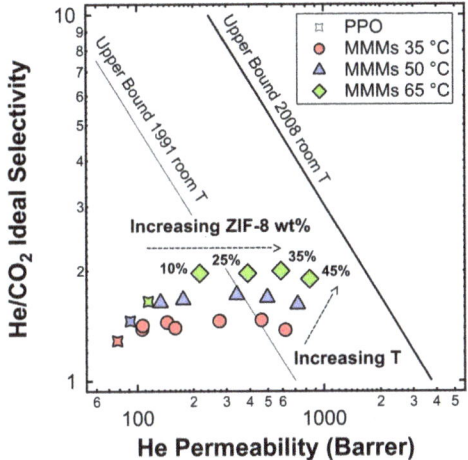

Figure 24. Positioning of the MMMs studied in this work in a Robeson plot for He/CO_2 separation. For data collected at 35 °C, filler loadings of 0%, 3%, 6%, 10%, 15%, 25%, 35%, and 45% in weight of ZIF-8 were tested. For data collected at 50 °C, filler loadings are equal to 0wt%, 3 wt%, 10 wt%, 25 wt%, 35 wt%, and 45 wt%, while for data collected at 65 °C, only data at wt %, 10 wt%, 25 wt%, 35 wt%, and 45 wt% were collected. The effect of the temperature is also shown.

3.9. Comparison with Other MMMs

In the present section, the results are compared in terms of the relative permeability and selectivity enhancement with respect to He/CO_2 or H_2/CO_2 separation, obtained by adding H_2-selective fillers to glassy polymers. The data shown in Figure 25 refer to MMMs formed by ZIF-7/Polybenzimidazole (PBI) [56]; Zeolite 3A/Polysulfone [54]; ZIF-8 blended with different polymer matrices such as Matrimid® [48], PIM-1 [61], and PPEES [60]; Cu-BPY-HSF/Matrimid® [90]; and $Cu_3(BTC)_2$/PMDA-ODA [91]. ZIF-8/PIM-1 MMMs showed a significant and balanced increase in the selectivity and permeability. In particular, He and H_2 permeabilities were improved by a factor of ~4, while He/CO_2 and H_2/CO_2 selectivity exhibited a three-fold increase. However, PIM-1 is intrinsically different from PPO in that it is CO_2-selective over H_2 in its pure unfilled state, as well as Matrimid®. The permeability enhancement observed by adding ZIF-8 to PPO, as assessed in this work, is among the highest ever recorded, second to only the 30 wt% PPEES/ZIF-8 membrane produced by Díaz et al. [60]. The addition of the hydrophilic Zeolite 3A to PSf, characterized by a smaller pore size, on the other hand, appreciably enhances the selectivity, but only slightly increases the permeability [54]. This effect is most likely due to the fact that Zeolite 3A has a higher selectivity than ZIF-8 given the rigidity of the inorganic cage. It can be said that the behavior of the mixed-matrix membranes inspected in this work resembles that of other materials obtained by mixing the same filler, ZIF-8, with a polymer that features similar initial properties in terms of the permeability and selectivity with respect to PPO. The data indeed indicate that the addition of ZIF-8 to a size-selective polymer can remarkably enhance the permeability, but has a limited effect on the selectivity, due to the intrinsically moderate selectivity of ZIF-8.

Finally, it can be noticed that the behavior of the mixed-matrix membranes obtained here is, at least qualitatively, intermediate between that of the pure polymer and that of the pure filler. This is strictly related to the optimization of a casting protocol that allows filler dispersion in solution and defect-free film formation. Such a result is in agreement with the fact that, according to SEM analysis and density tests, the composite materials show good adhesion between the polymer and the metalorganic phases, indicating that the properties of the composite do not significantly deviate from ideality. This can be considered, by all means, a qualitative structure-property correlation which can guide the design of novel mixed-matrix materials with targeted properties for specific separation applications by appropriate combinations of polymers and fillers with known initial properties.

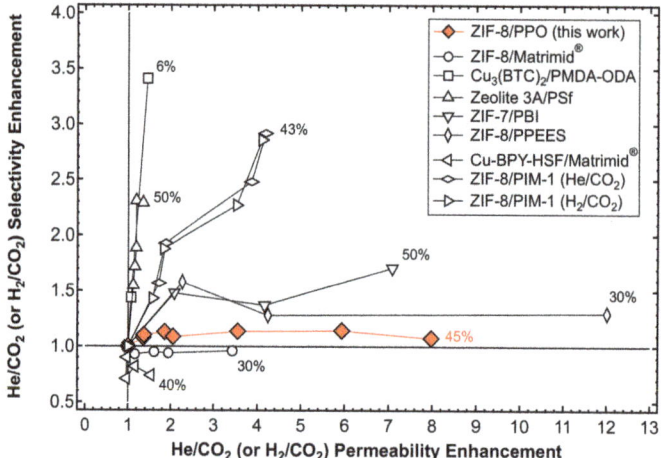

Figure 25. Effect of the addition of size-selective fillers on the He (or H_2) permeability and He(or H_2)/CO_2 selectivity of a series of glassy polymers with respect to the pure polymer matrix. The maximum loading achieved by each MMM is represented next to each series of data. Results from references [48,54,56,60,61,90,91].

4. Conclusions

In this work, we fabricated mixed-matrix membranes based on PPO and variable amounts of ZIF-8, with the goal of enhancing the sieving ability to achieve better H_2/CO_2 separation. A preparation protocol that allowed us to obtain membranes with loadings of filler as high as 45 wt% was defined. Morphological analysis highlighted the presence of generally well-dispersed filler particles with generally good interface adhesion between the polymer and filler, but also the formation of some aggregates at high loading. Thermal analysis showed the absence of residual solvents or moisture, the amorphous structure of the membranes, and the linear increase of the glass transition temperature with an increasing filler content in the polymer matrix. The buoyancy tests performed allowed us to estimate that the density of the composite membranes follows an additive behavior, based on the pure polymer and the theoretical density of the ZIF-8 crystal, which indicates good adhesion of the polymer and filler phase and no formation of voids at the interface.

Permeability tests were performed on membranes containing various contents of ZIF-8 at temperatures between 35 and 65 °C. The addition of ZIF-8 to the polymer produced a monotonous increase of permeability among all four gases tested, with factors as high as 8. This trend was obeyed at all temperatures. He/CO_2 selectivity, on the other hand, increased, to a smaller extent, up to a loading of ZIF-8 of about 35 wt%, due to the fact that the filler has a moderate selectivity with respect to this gas pair. The gas diffusivity also monotonously increased with the ZIF-8 content, for all gases and at each temperature. However, the enhancement of diffusivity alone does not justify the observed enhancement of permeability and, according to the solution diffusion-model, we were able to assess that there was also a beneficial effect of ZIF-8 addition on the gas solubility in the MMMs.

An analysis of the activation energies showed that the presence of filler in the polymer matrix made transport in the composite films easier. Indeed, $E_\mathcal{P}$ and $E_\mathcal{D}$ decreased with an increasing ZIF-8 content in the membrane. This behavior is compatible with the availability of a higher free volume for gas transport in the presence of filler particles.

The Robeson plot for He/CO$_2$ separation indicates that the addition of ZIF-8 pushes the MMM performance above the room temperature 1991 upper bound, and close to the room temperature 2008 upper bound. The temperature increase also yielded a simultaneous increase of permeability and selectivity, indicating that such membranes can have potential for applications at high temperatures. However, further optimization is required, especially in terms of the selectivity, before proceeding to produce thin film membranes required for upscaling of the materials.

The present results, together with those obtained previously on materials characterized by properties similar to PPO, i.e., a moderate selectivity for the H$_2$/CO$_2$ and He/CO$_2$ gas pairs, indicate that the addition of ZIF-8 to such materials leads to mixed-matrix membranes which are defect-free and can show remarkably higher permeabilities than initial values, thanks to the intrinsic filler permeability. However, the increase of selectivity achievable is not significant, due to the intrinsically moderate selectivity of ZIF-8 for such gas pairs. However, the properties of the mixed-matrix membranes inspected here are intermediate in relation to those of the pure polymer and the pure filler, as expected from defect-free composite materials: such a result can guide the design of novel composite materials with tunable separation properties by appropriate combinations of polymer and filler.

Supplementary Materials: The following are available online at http://www.mdpi.com/2077-0375/10/4/56/s1: Figure S1: TGA analysis results of PPO (red), ZIF-8 (black), 15 wt% ZIF-8/PPO (blue), and 25 wt% ZIF-8/PPO (green). Tests were performed in nitrogen atmosphere. PPO and MMMs were pre-treated at 200 °C overnight and then normally exposed to air for days/weeks, while ZIF-8 powder was tested as received; Figure S2: First DSC scan of a sample of PPO casted by inducing slow solvent evaporation, with subsequent formation of crystal domains and rupture of the membrane. Before performing DSC, the film was not pre-treated; Figure S3: SEM images of the cross-section of PPO/ZIF-8 mixed-matrix membranes at different loadings and magnitudes: (a) 25 wt.%, (b) 25 wt.%, (c) 25 wt.%, (d) 10 wt.%, (e) 6 wt.%, (f) 45 wt.%, (g) 25 wt.%, (h) 10 wt.%, (i) 25 wt.%, and (j) 25 wt.%; Figure S4: Positioning of the MMMs studied in this work in Robeson plots for (a) He/N$_2$, (b) He/CH$_4$, (c) CO$_2$/N$_2$, and (d) CO$_2$/CH$_4$ separations. For data collected at 35 °C, filler loadings of 0%, 3%, 6%, 10%, 15%, 25%, 35%, and 45% in weight of ZIF-8 were tested. For data collected at 50 °C, filler loadings were equal to 0 wt%, 3 wt%, 10 wt%, 25 wt%, 35 wt%, and 45 wt%, while for data collected at 65 °C, only data at 0 wt%, 10 wt%, 25 wt%, 35 wt%, and 45 wt% were collected. The effect of the temperature is also shown; Figure S5: Parity plot of diffusivity enhancement and permeability enhancement due to the addition of ZIF-8 to PPO at 35 °C for (a) He, (b) CO$_2$, (c) N$_2$, and (d) CH$_4$; Table S1: Pure gas permeability and ideal selectivity in PPO and ZIF-8/PPO MMMs. Tests were performed at 50 °C and 1.3 bar was employed as the upstream pressure; Table S2: Pure gas permeability and ideal selectivity in PPO and ZIF-8/PPO MMMs. Tests were performed at 65 °C and 1.3 bar was employed as the upstream pressure.

Author Contributions: Conceptualization, F.M.B. and M.G.D.A.; methodology, F.M.B., M.D.E. and P.F.; investigation, F.M.B. and M.D.E; data curation, F.M.B., M.D.E. and P.F.; writing—original draft preparation, F.M.B.; writing—review and editing, M.G.D.A., P.F. and A.P.; supervision, M.G.D.A., A.P. and A.O.; project administration, F.M.B., M.G.D.A., A.P., A.O. and A.M.; funding acquisition, M.G.D.A., A.P., A.O. and A.M. All authors have read and agreed to the published version of the manuscript.

Funding: This research was funded by the "Centre of Excellence on Clean Energy" project (CUP: D82I13000250001 and D83C17000370002), managed by Sotacarbo S.p.A. and funded by the Regional Government of Sardinia. The work was carried out in the framework of the project "Development and study of a system based on composite membranes for the purification of syngas from gasification process", INSTM project INDBO01183.

Acknowledgments: We are grateful to Giorgio Cucca, Andrea Sardano, Farzaneh Torabi Nabil, and Vahid Alizadeh for providing support during the experimental campaign; Mauro Zapparoli for the experienced support offered when performing the SEM experiments at Centro Grandi Strumenti (Modena, Italy); and Moon Joo Lee for the graphical representation of the structure of ZIF-8.

Conflicts of Interest: The authors declare no conflicts of interest. The funders had no role in the design of the study; in the collection, analyses, or interpretation of data; in the writing of the manuscript; or in the decision to publish the results.

References

1. Ockwig, N.W.; Nenoff, T.M. Membranes for hydrogen separation. *Chem. Rev.* **2007**, *107*, 4078–4110. [CrossRef] [PubMed]
2. Koros, W.J.; Mahajan, R. Pushing the limits on possibilities for large scale gas separation: Which strategies? *J. Membr. Sci.* **2000**, *175*, 181–196. [CrossRef]
3. Bernardo, P.; Drioli, E.; Golemme, G. Membrane Gas Separation: A Review/State of the Art. *Ind. Eng. Chem. Res.* **2009**, *48*, 4638–4663. [CrossRef]

4. Henis, J.M.S.; Tripodi, M.K. A Novel Approach to Gas Separations Using Composite Hollow Fiber Membranes. *Sep. Sci. Technol.* **1980**, *15*, 1059–1068. [CrossRef]
5. Galizia, M.; Chi, W.S.; Smith, Z.P.; Merkel, T.C.; Baker, R.W.; Freeman, B.D. 50th Anniversary Perspective: Polymers and Mixed Matrix Membranes for Gas and Vapor Separation: A Review and Prospective Opportunities. *Macromolecules* **2017**, *50*, 7809–7843. [CrossRef]
6. Robeson, L.M. Correlation of separation factor versus permeability for polymeric membranes. *J. Membr. Sci.* **1991**, *62*, 165–185. [CrossRef]
7. Robeson, L.M. The upper bound revisited. *J. Membr. Sci.* **2008**, *320*, 390–400. [CrossRef]
8. Rose, I.; Bezzu, C.G.; Carta, M.; Comesañā-Gándara, B.; Lasseuguette, E.; Ferrari, M.C.; Bernardo, P.; Clarizia, G.; Fuoco, A.; Jansen, J.C.; et al. Polymer ultrapermeability from the inefficient packing of 2D chains. *Nat. Mater.* **2017**, *16*, 932–937. [CrossRef]
9. Carta, M.; Malpass-Evans, R.; Croad, M.; Rogan, Y.; Jansen, J.C.; Bernardo, P.; Bazzarelli, F.; McKeown, N.B. An efficient polymer molecular sieve for membrane gas separations. *Science* **2013**, *339*, 303–307. [CrossRef]
10. Budd, P.M.; Elabas, E.S.; Ghanem, B.S.; Makhseed, S.; McKeown, N.B.; Msayib, K.J.; Tattershall, C.E.; Wang, D. Solution-Processed, Organophilic Membrane Derived from a Polymer of Intrinsic Microporosity. *Adv. Mater.* **2004**, *16*, 456–459. [CrossRef]
11. Lai, H.W.H.; Benedetti, F.M.; Jin, Z.; Teo, Y.C.; Wu, A.X.; De Angelis, M.G.; Smith, Z.P.; Xia, Y. Tuning the Molecular Weights, Chain Packing, and Gas-Transport Properties of CANAL Ladder Polymers by Short Alkyl Substitutions. *Macromolecules* **2019**, *52*, 6294–6302. [CrossRef]
12. He, Y.; Benedetti, F.M.; Lin, S.; Liu, C.; Zhao, Y.; Ye, H.; Van Voorhis, T.; De Angelis, M.G.; Swager, T.M.; Smith, Z.P. Polymers with Side Chain Porosity for Ultrapermeable and Plasticization Resistant Materials for Gas Separations. *Adv. Mater.* **2019**, *31*, 1807871. [CrossRef] [PubMed]
13. Kang, Z.; Peng, Y.; Qian, Y.; Yuan, D.; Addicoat, M.A.; Heine, T.; Hu, Z.; Tee, L.; Guo, Z.; Zhao, D. Mixed Matrix Membranes (MMMs) Comprising Exfoliated 2D Covalent Organic Frameworks (COFs) for Efficient CO_2 Separation. *Chem. Mater.* **2016**, *28*, 1277–1285. [CrossRef]
14. Biswal, B.P.; Chaudhari, H.D.; Banerjee, R.; Kharul, U.K. Chemically Stable Covalent Organic Framework (COF)-Polybenzimidazole Hybrid Membranes: Enhanced Gas Separation through Pore Modulation. *Chem. A Eur. J.* **2016**, *22*, 4695–4699. [CrossRef] [PubMed]
15. Dechnik, J.; Gascon, J.; Doonan, C.J.; Janiak, C.; Sumby, C.J. Mixed-Matrix Membranes. *Angew. Chem. Int. Ed.* **2017**, *56*, 9292–9310. [CrossRef]
16. Dong, G.; Li, H.; Chen, V. Challenges and opportunities for mixed-matrix membranes for gas separation. *J. Mater. Chem. A* **2013**, *1*, 4610–4630. [CrossRef]
17. Zornoza, B.; Tellez, C.; Coronas, J.; Gascon, J.; Kapteijn, F. Metal organic framework based mixed matrix membranes: An increasingly important field of research with a large application potential. *Microporous Mesoporous Mater.* **2013**, *166*, 67–78. [CrossRef]
18. Chung, T.S.; Jiang, L.Y.; Li, Y.; Kulprathipanja, S. Mixed matrix membranes (MMMs) comprising organic polymers with dispersed inorganic fillers for gas separation. *Prog. Polym. Sci.* **2007**, *32*, 483–507. [CrossRef]
19. Olivieri, L.; Ligi, S.; De Angelis, M.G.; Cucca, G.; Pettinau, A. Effect of Graphene and Graphene Oxide Nanoplatelets on the Gas Permselectivity and Aging Behavior of Poly(trimethylsilyl propyne) (PTMSP). *Ind. Eng. Chem. Res.* **2015**, *54*, 11199–11211. [CrossRef]
20. Althumayri, K.; Harrison, W.J.; Shin, Y.; Gardiner, J.M.; Casiraghi, C.; Budd, P.M.; Bernardo, P.; Clarizia, G.; Jansen, J.C. The influence of few-layer graphene on the gas permeability of the high-free-volume polymer PIM-1. *Philos. Trans. R. Soc. A Math. Phys. Eng. Sci.* **2016**, *374*. [CrossRef]
21. Merkel, T.C.; He, Z.; Pinnau, I.; Freeman, B.D.; Meakin, P.; Hill, A.J. Sorption and Transport in Poly(2,2-bis(trifluoromethyl)-4,5-difluoro-1,3-dioxole-co-tetrafluoroethylene) Containing Nanoscale Fumed Silica. *Macromolecules* **2003**, *36*, 8406–8414. [CrossRef]
22. Chi, W.S.; Sundell, B.J.; Zhang, K.; Harrigan, D.J.; Hayden, S.C.; Smith, Z.P. Mixed-Matrix Membranes Formed from Multi-Dimensional Metal–Organic Frameworks for Enhanced Gas Transport and Plasticization Resistance. *ChemSusChem* **2019**, *12*, 2355–2360. [CrossRef] [PubMed]
23. Dong, G.; Li, H.; Chen, V. Plasticization mechanisms and effects of thermal annealing of Matrimid hollow fiber membranes for CO_2 removal. *J. Membr. Sci.* **2011**, *369*, 206–220. [CrossRef]
24. Li, J.; Sculley, J.; Zhou, H. Metal–Organic Frameworks for Separations. *Chem. Rev.* **2012**, *112*, 869–932. [CrossRef] [PubMed]

25. Yaghi, O.M.; O'Keeffe, M.; Ockwig, N.W.; Chae, H.K.; Eddaoudi, M.; Kim, J. Reticular synthesis and the design of new materials. *Nature* **2003**, *423*, 705–714. [CrossRef]
26. Trickett, C.A.; Helal, A.; Al-Maythalony, B.A.; Yamani, Z.H.; Cordova, K.E.; Yaghi, O.M. The chemistry of metal–organic frameworks for CO_2 capture, regeneration and conversion. *Nat. Rev. Mater.* **2017**, *2*, 17045. [CrossRef]
27. Yehia, H.; Pisklak, T.J.; Balkus, K.J.; Musselman, I.H. Methane Facilitated Transport Using Copper(II) Biphenyl Dicarboxylate-Triethylenediamine/Poly (3-Acetoxyethylthiophene) Mixed Matrix Membranes. In *Abstracts of Papers of the American Chemical Society*; American Chemical Society: Washington, DC, USA, 2004.
28. Mahajan, R.; Koros, W.J. Factors Controlling Successful Formation of Mixed-Matrix Gas Separation Materials. *Ind. Eng. Chem. Res.* **2000**, *39*, 2692–2696. [CrossRef]
29. Mahajan, R.; Koros, W.J. Mixed matrix membrane materials with glassy polymers. Part 1. *Polym. Eng. Sci.* **2002**, *42*, 1420–1431. [CrossRef]
30. Caro, J. Are MOF membranes better in gas separation than those made of zeolites? *Curr. Opin. Chem. Eng.* **2011**, *1*, 77–83. [CrossRef]
31. Wang, Z.; Wang, D.; Zhang, S.; Hu, L.; Jin, J. Interfacial Design of Mixed Matrix Membranes for Improved Gas Separation Performance. *Adv. Mater.* **2016**, *28*, 3399–3405. [CrossRef]
32. Xin, Q.; Ouyang, J.; Liu, T.; Li, Z.; Li, Z.; Liu, Y.; Wang, S.; Wu, H.; Jiang, Z.; Cao, X. Enhanced Interfacial Interaction and CO_2 Separation Performance of Mixed Matrix Membrane by Incorporating Polyethylenimine-Decorated Metal–Organic Frameworks. *ACS Appl. Mater. Interfaces* **2015**, *7*, 1065–1077. [CrossRef] [PubMed]
33. Qian, Q.; Wu, A.X.; Chi, W.S.; Asinger, P.A.; Lin, S.; Hypsher, A.; Smith, Z.P. Mixed-Matrix Membranes Formed from Imide-Functionalized UiO-66-NH 2 for Improved Interfacial Compatibility. *ACS Appl. Mater. Interfaces* **2019**, *11*, 31257–31269. [CrossRef] [PubMed]
34. Banerjee, R.; Phan, A.; Wang, B.; Knobler, C.; Furukawa, H.; O'Keeffe, M.; Yaghi, O.M. High-Throughput Synthesis of Zeolitic Imidazolate Frameworks and Application to CO_2 Capture. *Science* **2008**, *319*, 939–943. [CrossRef] [PubMed]
35. Li, Y.; Liang, F.; Bux, H.; Yang, W.; Caro, J. Zeolitic imidazolate framework ZIF-7 based molecular sieve membrane for hydrogen separation. *J. Membr. Sci.* **2010**, *354*, 48–54. [CrossRef]
36. Melgar, V.M.A.; Kim, J.; Othman, M.R. Zeolitic imidazolate framework membranes for gas separation: A review of synthesis methods and gas separation performance. *J. Ind. Eng. Chem.* **2015**, *28*, 1–15. [CrossRef]
37. Parent, L.R.; Pham, C.H.; Patterson, J.P.; Denny, M.S.; Cohen, S.M.; Gianneschi, N.C.; Paesani, F. Pore Breathing of Metal-Organic Frameworks by Environmental Transmission Electron Microscopy. *J. Am. Chem. Soc.* **2017**, *139*, 13973–13976. [CrossRef]
38. Hyun, S.; Lee, J.H.; Jung, G.Y.; Kim, Y.K.; Kim, T.K.; Jeoung, S.; Kwak, S.K.; Moon, D.; Moon, H.R. Exploration of Gate-Opening and Breathing Phenomena in a Tailored Flexible Metal–Organic Framework. *Inorg. Chem.* **2016**, *55*, 1920–1925. [CrossRef]
39. Shieh, J.-J.; Chung, T.S. Gas permeability, diffusivity, and solubility of poly(4-vinylpyridine) film. *J. Polym. Sci. Part B Polym. Phys.* **1999**, *37*, 2851–2861. [CrossRef]
40. Bux, H.; Liang, F.; Li, Y.; Cravillon, J.; Wiebcke, M.; Caro, J. Zeolitic Imidazolate Framework Membrane with Molecular Sieving Properties by Microwave-Assisted Solvothermal Synthesis. *J. Am. Chem. Soc.* **2009**, *131*, 16000–16001. [CrossRef]
41. McCarthy, M.C.; Varela-Guerrero, V.; Barnett, G.V.; Jeong, H.K. Synthesis of zeolitic imidazolate framework films and membranes with controlled microstructures. *Langmuir* **2010**, *26*, 14636–14641. [CrossRef]
42. Bux, H.; Feldhoff, A.; Cravillon, J.; Wiebcke, M.; Li, Y.-S.; Caro, J. Oriented Zeolitic Imidazolate Framework-8 Membrane with Sharp H2/C3H8 Molecular Sieve Separation. *Chem. Mater.* **2011**, *23*, 2262–2269. [CrossRef]
43. Melgar, V.M.A.; Ahn, H.; Kim, J.; Othman, M.R. Highly selective micro-porous ZIF-8 membranes prepared by rapid electrospray deposition. *J. Ind. Eng. Chem.* **2015**, *21*, 575–579. [CrossRef]
44. Hara, N.; Yoshimune, M.; Negishi, H.; Haraya, K.; Hara, S.; Yamaguchi, T. Diffusive separation of propylene/propane with ZIF-8 membranes. *J. Membr. Sci.* **2014**, *450*, 215–223. [CrossRef]
45. Baker, R.W. Future Directions of Membrane Gas Separation Technology. *Ind. Eng. Chem. Res.* **2002**, *41*, 1393–1411. [CrossRef]
46. Merkel, T.C.; Zhou, M.; Baker, R.W. Carbon dioxide capture with membranes at an IGCC power plant. *J. Membr. Sci.* **2012**, *389*, 441–450. [CrossRef]

47. Marano, J.J.; Ciferino, J.P. Integration of Gas Separation Membranes with IGCC Identifying the right membrane for the right job. *Energy Procedia* **2009**, *1*, 361–368. [CrossRef]
48. Ordoñez, M.J.C.; Balkus, K.J.; Ferraris, J.P.; Musselman, I.H. Molecular sieving realized with ZIF-8/Matrimid®mixed-matrix membranes. *J. Membr. Sci.* **2010**, *361*, 28–37. [CrossRef]
49. Park, K.S.; Ni, Z.; Cote, A.P.; Choi, J.Y.; Huang, R.; Uribe-Romo, F.J.; Chae, H.K.; O'Keeffe, M.; Yaghi, O.M. Exceptional chemical and thermal stability of zeolitic imidazolate frameworks. *Proc. Natl. Acad. Sci. USA* **2006**, *103*, 10186–10191. [CrossRef]
50. Aguilar-Vega, M.; Paul, D.R. Gas transport properties of polyphenylene ethers. *J. Polym. Sci. Part B Polym. Phys.* **1993**, *31*, 1577–1589. [CrossRef]
51. Alentiev, A.; Drioli, E.; Gokzhaev, M.; Golemme, G.; Ilinich, O.; Lapkin, A.; Volkov, V.; Yampolskii, Y. Gas permeation properties of phenylene oxide polymers. *J. Membr. Sci.* **1998**, *138*, 99–107. [CrossRef]
52. Huang, Y.; Paul, D.R. Effect of Molecular Weight and Temperature on Physical Aging of ThinGlassy Poly(2,6-dimethyl-1,4-phenylene oxide) Films. *J. Polym. Sci. Part B Polym. Phys.* **2007**, *45*, 1390–1398. [CrossRef]
53. Galizia, M.; Daniel, C.; Fasano, G.; Guerra, G.; Mensitieri, G. Gas Sorption and Diffusion in Amorphous and Semicrystalline Nanoporous Poly(2,6-dimethyl-1,4-phenylene)oxide. *Macromolecules* **2012**, *45*, 3604–3615. [CrossRef]
54. Khan, A.L.; Cano-Odena, A.; Gutiérrez, B.; Minguillón, C.; Vankelecom, I.F.J. Hydrogen separation and purification using polysulfone acrylate–zeolite mixed matrix membranes. *J. Membr. Sci.* **2010**, *350*, 340–346. [CrossRef]
55. Song, Q.; Nataraj, S.K.; Roussenova, M.V.; Tan, J.C.; Hughes, D.J.; Li, W.; Bourgoin, P.; Alam, M.A.; Cheetham, A.K.; Al-Muhtaseb, S.A.; et al. Zeolitic imidazolate framework (ZIF-8) based polymer nanocomposite membranes for gas separation. *Energy Environ. Sci.* **2012**, *5*, 8359. [CrossRef]
56. Yang, T.; Xiao, Y.; Chung, T.-S. Poly-/metal-benzimidazole nano-composite membranes for hydrogen purification. *Energy Environ. Sci.* **2011**, *4*, 4171. [CrossRef]
57. Yang, T.; Chung, T.-S. High performance ZIF-8/PBI nano-composite membranes for high temperature hydrogen separation consisting of carbon monoxide and water vapor. *Int. J. Hydrog. Energy* **2013**, *38*, 229–239. [CrossRef]
58. Yang, T.; Shi, G.M.; Chung, T.-S. Symmetric and Asymmetric Zeolitic Imidazolate Frameworks (ZIFs)/Polybenzimidazole (PBI) Nanocomposite Membranes for Hydrogen Purification at High Temperatures. *Adv. Energy Mater.* **2012**, *2*, 1358–1367. [CrossRef]
59. Zhang, C.; Dai, Y.; Johnson, J.R.; Karvan, O.; Koros, W.J. High performance ZIF-8/6FDA-DAM mixed matrix membrane for propylene/propane separations. *J. Membr. Sci.* **2012**, *389*, 34–42. [CrossRef]
60. Díaz, K.; López-González, M.; Del Castillo, L.F.; Riande, E. Effect of zeolitic imidazolate frameworks on the gas transport performance of ZIF8-poly(1,4-phenylene ether-ether-sulfone) hybrid membranes. *J. Membr. Sci.* **2011**, *383*, 206–213. [CrossRef]
61. Bushell, A.F.; Attfield, M.P.; Mason, C.R.; Budd, P.M.; Yampolskii, Y.; Starannikova, L.; Rebrov, A.; Bazzarelli, F.; Bernardo, P.; Jansen, J.C.; et al. Gas permeation parameters of mixed matrix membranes based on the polymer of intrinsic microporosity PIM-1 and the zeolitic imidazolate framework ZIF-8. *J. Membr. Sci.* **2013**, *427*, 48–62. [CrossRef]
62. Sutrisna, P.D.; Hou, J.; Li, H.; Zhang, Y.; Chen, V. Improved operational stability of Pebax-based gas separation membranes with ZIF-8: A comparative study of flat sheet and composite hollow fibre membranes. *J. Membr. Sci.* **2017**, *524*, 266–279. [CrossRef]
63. Ma, X.; Swaidan, R.J.; Wang, Y.; Hsiung, C.; Han, Y.; Pinnau, I. Highly Compatible Hydroxyl-Functionalized Microporous Polyimide-ZIF-8 Mixed Matrix Membranes for Energy Efficient Propylene/Propane Separation. *ACS Appl. Nano Mater.* **2018**, *1*, 3541–3547. [CrossRef]
64. Küsgens, P.; Rose, M.; Senkovska, I.; Fröde, H.; Henschel, A.; Siegle, S.; Kaskel, S. Characterization of metal-organic frameworks by water adsorption. *Microporous Mesoporous Mater.* **2009**, *120*, 325–330. [CrossRef]
65. Ortiz, A.U.; Freitas, A.P.; Boutin, A.; Fuchs, A.H.; Coudert, F.X. What makes zeolitic imidazolate frameworks hydrophobic or hydrophilic? the impact of geometry and functionalization on water adsorption. *Phys. Chem. Chem. Phys.* **2014**, *16*, 9940–9949. [CrossRef] [PubMed]
66. Khulbe, K.C.; Matsuura, T.; Lamarche, G.; Kim, H.J. The morphology characterisation and performance of dense PPO membranes for gas separation. *J. Membr. Sci.* **1997**, *135*, 211–223. [CrossRef]

67. Khayet, M.; Villaluenga, J.P.G.; Godino, M.P.; Mengual, J.I.; Seoane, B.; Khulbe, K.C.; Matsuura, T. Preparation and application of dense poly(phenylene oxide) membranes in pervaporation. *J. Colloid Interface Sci.* **2004**, *278*, 410–422. [CrossRef]
68. Das, M.; Perry, J.D.; Koros, W.J. Gas-Transport-Property Performance of Hybrid Carbon Molecular Sieve–Polymer Materials. *Ind. Eng. Chem. Res.* **2010**, *49*, 9310–9321. [CrossRef]
69. Perry, R.H.; Green, D.W. *Perry's Chemical Engineers' Handbook*, 7th ed.; McGraw-Hill: New York, NY, USA, 1999; pp. 291–293.
70. Horn, N.R.; Paul, D.R. Carbon dioxide plasticization and conditioning effects in thick vs. thin glassy polymer films. *Polymer* **2011**, *52*, 1619–1627. [CrossRef]
71. Bos, A.; Pünt, I.G.M.; Wessling, M.; Strathmann, H. CO_2-induced plasticization phenomena in glassy polymers. *J. Membr. Sci.* **1999**, *155*, 67–78. [CrossRef]
72. Handa, Y.P.; Lampron, S.; O'neill, M.L. On the plasticization of poly(2,6-dimethyl phenylene oxide) by CO_2. *J. Polym. Sci. Part B Polym. Phys.* **1994**, *32*, 2549–2553. [CrossRef]
73. Minelli, M.; Deangelis, M.; Doghieri, F.; Marini, M.; Toselli, M.; Pilati, F. Oxygen permeability of novel organic–inorganic coatings: I. Effects of organic–inorganic ratio and molecular weight of the organic component. *Eur. Polym. J.* **2008**, *44*, 2581–2588. [CrossRef]
74. Bevington, P.R.; Robinson, D.K. *Error Data Reduction and Error Analysis for the Physical Sciences*, 3rd ed.; AIP Publishing LLC: Melville, NY, USA, 1992. [CrossRef]
75. Daynes, H.A. The Process of Diffusion through a Rubber Membrane. *Proc. R. Soc. A Math. Phys. Eng. Sci.* **1920**, *97*, 286–307. [CrossRef]
76. Crank, J. *The Mathematics of Diffusion*, 2nd ed.; Oxford University Press: Oxford, UK, 1975.
77. Moaddeb, M.; Koros, W.J. Gas transport properties of thin polymeric membranes in the presence of silicon dioxide particles. *J. Membr. Sci.* **1997**, *125*, 143–163. [CrossRef]
78. Muller, J.C.M.; Hakvoort, G.; Jansen, J.C. DSC and TG study of water adsorption and desorption on zeolite NaA: Powder and attached as layer on metal. *J. Therm. Anal. Calorim.* **1998**, *53*, 449–466. [CrossRef]
79. Story, B.J.; Koros, W.J. Sorption of CO_2/CH_4 mixtures in poly(phenylene oxide) and a carboxylated derivative. *J. Appl. Polym. Sci.* **1991**, *42*, 2613–2626. [CrossRef]
80. Toi, K.; Morel, G.; Paul, D.R. Gas sorption and transport in poly(phenylene oxide) and comparisons with other glassy polymers. *J. Appl. Polym. Sci.* **1982**, *27*, 2997–3005. [CrossRef]
81. Le Roux, J.D.; Paul, D.R.; Kampa, J.; Lagow, R.J. Surface fluorination of poly (phenylene oxide) composite membranes Part I. Transport properties. *J. Membr. Sci.* **1994**, *90*, 21–35. [CrossRef]
82. Yasuda, H.; Rosengren, K. Isobaric measurement of gas permeability of polymers. *J. Appl. Polym. Sci.* **1970**, *14*, 2839–2877. [CrossRef]
83. Ansaloni, L.; Minelli, M.; Baschetti, M.G.; Sarti, G.C. Effects of Thermal Treatment and Physical Aging on the Gas Transport Properties in Matrimid®. *Oil Gas Sci. Technol. Rev.* **2015**, *70*, 367–379. [CrossRef]
84. Savoca, A.C.; Surnamer, A.D.; Tien, C. Gas Transport in Poly(sily1propynes): The Chemical Structure Point. *Macromolecules* **1993**, *26*, 6211–6216. [CrossRef]
85. Huang, Y.; Paul, D.R. Physical aging of thin glassy polymer films monitored by gas permeability. *Polymer* **2004**, *45*, 8377–8393. [CrossRef]
86. Huang, Y.; Paul, D.R. Effect of Temperature on Physical Aging of Thin Glassy Polymer Films. *Macromolecules* **2005**, *38*, 10148–10154. [CrossRef]
87. Perez, E.V.; Balkus, K.J.; Ferraris, J.P.; Musselman, I.H. Mixed-matrix membranes containing MOF-5 for gas separations. *J. Membr. Sci.* **2009**, *328*, 165–173. [CrossRef]
88. Rea, R.; Ligi, S.; Christian, M.; Morandi, V.; Baschetti, M.G.; De Angelis, M. Permeability and Selectivity of PPO/Graphene Composites as Mixed Matrix Membranes for CO_2 Capture and Gas Separation. *Polymers* **2018**, *10*, 129. [CrossRef] [PubMed]
89. Rowe, B.W.; Robeson, L.M.; Freeman, B.D.; Paul, D.R. Influence of temperature on the upper bound: Theoretical considerations and comparison with experimental results. *J. Membr. Sci.* **2010**, *360*, 58–69. [CrossRef]

90. Zhang, Y.; Musselman, I.H.; Ferraris, J.P.; Balkus, K.J. Gas permeability properties of Matrimid®membranes containing the metal-organic framework Cu–BPY–HFS. *J. Membr. Sci.* **2008**, *313*, 170–181. [CrossRef]
91. Hu, J.; Cai, H.; Ren, H.; Wei, Y.; Xu, Z.; Liu, H.; Hu, Y. Mixed-Matrix Membrane Hollow Fibers of Cu3(BTC)2 MOF and Polyimide for Gas Separation and Adsorption. *Ind. Eng. Chem. Res.* **2010**, *49*, 12605–12612. [CrossRef]

 © 2020 by the authors. Licensee MDPI, Basel, Switzerland. This article is an open access article distributed under the terms and conditions of the Creative Commons Attribution (CC BY) license (http://creativecommons.org/licenses/by/4.0/).

Review

Recent Advances in Membrane-Based Electrochemical Hydrogen Separation: A Review

Leandri Vermaak [1,*], Hein W. J. P. Neomagus [2] and Dmitri G. Bessarabov [1,*]

[1] HySA Infrastructure Centre of Competence, Faculty of Engineering, Potchefstroom Campus, North-West University, Potchefstroom 2520, South Africa

[2] Centre of Excellence in Carbon Based Fuels, Faculty of Engineering, Potchefstroom Campus, School of Chemical and Minerals Engineering, North-West University, Potchefstroom 2520, South Africa; Hein.Neomagus@nwu.ac.za

* Correspondence: 24088633@nwu.ac.za (L.V.); Dmitri.Bessarabov@nwu.ac.za (D.G.B.)

Citation: Vermaak, L.; Neomagus, H.W.J.P.; Bessarabov, D.G. Recent Advances in Membrane-Based Electrochemical Hydrogen Separation: A Review. *Membranes* **2021**, *11*, 127. https://doi.org/10.3390/membranes11020127

Academic Editor: Yuri Yampolskii

Received: 15 December 2020
Accepted: 25 January 2021
Published: 13 February 2021

Publisher's Note: MDPI stays neutral with regard to jurisdictional claims in published maps and institutional affiliations.

Copyright: © 2021 by the authors. Licensee MDPI, Basel, Switzerland. This article is an open access article distributed under the terms and conditions of the Creative Commons Attribution (CC BY) license (https://creativecommons.org/licenses/by/4.0/).

Abstract: In this paper an overview of commercial hydrogen separation technologies is given. These technologies are discussed and compared—with a detailed discussion on membrane-based technologies. An emerging and promising novel hydrogen separation technology, namely, electrochemical hydrogen separation (EHS) is reviewed in detail. EHS has many advantages over conventional separation systems (e.g., it is not energy intensive, it is environmentally-friendly with near-zero pollutants, it is known for its silent operation, and, the greatest advantage, simultaneous compression and purification can be achieved in a one-step operation). Therefore, the focus of this review is to survey open literature and research conducted to date on EHS. Current technological advances in the field of EHS that have been made are highlighted. In the conclusion, literature gaps and aspects of electrochemical hydrogen separation, that require further research, are also highlighted. Currently, the cost factor, lack of adequate understanding of the degradation mechanisms related to this technology, and the fact that certain aspects of this technology are as yet unexplored (e.g., simultaneous hydrogen separation and compression) all hinder its widespread application. In future research, some attention could be given to the aforementioned factors and emerging technologies, such as ceramic proton conductors and solid acids.

Keywords: electrochemical hydrogen separation; electrochemical hydrogen pump; proton exchange membrane (PEM); hydrogen purification/separation

1. Introduction

The continued expansion of the commercial and industrial sectors, such as heavy-duty mobility/shipping and manufacturing, raises concerns regarding the supply capacity of existing energy resources [1,2]. Currently the global energy demand is primarily met by fossil fuel utilisation methods, which raises environmental concerns related to CO_2 emissions and other greenhouse gas emissions [2,3]. Subsequently, the energy sector faces a major challenge—to decarbonise energy supply and to find new reliable, sustainable, and environmentally friendly energy alternatives [4,5]. Renewable energy (RE) is expected to play a key role in future energy systems as it is clean and sustainable [2]. However, some key challenges, such as its variable and intermittent nature [3,6,7] remain to be addressed before completely transitioning towards RE [2]. The solution to this problem lies in adequate large-scale energy storage, which would increase energy supply reliability [8]. Energy storage can provide energy flexibility and will reduce the global dependence on fossil fuel backup power. Various types of energy storage systems exist [6,9,10]. These can be broadly categorized [8] as electrochemical (batteries) [11], chemical (hydrogen systems: Fuel cells/electrolyses [12–14]), electrical (capacitors, super capacitors and ultra-capacitors) [15,16], mechanical (flywheels [17–19], compressed air [20,21] and pumped hydro-storage [22,23]) and thermal (hot water, sensible/latent heat storage, solar energy storage) [24–28], and magnetic (superconducting energy storage) [29]. Alternative methods for large-scale energy storage are being researched, including renewable hydrogen and synthetic natural gas [6].

Hydrogen is part of an industrial concept known as power-to-gas (P2G) technology, which is a power grid balancing mechanism used to capture and store surplus energy to use at times of limited supply (e.g., night-time or at times of low wind speed when solar and wind is used as energy source) [7]. In principle, P2G converts excess RE into a chemical carrier such as hydrogen or methane [7]. Hydrogen, in particular, is attracting great interest as an energy carrier, with many unique properties to commend itself [30,31]. It can be produced/converted into electricity by means of electrochemical devices (e.g., electrolysers and fuel cells) with relatively high energy conversion efficiencies and can be stored using a variety of methods [3,31–35], such as compressed gas, cryogenic liquid [36,37], chemical compounds (e.g., liquid organic hydrogen carriers (LOHC) [38–40], ammonia) [41] or it can be adsorbed/absorbed on special materials (e.g., metal hybrids [42], chemical hybrids, carbon nanostructures). Long-distance hydrogen transportation can be achieved through pipelines [43,44] or via tanker trucks [45], and can be converted into various forms of energy more efficiently than other fuels [30,31]. Furthermore, hydrogen can be generated in an environmentally friendly manner with no greenhouse gas pollutants [30,31]. Hydrogen also has potential to provide energy to the main sectors of the economy, including transportation, buildings, and industry [46,47]. This, in turn, may lead to a low-carbon energy system known as the "hydrogen economy", which was introduced in 1972 [48].

Currently, hydrogen is a very important industrial commodity, as it is a key reactant and/or by-product of several industrial processes, including the food industry, petrochemical and petroleum refining, ammonia production, methanol production, hydrogenation processes, hydrometallurgical processes, and metal refining (mainly nickel, tungsten, molybdenum, copper, zinc, uranium, and lead) [3,46,49–51]. It can also be employed for application in electricity production from fuel cells, transportation, and energy storage [51].

Hydrogen is not widely available in gaseous state, but rather in a form of chemical compounds in natural sources, such as natural gas, water, coal and biomass (after gasification), which are all major feedstocks for hydrogen production [52]. Many hydrogen production pathways can be found in literature [45,53–55] and the selection thereof is mainly dependent on the feedstock used to produce hydrogen, the scale of production, and the available energy sources [2]. These pathways can be classified in various ways: The hydrogen source/feedstock (hydrocarbons or non-hydrocarbons) [2,3,56,57], the chemical nature and/or energy input [2,3,56] (thermochemical, electrochemical and biological [46]), the production method used [3,52,56,57] (its maturity level and efficiency) [2], the catalyst material [52], storage [51], the distribution mechanism (i.e., on-site generation or delivered) [53,56] and end use (e.g., hydrogen purity required) [50]. The choice of the hydrogen production pathway should take into account, (i) the hydrogen fuel quality grade required for end-use application and (ii) purification technology feasibility [2]. Separation processes, such as pressure swing adsorption (PSA), are applied to improve the economics of the conventional hydrogen production methods [58].

Table 1 summarizes the state-of-the-art hydrogen production technologies based on their advantages and disadvantaged, the technology maturity level (TML), the process efficiency, cleanness of the hydrogen, and the impurities commonly contained in the product streams [2].

For hydrogen production, fossil fuels are currently the main source [59]. Fossil fuel-based hydrogen production technologies are already developed and mature industrial technologies [60], capable of producing high grade hydrogen at relatively lower costs compared to some alternatives [59]. Therefore, of the over 50 million tons of hydrogen produced annually, fossil fuel-based hydrogen production constitutes an estimated 95% [3,61]. There are a number of feedstocks used to produce industrial hydrogen, but the most favoured feedstock is natural gas due to it being abundantly available and cost efficient [52,62].

The two main methods used in industry to produce hydrogen from fossil fuels are reforming processes and gasification [53]. These two methods are distinguished by the nature of the incoming fuel [3]. Gasification processes use solid fuel, such as coal, biomass and solid waste to produce hydrogen or syngas (a mixture of mainly H_2, CO [63] and, in some instances, CO_2 [46,64,65]), while reforming processes make use of fluid fuel, either in gas or liquid form, for syngas production [3]. Three reforming processes can be differentiated to produce hydrogen from hydrocarbons: (i) Steam reforming (particularly steam methane reforming (SMR)) [55,61,65–68], (ii) partial oxidation (POX) [3,46,66], and (iii) auto-thermal reforming (ATR) [3,55,65]. These processes are distinguished by the reactants involved and the thermodynamic nature of the reactions taking place [3,55]. For example, in SMR and steam-gasification, steam (water) reacts with hydrocarbons to produce hydrogen. This reaction is endothermic. In the case of POX and gasification, oxygen reacts with the hydrocarbons to produce hydrogen and results in an exothermic reaction. When these two reactions are combined (SMR and POX), the process is termed as ATR [3,55,67]. In addition to H_2, CO_2 and CO are emitted by reforming processes [46,65]. Other hydrogen reforming technologies can also be found in literature, such as hydrocarbon pyrolysis, plasma reforming, ammonia reforming and aqueous phase reforming [67,69]; however, they are not as common as SMR and coal gasification. In the majority of the processes listed above, CO_2 and/or CO is produced. One of the promising technologies that receive significant attention is the utilization of CO_2 by reacting it with H_2 to produce valuable chemicals, such as methane and methanol (e.g., through the Sabatier process) [70–73]. Similarly, CO can be converted through the water–gas shift (WGS) reaction [3,55,60,67,74].

Table 1. Summary of hydrogen production processes, their advantages and disadvantages, and technological status.

Method	Advantages	Disadvantages	TML *	PE ** (%)	Cleanness ***	Impurities	References
Reforming:							
SMR [a]	Most developed industrial process, lowest cost, existing infrastructure, high efficiency, best H_2/CO ratio	Highest air emissions, system is complex, system is sensitive to natural gas quantities. Capital, operation, and maintenance cost. Fossil fuel feedstock.	10	65–75	NC/CCS	CO_2, CO, CH_4, N_2	[2,3,46,55,59–61,65,68,75–79]
POX [c]	Well-established. Variety of fuels, reduced desulphurization requirement, no catalyst required	Complex handling process, high operating temperature, low H_2/CO ratio. Fossil fuel feedstock	7–9	50	NC	CO, CO_2, H_2O, CH_4, H_2S, COS and sometimes CH_4	[46,60,61,65,66,78,80]
ATR [b]	Lower temperatures than POX [c], Requires less oxygen than POX [c]	Limited commercial application, required air or oxygen. Fossil fuel feedstock.	6–8	60–75	NC	CO, CO_2, N_2, CH_4 and sometimes Ar	[60,81]
Gasification:							
Coal	Abundant and affordable, Low-cost synthetic fuel in addition to H_2	Reactor costs, system efficiency, feedstock impurities, significant carbon footprint unless CCS is used. Separation and purification of gas products are difficult [82]. Fossil fuel feedstock (coal gasification). Season limitations and heterogeneity (biomass)	10	74–85	NC/CCS	N_2, CO_2, CO, CH_4, H_2S	[79,83–85]
Biomass			3 (R&D)	35–50	NC/CCS	CO_x, SO_x and CH_4	[2,78,84,86,87]
Electrolysis:							
Water electrolysis	Simplicity of process design, compactness, renewable feedstock, cost effective way to produce hydrogen locally. Does not involve moving parts. Silent operation.	Energy input is required and it is more costly than fossil-fuel alternatives.	9–10	62–82	C	H_2O	[2,66,67]

* Technology maturity level (TML) is defined by a rating scale (1–10) used to indicate the commercial readiness of the technology. Level 1 refers to initial research stages, whilst level 10 refers to well-established mature commercial technologies [2]. ** Process efficiency (PE). *** C = clean without emissions, NC = not clean with emissions, CCS = quasi-clean using carbon capture and storage (CCS). Abbreviations: [a] Steam methane reforming (SMR). [b] Auto-thermal reforming (ATR). [c] Partial oxidation (POX).

Although fossil fuels are currently the main feedstock used to produce hydrogen, renewable integrated technologies are unavoidable for the global energy future [67]. Several processes have been proposed for hydrogen production from renewables [87]. Though not widely implemented, hydrogen can be produced from biomass using processes such as pyrolysis/gasification, but this is commonly accompanied by large amounts of impurities [31,61,65,66,88]. Several methods of hydrogen production from water are also available, including electrolysis, thermochemical processes, photolysis, and direct thermal decomposition or thermolysis [61]. Water electrolysis is a common method used to produce hydrogen; furthermore, it is the only method, at present, that can be used for large-scale hydrogen production without fossil fuel utilization [31]. One major advantage of water electrolysis is that no-carbon containing compounds are present in the exhaust, only water [2,66,67]. Hydrogen is obtained by splitting water into oxygen and hydrogen, achieved by an electrical current [89]. The electricity required for electrolysis can be generated from renewable sources (e.g., solar, wind and hydropower) or non-renewable sources (fossil fuel or nuclear-based) [90].

The benefits of hydrogen as a fuel, which is clean and efficient, can only be fully recognized when hydrogen is produced from renewable energy sources [6]. Most of the current hydrogen production methods yield hydrogen-rich streams, but are commonly accompanied by contaminant gases including CO_2, CO, sulphur-containing components, CH_4, and N_2 (see Table 1). Shalygin et al. [91] gives a detailed composition of all small to large-scale hydrogen production process streams. High-purity hydrogen (>99.97%) is required for fuel cells (according to SAE J2719—see Table 2), the chemical industry and stationary power production. The hydrogen produced from commercial processes should, therefore be purified after production, based on end-use application, e.g., fuel cells.

Table 2. Hydrogen fuel quality specifications.

Constituent	Limits ($\mu mol \cdot mol^{-1}$) Unless Stated Otherwise)	Minimum Analytical Detection Limit
Hydrogen fuel index	>99.97%	
Water [a]	5	0.12
Total hydrocarbons [b] (C_1 basis)	2	0.1
Oxygen	5	1
Helium	300	100
Nitrogen, Argon	100	5
Carbon dioxide	2	0.1
Carbon monoxide	0.2	0.01
Total sulphur [c]	0.004	0.00002
Formaldehyde	0.01	0.01
Formic acid	0.2	0.02
Ammonia	0.1	0.02
Total halogenates [d]	0.05	0.01
Particulate concentration	1 $mg \cdot kg^{-1}$	0.005 $mg \cdot kg^{-1}$

[a] Due to the water threshold level, the following should be tested for, if there are concerns regarding the water content: (1) Sodium (Na^+) < 0.05 $\mu mol \cdot mol^{-1}$ H_2 or < 0.05 $\mu g \cdot L^{-1}$; (2) Potassium (K^+) < 0.05 $\mu mol \cdot mol^{-1}$ H_2 or < 0.08 $\mu g \cdot L^{-1}$ (3) Pottasium hydroxide (KOH) < 0.05 $\mu mol \cdot mol^{-1}$ H_2 or < 0.12 $\mu g \cdot L^{-1}$; [b] e.g., ethylene, propylene, acetylene, benzene, phenol (paraffins, olefins, aromatic compounds, alcohols, aldehydes). The summation of methane, nitrogen and argon is not to exceed 100 ppm. [c] e.g., hydrogen sulphide (H_2S), carbonyl sulphide (COS), carbon disulphide (CS_2) and mercaptans. [d] e.g., hydrogen bromide (HBr), hydrogen chloride (HCl), Chlorine (Cl_2) and organic halide (R-X).

The purpose of this review is twofold: (i) To provide a general overview and comparison of the commercially available hydrogen separation technologies, and (ii) to survey open literature, on the status, and advances made, in the field of electrochemical hydrogen separation (EHS). With the former, special emphasis is given to membrane technologies, especially the membrane materials/types and their performance properties. With the latter, all available literature on EHS are summarized and classified based on the type of article (experimental, modelling, case study or review) and the year it has been published, the operating temperature range, the membrane materials, the type of electrocatalyst and the type of impurities contained in the feed. In the conclusion, literature gaps in the field of EHS is identified for further (future) research.

2. Hydrogen Separation/Purification Technologies

For hydrogen to be realized as a widespread energy carrier (especially as a carrier of RE), its purification and compression are unavoidable industrial processes. Several technological approaches are used to extract hydrogen from gas mixtures, utilizing various characteristics of hydrogen, under different industrial conditions. Common approaches for hydrogen recovery include the following: Adsorbing the impurities (pressure swing adsorption, PSA), condensing the impurities (cryogenic distillation) or by using permselective membranes [49,92,93]. Although PSA and cryogenic distillation processes are both commercial processes used in hydrogen separation, pressure-driven membranes are considered a better candidate for hydrogen production because they are not as energy intensive and they yield high-purity hydrogen [92]. Moreover, PSA and cryogenic distillation technologies all require multiple units and, in some instances, may involve supplementary wash columns to remove CO and CO_2 [49]. Additional advantages offered by membrane technologies include the ease of operation, low energy consumption, possibility of continuous operation, cost effectiveness, low maintenance and compactness [86,92,94–96]. Nonetheless, despite their numerous advantages, membrane systems commonly depend on high-pressure feed streams and hydrogen embrittlement is often experienced [49]. As a rule, hydrogen appears in the permeate (low pressure stream after membrane), and additional compression is required after hydrogen purification for transport and storage purposes [97]—including expenses of energy, additional equipment, etc. Properties of the various hydrogen purification technologies are summarized in Table 3. A brief discussion of each technology then follows.

Table 3. Properties of different hydrogen purification processes (adapted from Refs [2,95,98,99]).

Properties	PSA	Membranes	Cryogenic
Min. feed purity (vol.%)	>40	>25	15–80
Product purity (vol.%)	98–99.999	>98	95–99.8
Hydrogen recovery (%)	Up to 90	Up to 99	Up to 98

2.1. Pressure Swing Adsorption

PSA is the most extensively used state-of-the-art industrial process for hydrogen separation [100]; it is capable of yielding hydrogen with a purity ranging from 98–99.999% [2,81,97]. It is most frequently used in the chemical/petrochemical industry, as well as to recover hydrogen from industrial-rich exhaust gases, including reforming off-gases, coke oven gases, and pyrolysis effluent gases [92,94,101]. Currently, about 85% of the produced hydrogen is purified by PSA [102]. Although the system is mainly classified as a batch system, continuous operation can be achieved by implementing multiple adsorbers [92–94], creating a cyclic process [103,104]. The system can be divided into five primary steps: (i) Adsorption, (ii) concurrent depressurization, (iii) counter-current depressurization, (iv) purge and (v) counter-current pressurization [94,100].

In PSA, a hydrogen-rich gas mixture is passed through a high-surface-area adsorber, capable of adsorbing the impurities (e.g., CO, CO_2, CH_4, H_2O and N_2 [105]), whilst allowing hydrogen to permeate through the material [103]. The impurities are removed by swinging the system pressure from the feed to the exhaust pressure, coupled with a high-purity hydrogen purge. The driving force of PSA is the difference in the impurities' gases' partial pressure of the feed gas and the exhaust. Generally, hydrogen separation requires a pressure ratio of 4:1 between the feed and the exhaust [103]. In the initial layer H_2O, CO_2 and CH_4 is removed, whilst a second layer removes other components until the levels of CO is <10 ppm [105]. The reaction takes place at room temperature and at pressures of 20–25 bar [104]. Zeolites are commonly used as adsorbent materials [103].

To increase the hydrogen recovery, a complex arrangement of the columns is required; generally, more than eight columns are required [105]. The quantity of recovered hydrogen is dependent on the feed and purge gas pressures and hydrogen-to-impurity ratio [92]. The hydrogen recovery is typically in the range between 60% and 90% [97].

2.2. Cryogenic Distillation

Cryogenic distillation is a widely used separation process at low temperature (LT). It is used to separate gas components based on differences in their boiling temperatures [92–94]. Hydrogen's low boiling point of $-252.9\ °C$ (below that of almost all other substances) is used as a measure to separate it from other components, where the collected hydrogen can be stored as a liquid [106]. The gas needs to be cooled down, to condense, resulting in large energy consumption [92,94].

If significant amounts of CO, CO_2 and N_2 are found in the feed stream, a methane wash column is required to reduce the concentrations of these gases [92,93]. The feed gas also requires pretreatment to remove the components that might freeze; therefore, water should be reduced to <1 ppm and CO_2 to <100 ppm [94]. It is not practical to use this method to obtain high-purity hydrogen, however, higher hydrogen recovery can be achieved at moderate hydrogen purity yields ($\leq$95%) [92,94]. Similar to PSA, cryogenic distillation is perfect for large industrial scales, but unsuitable for small portable applications [103].

2.3. Membrane Technologies

Besides PSA and cryogenic distillation, membrane separation has attracted the widest interest. A membrane is a selective barrier between two phases [107] that allows mass transfer under the action of a driving force [108] (e.g., gradients in pressure, temperature, concentration or electrical potential [107]). This allows for preferential permeation of some components of the feed stream, with retention of the other components [109]. See Figure 1.

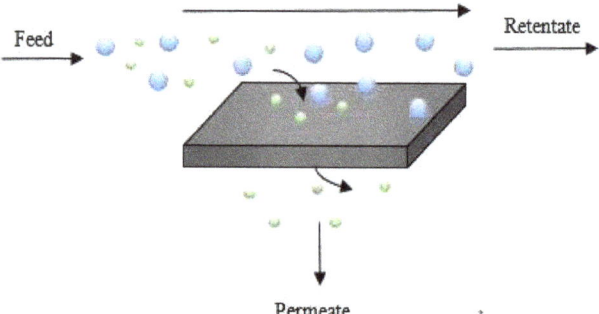

Figure 1. Simplified schematic of membrane separation.

Membranes for hydrogen separation can be divided into organic (polymeric), inorganic and mixed-matrix (hybrid). See membrane classification scheme in Figure 2, which is based on the nature of the material of the membranes. Currently, industrial processes mainly use polymer membranes (glassy or rubbery [110]) due to their capability to cope with high pressure drops, their low cost and good scalability [108]. Though organic/polymeric membranes are temperature limited (363–373 K) [108,111], recent progress have been made with thermally rearranged polymers—which show good separation at high temperatures [112,113]. Inorganic membranes provide several advantages, including mechanical, thermal and chemical stability [114], it's typically not subject to dimensional changes, such as plasticization, or swelling of the membrane upon adsorption of the components of the feed gas and controllable pore size distribution allowing for better control over selectivity and permeability [108]. To take advantage of the capabilities of inorganic membranes combined with the low manufacturing costs of the organic membranes, mixed-matrix membranes have been developed (composite materials that consist of continuous polymeric matrix and imbedded, mainly inorganic, particles) [115]. According to their characteristics, membranes can be classified as either dense (non-porous) or porous [92,96]. Porous membranes can be divided into glasses, organic polymer, ceramic- and carbon-based membranes (carbon molecular sieves) [116]. A particular important type of membranes is metallic or metal-alloys membranes (mainly Pd and/or its alloys [110]). Ceramic proton-conducting membranes are also known, but are still at the earlier development stage [108]. According to the IUPAC classification of pores sizes, porous membranes can be either microporous ($d_p < 2$ nm), mesoporous ($2 < d_p < 50$ nm), or macroporous ($d_p > 50$ nm), where d_p is the average pore diameter [108].

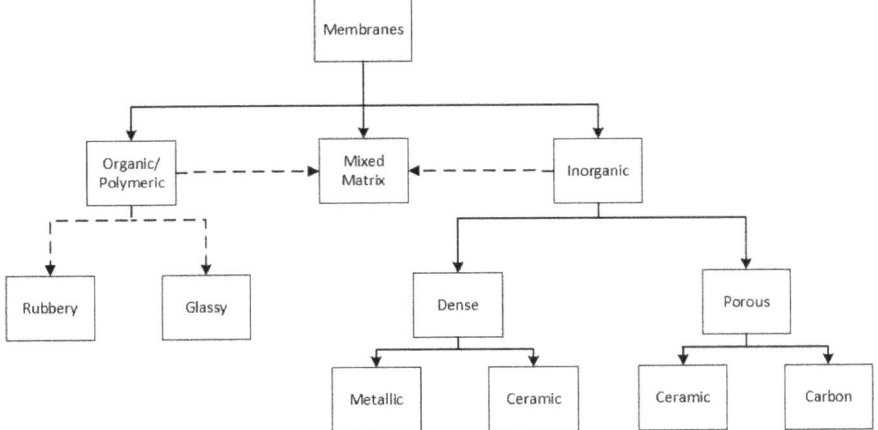

Figure 2. Schematic of membrane classification.

Gas transport through membranes can occur through a number of mechanisms [109]. A comprehensive review of the permeation mechanisms of inorganic membranes can be found in Oyama et al. [117]. Porous membranes achieve fractionation based on differences in size, shape and/or affinity between the permeating molecules and the membrane [108]. Depending on the pore size, different mechanisms will dominate molecular transport [118]. Mechanisms associated with porous membranes include, broadly: (i) Knudsen diffusion, (ii) surface diffusion, (iii) capillary condensation and (iv) molecular sieving (MS) [79,92,103,109,119]. See Figure 3. MS is an activated process that will be dominant for the smallest pores (pore diameter and diameter of diffusing molecules are approximately the same). Whereas surface diffusion will dominate the some somewhat larger pores, and Knudsen diffusion will dominate the even larger pores [118]. Separation based on Knudsen diffusion is mainly determined by the pore size and it occurs when the pore diameter of the membrane is smaller than the mean free path of the gas being separated [107]. Separation based on MS operate on size-exclusion principal [92]. In other words, only molecules with small enough kinetic diameters can permeate through the membrane [103]. Separation based on capillary condensation takes place when a partially condensed gas phase occupies a pore. Only gases that are soluble in this condensed phase can permeate through the pores when the pores are completely filled [103]. Surface diffusion can occur alongside Knudson diffusion. It involves the adsorption of gas molecules onto the pore walls of the membrane and spread along the surface [92,103]. Permeability will be high for the molecules that are readily adsorbed onto the pore walls. However, this form of diffusion is limited to certain temperatures and pore diameters [103]. In the case of dense membranes, molecular transport occurs through a solution–diffusion (Sol–D) mechanism [120]. Thus, for the separation of hydrogen from other components in a gas mixture, both size (diffusivity) and the condensability (solubility) of the gases to be separated play important roles [92,108,111,120,121].

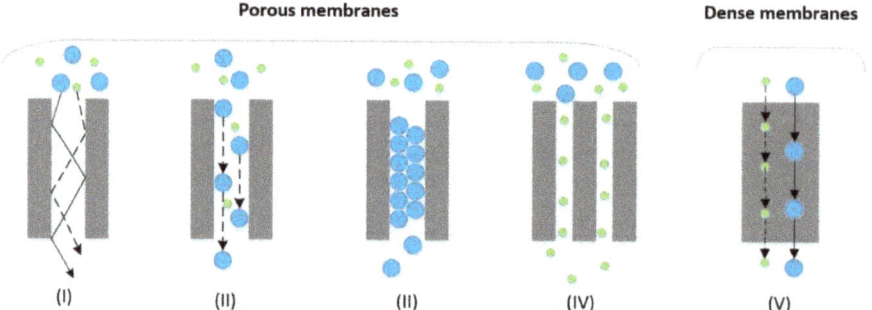

Figure 3. Transport mechanism for porous membranes ((**I**) Knudson diffusion, (**II**) surface diffusion, (**III**) capillary condensation, and (**IV**) molecular sieving) and dense membranes ((**V**) solution–diffusion).

Hydrogen purification typically involves the separation of H_2 from light gas molecules, such as CO_2, CO, CH_4, H_2O and impurities, e.g., H_2S. Physical adsorption becomes negligible at temperatures >400 °C, therefore hydrogen separation is mainly based on MS (hydrogen kinetic diameter = 2.89 Å) and differences in molecular diffusivity [114,122]. Gas transport through non-porous metals (Pd) involves dissociation of H_2 molecules into atomic hydrogen and its transport through the films [123]. Knudsen diffusion or a combination of Knudsen diffusion and surface diffusion are characteristic for porous metal membranes, e.g., porous stainless steel [124,125]. Likewise, hydrogen transport through ceramic membranes is based on Sol–D (dense ceramic) and MS (microporous). Gas transport in polymeric membranes and carbon membranes, on the other hand, occurs based on the Sol–D mechanism and MS, respectively. The kinetic diameters of light gases commonly found in produced H_2 streams, such as He, H_2, NO, CO_2, Ar, O_2, N_2, CO and CH_4, can be found in Teplyakov and Meares [126] and Nenoff et al. [127].

Two basic properties are used to evaluate the performance of a membrane: Permeance (also commonly referred to as the flux or permeation rate [107]) and selectivity [92,108,109,111]. The higher the selectivity, the lower the driving force required to achieve a certain separation—thereby reducing the cost of operation of the system. Conversely, the higher the flux, the smaller the required membrane area—thereby reducing the capital cost of the system [111]. Permeance is defined as the net transport of constituents through the membrane and it can be expressed as either mass per unit time and unit area, or mole per unit time and unit area [79]. It is usually used to assess the permeability for composite and asymmetric membranes [109]. The permeance in $mol \cdot m^{-2} \cdot s^{-1} \cdot Pa^{-1}$ (SI) (or $cm^3(STP) \cdot cm^{-2} \cdot s^{-1} \cdot cm^{-1}$ Hg)), is the permeability coefficient P_i divided by the membrane thickness (cm, or sometimes reported in m) [92,117,128]:

$$Q_i = \frac{P_i}{\delta} \quad (1)$$

Permeability coefficient, or permeability, is typically used to define the membrane's capacity to perform gas transport. In other words, permeability is the measure of the ability of certain gas components to diffuse through a membrane [103], where high permeability indicates a high throughput [111]. It denotes the amount of hydrogen (molar/volume) diffusing through the membrane per unit area and time at a given pressure gradient [103,111]. Several units are commonly used for permeability, including Barrer (10^{-10} $cm^3(STP) \cdot cm \cdot cm^{-2} \cdot s^{-1} \cdot cm^{-1}$ Hg = 3.347×10^{-16} $mol \cdot m^{-1} \cdot s^{-1} \cdot Pa^{-1}$ (SI units)), gas permeability units (GPU = 10^{-6} $cm^3(STP) \cdot cm^{-2} \cdot s^{-1} \cdot cm^{-1}$ Hg) or molar permeability ($mol \cdot m \cdot m^{-2} \cdot s^{-1} \cdot Pa^{-1}$) for characterization of the permeance of membranes.

The selectivity of a membrane measures the membrane's ability to separate a desired component from the feed mixture [108]. Membrane selectivity towards gas mixtures and mixtures of organic liquids is usually expressed by a separation factor α. The ideal selectivity/permselectivity [111] is defined as the permeability or permeance ratio of two pure gases:

$$\alpha_{i/j} = \frac{P_i}{P_j} = \frac{Q_i}{Q_j} \qquad (2)$$

where, P_i and P_j, and Q_i and Q_j are the permeability coefficients and permeance as gases i and j, respectively.

Typically, the more permeable gas is taken as i, so that $\alpha_{i/j}$ >1 [129]. Ideal selectivity is very useful for the description of the separation of mixtures of light gases, such as H_2 and N_2, which have low solubility in membrane materials. Thus, the gases only weakly affect the property and behaviour of the polymer, they do not affect the mutual diffusion and sorption parameters in the process of simultaneous transport of gases in mixture separation [128]. Whereas, in the case of heavy vapours/gasses with great solubility, the applicability of this parameter is not as predictable [128]. The following equation is used for mixture separation [128]:

$$\alpha_{i/j}^* = \frac{y_i/y_j}{x_i/x_j} \qquad (3)$$

where, y_i and y_j depict the concentration of components i and j in the permeate. Mass and molar concentrations (kg·m^{-3}; mol·m^{-3}), as well as weight fraction, mole fraction, and volume fraction are frequently used. As for x_i and x_j, two definitions are found in literature and are considered. According to Koros et al. [130], if x_i and x_j characterize the composition of the feed stream, this ratio (Equation (3) is termed the *separation coefficient*, similarly, if x_i and x_j refer to the composition in the retentate, the ratio is termed the *separation factor* [128]. The former is also sometimes referred to as the ideal/"actual" selectivity [107]. If $\alpha_{A/B} = \alpha_{B/A} = 1$, no separation is achieved [107]. In the case where the feed pressure is significantly larger than the permeate pressure, and the permeate pressure approaches zero, the ideal selectivity and real selectivity will be in equal [109].

A fundamental expression for transport in membranes is derived from Fick's first law, which relates the flux of species i to the concentration gradient. Under steady state conditions, this equation can be integrated to give:

$$J = -D\frac{dC}{dx} \qquad (4)$$

$$J_i = \frac{D_i(C_{i,0} - C_{i,\delta})}{\delta} \qquad (5)$$

where, J_i is the flux (cm^3(STP)·cm^{-2}·s^{-1} or mol·m^{-2}·s^{-1}), D_i is the diffusion coefficient (m^2·s^{-1}), $C_{i,0}$ and $C_{i,\delta}$ depicts the inlet and outlet concentrations (usually mol·m^3), and δ is the membrane thickness (m) [98,131].

2.3.1. Porous Membranes

The transport mechanism for each mechanism (porous membranes) can be summarized as follows [118,132]. The molar flux through the pores can be defined by Fick's law Equation (4). According to Burggraaf et al. [133] and Bhandarkar et al. [134], the Fickian diffusion coefficient (D_i) is composed of a corrected diffusion coefficient together with a thermodynamic correction factor $\Gamma(T,P)$:

$$D_i = D_{i,c}\Gamma, \text{ where } \Gamma \equiv \frac{\partial \ln p}{\partial \ln C} \qquad (6)$$

The diffusion coefficients are defined as:

$$D_{K,c} = \frac{d_p}{3}\sqrt{\frac{8RT}{\pi M}} \quad \text{(Knudsen)} \qquad (7)$$

$$D_{S,c} = D_S^0 exp\left(\frac{-E_{a,S}}{RT}\right) \quad \text{(Surface diffusion)} \qquad (8)$$

$$D_{MS,c} = D_{MS}^0 exp\left(\frac{-E_{a,MS}}{RT}\right) \quad \text{(Molecular sieving)} \qquad (9)$$

where, E_a denotes the activation energy (kJ·mol^{-1}), R is the universal gas constant and T is the temperature (K). Different methods for obtaining $D_{K,c}$ have been reported [98,117].

If ideal gas behaviour is assumed, $\Gamma = 1$ and Equations (4) and (7) can be combined and integrated to give:

$$J_K = \frac{D_K \Delta p}{RT\delta} \qquad (10)$$

where, δ is the thickness of the membrane and Δp is the difference in the outlet and inlet partial pressures. Similarly, the integrated flux equations for activated processes, assuming Henry's law for adsorption, can be written as [132,133,135]:

$$J_i = \frac{\Delta p}{RT\delta} D_0(T) exp\left\{\frac{-(E_a - E_{ads})}{RT}\right\} \qquad (11)$$

where, E_{ads} denotes the adsorption energy (kJ·mol^{-1}).

Substituting these expressions into the flux equation and defining the permeability, P_i, and permeance, Q_i, due to a particular transport mechanism as $Q_i = \frac{P_i}{\delta} = \frac{J_i}{\Delta p}$ the following temperature dependencies for the permeance due to the transport mechanism are obtained:

$$R_K \sim (MRT)^{-0.5} \quad \text{(Knudsen)} \qquad (12)$$

$$R_S \sim D_0(T) exp\left(\frac{-(E_{a,S} - E_{ads})}{RT}\right) \quad \text{(Surface diffusion)} \qquad (13)$$

$$R_{MS} \sim D_0(T) exp\left(\frac{-(E_{a,MS} - E_{ads})}{RT}\right) \quad \text{(Molecular sieving)} \qquad (14)$$

In the case of Knudsen diffusion, the selectivity can be calculated using Equation (15) [111]:

$$\alpha_{i/j} = \sqrt{\frac{M_j}{M_i}} \qquad (15)$$

2.3.2. Dense (Non-Porous) Membranes-Diffusion Mechanism

Solid-state diffusion occurs with a further decrease in the pore size, where the gas molecules interact strongly with the membrane material and its solubility needs to be considered [117]. In this case, the permeability can be determined using the equation that is based on the sol-D model [103,107]:

$$P = D \times S \qquad (16)$$

where, S is the gas solubility in mol·m^{-3}·Pa^{-1} (see Equation (23)). If Equation (16) is substituted into Equation (2), it is possible to speak of selectivity of diffusion and sorption [128]:

$$\alpha_{i/j} = (D_i/D_j)(S_i/S_j) \qquad (17)$$

There are three instances where this transport mechanism applies: (i) permeation through glassy membranes, (ii) metallic membranes, and (iii) polymeric membranes [117,136]. Overall, the permeability can be calculated using the following expression [98,128]:

$$P_i = \frac{J_i \delta}{\Delta p} \tag{18}$$

Steady-state flux of gases through a dense (non-porous) metallic membranes can be expressed by Equation (1) [137], which is derived by combining Fick's first law (driving force = concentration gradient) (Equation (4) [107,138] and Henry's law (Equation (20) [139]), as follows:

$$\text{Steady state permeability:} \quad J_i = \frac{P_i(p^n_{i,feed} - p^n_{i,perm})}{\delta} \tag{19}$$

$$\text{Henry's law:} \quad S_i = \frac{C_i}{p^n_i} \tag{20}$$

where, J_i is the flow rate (mol·m^{-2}) of the diffusing species, P_i is the permeability coefficient (mol·m^{-1}·s^{-1}·Pa^{-1}) of component i, δ is the thickness of the membrane (m), and $p^n_{i,feed}$ and $p^n_{i,perm}$ are the hydrogen partial pressures at the feed and permeate side, respectively. Furthermore, S_i and C_i, refers to the solubility coefficient (mol·m^{-3}·Pa^{-1}), and the concentration gradient across the membrane, respectively. The pressure component (n) is generally in the range of 0.5–1, depending on the limiting step of the hydrogen permeation mechanism [108,117,139]. The hydrogen permeation is controlled by the rates of adsorption/desorption and the diffusion through the lattice [117,140]. Generally, if diffusion through the metal is the rate limiting step, the pressure is relatively low, and the hydrogen atoms form an ideal solution in the metal, then n = 0.5 known as Sievert's law [108,111]. Higher values of n are expected when mass transport to/from the surface or dissociative/associative adsorption steps become rate determining [108]. Moreover n > 0.5 for defective membranes (e.g., pinhole formation) or when the rate is influenced by the membrane's porous support [108]. Furthermore, the thickness of the membrane can also influence the value of n, causing it to be >0.5 [141].

The relationship between the permeability and temperature follows Arrhenius behaviour [142]:

$$P = P_0 exp\left(\frac{-E_a}{RT}\right) \tag{21}$$

where, P^0_i is the maximum permeability at infinitely high temperature, E_a is the activation energy for permeation, R is the universal gas constant, and T the absolute temperature. Similarly, the temperature dependence of the gas diffusion coefficient and solution coefficient can be expressed as follows (Equations (22) and (23)) [121]:

$$D = D_0 exp\left(\frac{-E_d}{RT}\right) \tag{22}$$

$$S = S_0 exp\left(\frac{-\Delta H_s}{RT}\right) \tag{23}$$

where, E_d is the activation energy of diffusion, ΔH_s is the partial molar sorption, D_0 and S_0 are the diffusivity and solubility, respectively, at infinite temperature [103]. Consequently, when n is 0.5, the flux can be written in terms of the so-called Richardson equation, where Equation (21) is substituted into Equation (19), as follows:

$$J_i = P_0 exp\left(\frac{-E_a}{RT}\right) \frac{(p^n_{i,feed} - p^n_{i,perm})}{\delta} \tag{24}$$

2.3.3. Ceramic Proton-Conducting Membranes

Although proton-conducting ceramics are still at the early stages of development, several research efforts have been made [143–150]. Overall, the process of hydrogen permeation through a dense proton conducting membrane involves several steps [122,151]:

1. H_2 gas diffusion to reaction sites on the surface of the feed side;
2. H_2 adsorption, dissociation, and charge transfer at the membrane surface;

$$H_2 \rightarrow H_{2,ads} \; ; \; H_{2,ads} \rightarrow 2H_{ads} \; ; \; H_{ads} \rightarrow H^+ + e^-$$

3. Proton reduction and hydrogen re-association at the membrane surface

$$(H_{ads}^+ + e^-)_{S'} \xrightarrow{incorporation} (H_{ads}^+ + e^-)_{BM} \xrightarrow{diffusion} (H_{ads}^+ + e^-)_{S''} \xrightarrow{re-association} (H_2)_{S''} \xrightarrow{diffusion} (H_2)_G$$

where, S', BM, S'' and G is the membrane surface at the inlet, the bulk membrane, the membrane surface at the outlet, and the gas, respectively.

Two transport mechanisms have been proposed in literature: The vehicle mechanism and the Grotthus/proton hopping mechanism [151]. In the vehicle mechanism, protons bind to oxygen to form a hydroxide ion which diffuses through the lattice by vacancy or interstitial diffusion. Electroneutrality is achieved through the counter-diffusion of unprotonated vehicles or oxygen vacancies [108]. The Grotthuss mechanism instead assumes that protons jump between stationary oxygen ions, and that the diffusion of protons and electrons occur in the same direction, whilst maintaining electroneutrality and a zero net electric current [151,152]. For structural reasons, proton transport in oxides is better explained by the Grotthuss mechanism [153].

The proton flux (mol·cm^{-2}·s^{-1}) for a membrane containing only protonic-electronic conductors can be described by the Wagner equation [108,154]:

$$J_{H^+} = -\frac{RT}{2F^2\delta} \int_I^{II} (\sigma_{H^+} \times t_{e^-}) d\ln p_{H_2} \qquad (25)$$

where, R is the universal gas constant (8.314 J·mol^{-1}·K^{-1}), $T(K)$ is the absolute temperature, F is the Faraday constant (96 485 C·mol^{-1}), δ is the thickness of the ceramic membrane (m), σ_{H^+} (S·cm^{-1}) is the proton conductivity within the membrane, t_{e^-} is the electronic transport number, and p_{H_2} is the hydrogen partial pressure (Pa). In this equation, the pressure gradient serves as driving force and the flux is directed from $I \rightarrow II$ [108]. Since σ_{H^+} and t_{e^-} can be dependent on the pressure, Norby and Haugsrud [152] and Marie-Laure et al. [153] integrated Equation (20) for different limiting cases. For example, when $t_{e^-} \cong 1$ and σ_{H^+} is proportional to $p_{H_2}^n$ where (i) $n = 0.5$ when protons are minority defects, (ii) $n = 0.25$ when protons are majority defects compensated by electrons, and (iii) $n = 0$ when protons are majority defects compensated by acceptor dopants. In (i) and (ii), $J_{H^+} \sim \left[(p_{H_2}^n)^I - (p_{H_2}^n)^{II}\right]$ and, therefore, the partial pressure on the feed side (I) has a strong effect on the proton flux. In case (iii), $J_{H_2} \sim (p_{H_2}^{II}/p_{H_2}^I)$, and consequently the pressure on both sides are equally important [108,152].

A comparison of membrane properties, such as the temperature range, hydrogen selectivity and flux, and stability and poisoning issues are summarized in refs [79,92,103,155]. A more detailed quantification of hydrogen flux and selectivity for polymeric membranes can also be found in the following references [79,91,92,103,128,130,155]. Dense metal membranes (mainly palladium alloys) are seen as the most appropriate construction materials due to their high hydrogen selectivity [92,94,123,155]. However, Pt- and Pd-based membranes are known to have a high sensitivity to a variety of surface contaminants, such as H_2S, CO, thiophene, chlorine and iodine [79]. These contaminants severely influence the performance of these membranes [79]. This is explained by the favourable interaction between the membrane and the contaminants. Membrane degradation/hydrogen embrittlement is also a common issue in metal membranes with high diffusivity or solubility (e.g., Pd-membranes) [140], which make them less durable. However, the problem of hydrogen embrittlement in Pd-based can be minimized by alloying the membranes [140] or by controlling the operating conditions to avoid a two-phased region [156]. Besides dense metallic membranes, dense ceramic membranes are also seen as a favourable option for hydrogen separation, for the same reason [155].

3. Electrochemical Hydrogen Separation

Electrochemical membranes are seen as a promising alternative to pressure-driven membranes [94]. Electrochemical membranes are known to generate electricity (fuel cells) or to apply it (water electrolysis), and they are also used to purify/enrich and compress hydrogen streams [94]. EHS has the following advantages over other conventional separation systems [157,158]:

- Hydrogen, in the form of protons, is selectively transferred through the proton-conducting electrolyte;
- one-step operation provides pure hydrogen;
- the hydrogen separation rate can be controlled by the current (Faraday's Law);
- a high hydrogen collection rate is achieved;
- simultaneous purification and compression of hydrogen is possible, in principle;
- high hydrogen separation is achieved at low cell voltages, with a high separation efficiency [156] and
- high selectivity and low permeability results in pure hydrogen (up to 99.99 vol.%) [159].

An additional benefit of this method is that CO_2 is concentrated, and it can be captured and stored, without any further treatment—hence reducing greenhouse gas emissions [156,160]. In terms of application, electrochemical hydrogen separation could be beneficial in numerous industrial fields, including: (i) Hydrogen purification, for the use in fuel cell technologies, (ii) cooling agent in turbines, (iii) for the application in nuclear reactors (separation and concentration of hydrogen isotopes) and (iv) separation of hydrogen from natural gas mixtures (e.g., transportation of hythane through gas pipelines) [161]. Furthermore, compressed high-purity hydrogen is required for fuel cells (>99.97%) (SAE International, 2015) and hydrogen mobility (e.g., Hydrogen Mobility Europe (H2ME) [162,163]).

3.1. Working Principle

In principal, EHS is based on the following process (Figure 4). Molecular gaseous hydrogen (or a gas hydrogen-containing gas mixture) is injected at low pressure at the anode side of the proton exchange membrane (PEM) electrochemical cell. When the hydrogen makes contact with the anode electrode, hydrogen is oxidized into its constituents (protons and electrons), achieved with a Pt-based catalyst (Equation (26)). The electrons then travel through an electrical circuit, whilst the protons move through the PEM to the cathode compartment. Finally, the protons and electron reduce to H_2, at the cathode compartment (Equation (27)). This can be carried out at high pressure to improve the specific volumetric energy density, for storage.

$$\text{Anode reaction:} \quad H_2 \rightarrow 2H^+ + 2e^- \tag{26}$$

$$\text{Cathode reaction:} \quad 2H^+ + 2e^- \rightarrow H_2 \tag{27}$$

$$\text{Overall reaction:} \quad H_2(Anode) \rightarrow H_2(Cathode) \tag{28}$$

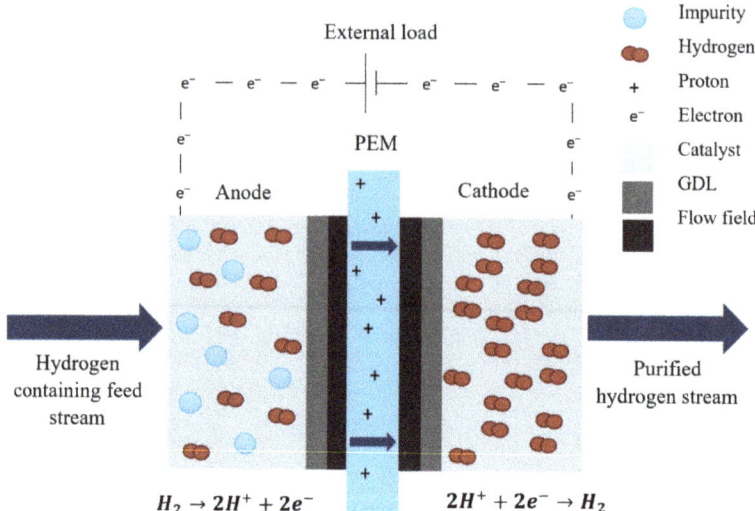

Figure 4. Schematic of the working principle of an ideal electrochemical hydrogen separator.

In the case of hydrogen purification and separation, a hydrogen-rich gas enters the anode compartment and hydrogen is selectively transferred through the PEM. This reaction does not occur spontaneously. An external voltage (DC power source) is required to drive the chemical reaction (electrolytic mode) [164]. According to literature, minimal power is required to operate the cell [49]. The oxidation and reduction reactions of hydrogen are facile and corresponds to Nernstian/Faradic theoretical values in their electrochemical behaviour [49].

Unlike conventional membrane systems that rely on pressure or concentration differentials, the hydrogen permeation of an electrochemical hydrogen separator is only dependent on the applied current as driving force, which is quantified by the law of Faraday:

$$I = nF\dot{n} \qquad (29)$$

where, $\dot{n}$ depicts the flow of hydrogen (mol·s^{-1}), I is the current (A), n depicts the number of electrons, and F is Faraday's constant (96,485 C·mol^{-1}). Faraday's law can also be rewritten in terms of the hydrogen mass flow rate ($\dot{m}$), which is proportional to the current [165]:

$$\dot{m} = \frac{MI}{nF} \qquad (30)$$

where, M is the molecular weight in kg·mol^{-1}.

Although the cell overvoltage in this process is normally quite low, high hydrogen separation efficiencies can be achieved [156]. The purity of the hydrogen that is produced at the cathode side is high. This is highly dependent on several factors, for example: (i) The permeability of the membrane with respect to the feed gas composition, (ii) the integrity of the membrane and (iii) in the case of LT gas separation (<100 °C), when Nafion is used, the water content of the membrane [49].

3.2. Current Status of Electrochemical Hydrogen Separation Technology

A summary of previously reported literature on electrochemical hydrogen separation technologies is presented in Table 4 (in chronological order). The type of article, membrane, and electro-catalyst, as well as the gas mixtures that was used in the listed articles are included. A brief discussion on each contribution follows. The articles are characterized as high-temperature and low-temperature, with subsections "active" and "passive" gas mixtures. "Active" gas mixtures in this paper refer to gases that have affinity to either the membrane or the catalyst, e.g., CO and Pt catalyst. "Passive" gasses refer to gas mixtures that do not react with the membrane or the catalyst, typically made of permanent gases, such as N_2 and Ar.

Table 4. Summary of published articles on electrochemical hydrogen separation.

Year	Type	Membrane	Catalyst *	Impurities	Temp. (°C)	Refs.
2004	Experimental	Nafion	Pt (B)	N_2, CO_2	30–70	[166]
2005	Review	N/A	N/A	N/A	N/A	[167]
2007	Experimental	Nafion	Pt (A), Ru (C)	CO, CO_2	20	[156]
	Experimental/modelling	Nafion	Pt/Pt-Ru (B)	Ar, CH_4	20–70	[168]
	Simulation	N/S **	N/S **	N_2	25, 60	[169]
2008	Experimental	Nafion	Pt (B)	N_2	25, 60	[170]
	Experimental	PBI	Pt (B)	CO, CO_2, N_2	120–160	[49]
2009	Experimental/modelling	Nafion	Pt (B)	Ar/C_2H_4	25	[171]
	Experimental	N/S **	N/S **	N_2/CO_2	60	[172]
2010	Experimental	PBI	N/S **	N_2; CO_2, CO, CH_4; N_2,CO_2, CO	160–180	[93]
2011	Experimental	Nafion	Pt/C (B)	CO_2, H_2O	50–70	[173]
	Experimental	Nafion	Pt (B)	N_2	35, 55, 75	[161]
	Experimental	Nafion	Pt (B)	Ar	20–70	[159]
2012	Experimental	Nafion	Pt/C, Pd/C (B)	CO_2 reformate	30–50	[160]
2013	Experimental	PBI	Pt(B)	CO_2	80, 160	[174]
2014	Experimental	PBI	Pt (B)	Simulated reformate: N_2, CO	140–160	[175]
	Experimental	Nafion	Ir/C (B)	Ar, CO_2	25, 70	[176]
2016	Experimental	Nafion	N/S **	CO_2, CO, CH_4	25–75	[177]
	Experimental	SPPESK, Nafion	Pt (B)	CO_2	20–60	[178]
2018	Experimental	Nafion	Pt-Ru (A), Pt (C)	CO_2	25, 50	[179]
2019	Experimental/modelling	Nafion	N/S **	N_2, CH_4, He, CO_2	≤28	[180]
	Experimental	Nafion	N/S **	N_2, CO_2 and air	15–22.5	[181]
	Experimental case study: MEMPHYS (Membrane based purification of hydrogen system) system	N/S **	N/S **	N_2	35	[97]
2020	Experimental	PBI	Pt (B)	N_2, CO	160–200	[182]
	Experimental/modelling	Nafion	Pt/C (B), Pt-Ru/C (A) and Pt/C (C), Pt-Ni/C (A) and Pt/C (C)	CO/Ar; N_2	35	[183]
	Experimental/modelling	Nafion	Pt/C (A), Pt-Ru/C (A) and Pt-Ru(C)	CO; CO_2, CH_4, CO, H_2S	35	[184]
	Case study/modelling	N/A	N/A	H_2,N_2 and CO_2	N/A	[185]
	Review	N/A	N/A	N/A	N/A	[186]
	Review	N/A	N/A	N/A	N/A	[187]
2021	Experimental	TPS	Pt-Co/C (A) and Pt/C (C)	CH_4, CO_2, and NH_3	120–160	[188]

* A = anode; C = cathode, B = both electrodes; ** N/S = Not specified.

3.2.1. Low-Temperature EHS
"Passive" Gas Mixtures

- **H_2/CH_4 Mixtures**

Ibeh et al. [168] investigated hydrogen separation from H_2/CH_4 mixtures with both Pt and Pt-Ru catalysts. Results indicated that little energy was consumed and CH_4 seemed to be inert at LTs, hence resulting in negligible fouling of the anode electrocatalyst. Moreover, it was found that if the hydrogen recovery is <80%, a small portion of the hydrogen energy is required for separation. A mathematical model was developed to simulate the electrochemical separation process. The model was adapted from the model used by Zhang [189]. This model takes into account the hydrogen adsorption and oxidation reaction, together with their reaction rates. Also, differential equations were used and solved using the ODE solver in Scilab. The differential equation included: The time dependence of the hydrogen surface coverage, the material balance for species i, the variation of the hydrogen pressure in the anode chamber, and the time dependence of the anode potential. There was a good agreement between the experimental data recorded and the simulated results.

- **H_2/Ar Mixtures**

Nguyen et al. [159] used electrochemical impedance spectroscopy (EIS) to characterise and measure the kinetics and efficiency of the PEM and electrochemical cell. The roles of the cell structure and the different operating parameters (cell temperature, relative humidity and the partial pressure of hydrogen) on the overall process efficiency were investigated. Both the cell conductivity and electrode activity were measured. The relative humidity played a key role in the performance of the cell; the best performance (lowest membrane electrode assembly (MEA) resistance) were recorded when the cell and gas humidifier temperatures were similar, and close to 70 °C. Under these conditions, catalytic layers showed the highest activity. Consequently, lower catalyst activity was recorded at lower temperatures. Furthermore, when the humidifier and cell temperature difference was >10 °C, the water content for membrane humidification decreased, resulting in high MEA resistance. Moreover, mass-transport limitations were observed when the hydrogen partial pressure of the feed decreased. They concluded that an electrochemical hydrogen pump is limited to the treatment of gases in which the hydrogen content is >50%.

- **H_2/N_2 Mixtures**

Casati et al. [170] investigated the concentration of hydrogen achieved electrochemically from a lean hydrogen-inert gas mixture under galvanostatic and potentiostatic conditions. Results of the experimental runs suggested that galvanostatic operating conditions were unstable, whereas operation at potentiostatic conditions appears to be more stable. They reported that the hydrogen recovery increased with applied potential and the coefficient of performance (defined as the ratio of the hydrogen produced over the hydrogen consumed) decreased with the applied potential difference. Furthermore, the feed flow rate had a noteworthy effect on the recovered hydrogen. They proposed an optimum energy efficiency for separation, but did not identify its dependency on the process parameters.

The current efficiencies for current densities between 0 and 2 $A \cdot cm^{-2}$ were >83% and >90% for current densities ≥ 0.4 $A \cdot cm^{-2}$. Premixed natural gas reformate (35.8% H_2, 1906 ppm CO, and 11.9% CO_2, with the balance N_2) and methanol reformate (1.03% CO and 29.8% CO_2, with the balance H_2) were used to investigate the CO tolerance and the cells' tolerance to impurities. Near Faradic flows were achieved, irrespective of the feed composition. The effect of CO on the Pt catalyst was completely reversible at 120–160 °C. Furthermore, the CO concentrations were reduced from 1906 ppm to ~12 ppm, and CO_2 concentrations were reduced from 11.9% to 0.37% at 0.4 $A \cdot cm^{-2}$ and 0.19% at 0.8 $A \cdot cm^{-2}$. Results of these experiments justified the use of polybenzimidazole (PBI) membrane-based hydrogen purification at high-temperature (HT) from feed streams containing relatively high CO and CO_2 concentrations and low concentrations of hydrogen—not observed at LT (<100 °C), due to catalyst poisoning and water management issues.

Grigoriev et al. [161] investigated the extraction and compression of hydrogen from a H_2/N_2 mixture. They found that a lower hydrogen content in mixtures resulted in high mass transport limitations and that the compression efficiency decreased as the hydrogen content decreased. The current density of the compressor must be reduced in order to match mass transport limitations that result from lower hydrogen concentrations in the feed, in order to avoid low concentration and compression efficiencies. They concluded that this system could not be used to effectively extract hydrogen from diluted hydrogen streams.

Schorer et al. [97] reports on the MEMPHYS (Membrane based purification of hydrogen system) project (including the project itself, project targets, and different work stages). Early measurements have been conducted and the results are discussed. The system was able to extract H_2 from H_2/N_2 gas mixtures (1:1 and 4:1). However, the project targets were not fully reached using this system, specifically with the 1:1 H_2/N_2 feed stream. If the MEMPHYS system is to be used on a feed stream with <50% hydrogen content, the electrical power demand or CAPEX will increase significantly. In their conclusion, the authors state some possible adjustments that could be made to the system (e.g., addition of an ozone generator) and their (future) intention to carry out further testing with other substances in the anode gas mixture.

- $H_2/CH_4/Ar$

Onda et al. [169] carried out preliminary investigations into the separation and compression efficiencies, and performance of a hydrogen pump and found that >98% of hydrogen in the feed could be separated, at cell voltage 0.06–0.15 V and current efficiencies ~100%. They developed a simulation code, based on a pseudo two-dimensional code for a proton exchange membrane fuel cell (PEM-FC). The current distribution along the flow direction calculated by their simulation agreed well with the measured distribution, except when the H_2 concentration is low and the H_2 transport rate is high.

- $H_2/CH_4/Ethylene$

Doucet et al. [171] investigates the feasibility of separating hydrogen from H_2/ethylene gas mixtures. Experimental results showed that a large amount of the ethylene reacted with the H_2 to form ethane. In spite of this reaction taking place, results still indicated that it is possible to obtain reasonable separation. Results indicated that if Pt is used as the electrocatalyst to separate alkenes and hydrogen, the consumption of hydrogen is somewhat less than the amount of alkenes in the inlet. The suggestion was made that another catalyst (which has a lower hydrogenation rate and different selectivities for H_2 and ethylene adsorption) should be used when alkenes are present, as it might improve the general efficiency of the process. The authors concluded that EHS by means of PEM-FC technology is not suitable for large quantities of ethylene.

"Active" Gas Mixtures

- CO and CO_2-Containing Mixtures

Lee et al. [166] investigated the electrochemical separation of hydrogen from a $H_2/N_2/CO_2$ mixture. They reported that an increase in cell temperature resulted in enhanced hydrogen purity and energy efficiency. Application of pressure on the feed gas side increased the performance and the amount of hydrogen produced but decreased the hydrogen purity, as the permeation flux of impurities increased. The performance decreased, with a higher concentration of impurities in the feed gas stream. The permeability of CO_2 is greater than that of N_2, therefore, the hydrogen purity is more dependent on CO_2. At a current density of 0.7 $A \cdot cm^{-2}$, a hydrogen purity of 98.6% and 99.73% can be achieved from a low-purity feed (30% H_2) using a one-stage and two-stage process, respectively.

Gardner and Turner [156] investigated hydrogen separation from hydrogen-rich streams obtained from steam reforming of a hydrocarbon or alcohol source (H_2/CO_2 and $H_2/CO_2/CO$ gas mixtures). The cell resistance was ~17 mΩ (using pure hydrogen to plot a polarization curve) and pure hydrogen yielded efficiencies of >80%. When extracting hydrogen from a H_2/CO_2 mixture, the extraction efficiency was ~80%. However, when the hydrogen feed stream contained CO (1000 ppm), the efficiency was relatively poor. CO severely contaminates the anode, raising the anode potential by up to 300 mV. In attempt to improve the efficiency in the presence of CO, periodic pulsing was examined. It was determined that pulsing can substantially mitigate the problem of CO contamination and reduce the anode potential, although the anode potential is still very high compared with that of an unpoisoned cell. The maximum current densities observed were ~ 0.2 $A \cdot cm^{-2}$.

Onda et al. [172] measured the voltage–current characteristics of an electrochemical hydrogen pump (EHP) and PEM-FCs; they changed the hydrogen concentration from 99.99% to 1%. Nearly all of the hydrogen could be separated and recovered, even when the hydrogen feed mixture flow rate or feed gas concentration was changed (impurity concentration in the treated gas stream: Max. 1000 ppm). The concentration of hydrogen released at the anode side was ~50 ppm, which is well below the permissible limit (1%). The cell voltage became unstable for CO_2 gas mixtures

Abdulla et al. [173] investigated the recovery and energy efficiency of hydrogen separation from mixtures of CO_2, H_2O and H_2 using an EHP. Measurements were recorded as functions of the operating parameters: Gas flow rate, gas composition, applied potential difference, and temperature. High-purity hydrogen (>99.99%) was recovered in single-stage EHP experiments, at energy efficiencies of 45%. Experimental analysis revealed that energy efficiencies of >90% with >98% hydrogen recovery is possible with a programmed voltage profile multistage EHP, and 90% hydrogen recovery is achievable with 75% energy efficiency if a fixed applied potential difference multistage EHP is used. The experimental results were used as a basis for predictive models in single-stage and multistage EHPs.

Wu et al. [160] compared carbon-supported Pt/C and Pd/C catalysts in the electrochemical recovery of hydrogen from H_2/CO_2 reformate mixtures. Electrochemical activity and separation efficiency were evaluated using cyclic voltammetry, polarization, and potentiostatic hydrogen pumping. Results revealed that the MEA resistance increased as the dilution of hydrogen in CO_2 increased. Furthermore, the effective resistance of the Pd/C catalyst was greater than that of the Pt/C catalyst. The CO_2 adsorbed more strongly to the surface of the Pd/C catalyst, thus impairing the electrochemical active surface area available for hydrogen oxidation/reduction—hence resulting in lower separation efficiencies. Furthermore, the Pd/C catalyst showed lower resistance to CO_2 poisoning, higher effective ohmic resistance, and a lower mass transport coefficient of hydrogen. The authors concluded that it would be possible to replace Pt with Pd as catalyst to reduce costs; however, this will be accompanied by reduced energy efficiencies.

Kim et al. [176] investigated the replacement of Pt electrocatalysts with Ir-based electrocatalysts, and carried out the EHS from H_2/CO_2 mixtures. The Ir-based catalysts were characterised using X-ray diffraction, X-ray photoelectron spectroscopy, transmission electron microscopy and thermogravimetric analysis. CO_2 stripping indicated that the Ir catalysts were unaffected by CO_2, unlike the Pt catalysts. Furthermore, the performance of the Ir catalysts were better than that of the Pt catalysts in terms of H_2/CO_2 separation. The operating voltage to separate hydrogen from the CO_2 mixture was less for Ir300 (0.18 V at a current density of 0.8 $A \cdot cm^{-2}$) than for Pt (0.20 V at a current density of 0.8 $A \cdot cm^{-2}$).

Bouwman [177] reported on electrochemical hydrogen compression and EHS. They used a single cell stack with a pumping capability of 2 kg H_2/day. Hydrogen was separated from a premixed gas mixture containing 70.05% H_2, 19.97% CO_2, 7.477% CO and 2.507% CH_4. Results revealed that successful hydrogen purification could be achieved with 188 ppm CO_2, 14 ppm CO, and no detectable CH_4 in the permeate. The estimated concentration reduction was 1000× for CO_2, 5000× for CO, and "infinite" for CH_4.

Chen et al. [178] investigated the possibility of coupling H_2/CO_2 separation with hydrogenation of biomass–derived butanone in an EHP, and replacing Nafion membranes with non-fluorinated SPPESK (sulphonated poly (phthalazinone ether sulphone ketone)) membranes. Due to higher resistance, caused by swelling, the SPPESK-based EHP reactor had exceptional reaction rates at elevated temperatures (60 °C). It also had a higher butanone flux of 270 $nmol \cdot cm^{-2} \cdot s^{-1}$, compared to that of a Nafion-based EHP reactor, which had a butanone flux of 240 nmol·cm^{-2}·s^{-1}. Furthermore, the SPPESK-based EHP exhibited better hydrogenation than an EHP with Nafion membranes due to lower CO_2 permeation; however, efficiencies of the former approximately 20% lower than those of the latter. With H_2/CO_2 as feed stream, an efficiency of 40% was reached with the EHP, with a power consumption of 0.3 kWh per Nm^3 hydrogen. This is superior to what is reached with alternative processes such as PSA and water electrolysis. The separation efficiencies of the membranes used in this study by Chen et al. [178] are tabulated in Table 5.

Table 5. Hydrogen purity of the product gas (inlet stream properties: 56.7 mA·cm^{-2}, 40 °C, 1 atm, 75 vol.% CO_2) [178].

Gas	Composition [%]			
	SPPESK-0.71	Nafion 115	Nafion 212	Nafion/PTFE
H_2	>99.99	>99.99	99.79	99.25
CO_2	<0.01	<0.01	0.21	0.75

Ru et al. [179] conducted a series of experiments at different currents (0–2.5 A), feed flow rates (90 mL·min^{-1}–300 mL·min^{-1}) and temperatures (25 °C, 50 °C) to determine the optimal operating conditions. A gas mixture of 50:50 H_2/CO_2 were fed to the cell at 1 atm and purified. Results showed that the optimal operating conditions were a feed flow rate of 90 mL·min^{-1} at 2.5 A, with 93.62% hydrogen purity and 60.45% hydrogen recovery. The hydrogen purity was enhanced by increasing the temperature from 25 to 50 °C. However, the hydrogen permeation was much lower at 50 °C compared to 25 °C.

Nordio et al. [180] investigated several parameters of an electrochemical hydrogen compressor/separator, including the hydrogen recovery factor (HRF), hydrogen purity and concentration, the mixture type, the total flow rate, and the temperature. An experimental and modelling (Matlab) approach was applied to study these parameters. The results were promising with regards to HRF and purity. The hydrogen purity, in the case of N_2 and CH_4, was ~100%, and more or less >98% in the case of He. Elevating the temperature had a positive effect on hydrogen separation due to increasing membrane conductivity and decreasing ohmic resistance. The model that was developed by these authors was able to effectively predict both the polarization curves and the product hydrogen purity. Finally, a case study in which 75% H_2:25% CH_4 and 30% H_2:70% CH_4 was used revealed that the EHS is more flexible regarding the hydrogen feed content (lower H_2 concentration) compared to PSA.

Nordio et al. [181] investigated the performance of an EHS/compressor for binary H_2/CO_2 mixtures. More specifically, they investigated whether the reverse water–gas shift (RWGS) or electrochemical CO_2 reduction is responsible for CO formation and subsequent catalyst poisoning. They demonstrated that the RWGS is largely responsible for the decreased performance. The lower hydrogen product purity, in comparison with other gas mixtures, was attributed to the extremely high CO_2 water solubility (in comparison with the other gases). Higher voltages, CO_2 concentrations, and temperatures have a more pronounced adverse influence on the cells' performance, together with increased catalyst inhibition. The hydrogen product purity increased with lower CO_2 concentration in the feed and higher applied voltage. Increasing the temperature resulted in in a promotion of the RWGS reaction, and subsequently led to a more rapid hindrance of the catalyst. Finally, the authors experimentally verified the fast performance of air bleeding, compared to temperature swing desorption—to generate the catalyst.

Jackson et al. [183] investigated the low-loading, high mass transport "floating electrode" technique for EHP catalyst characterisation. The feed streams were pure H_2 and 20 ppm CO in H_2. The cyclic voltammograms for Pt/C, Pt-Ru/C and Pt-Ni/C were compared, and the poisoning effects in both in situ and ex situ systems were discussed. Furthermore, the kinetic modelling of CO poisoning on the floating electrode was addressed. The model is based on the Langmuir–Hill isotherm, and takes into account the complex nature of CO_{ads} over time. Result showed that mass activities of 68–93 $A \cdot mg_{metal}^{-1}$ were achieved. This is significantly higher than that achieved with the EHP: 6–12 $A \cdot mg_{metal}^{-1}$. The authors attributed this to mass transport limitations (water electroosmotic drag). CO poisoning of the EHP and floating electrode systems showed the same tendency in terms of the CO tolerance of the catalyst—Pt-Ru/C>Pt/C>Pt-Ni/C. Finally, the kinetic model was well fitted to the experimental floating electrode data.

Jackson et al. [184] investigated the operation of an EHP under different EHP poisons. First, the effect of 20 ppm CO was investigated using Pt/C and Pt-Ru/C as anode catalysts. Secondly, a gas mixture, comparable to a product stream from steam methane reforming water–gas shift (SMR-WGS) reactor (78.6% H_2, 18.5% CO_2, 2.9% CH_4 and 20 ppm CO) was tested. Then, lastly, the effects of H_2S (100 ppb, 1 ppm and 5 ppm in H_2 and 100 ppb in the SMR-WGS H_2 feed) was investigated. Two poison mitigation strategies were introduced and investigated: "Online" and "offline" regeneration. This approach included O_3 in the O_2 bleed (online cleaning). The ozone was also used to recover the system after exposure to poisons (offline cleaning). It was determined that the gas compositions that contained O_3 were more effective in cleaning poisons (e.g., CO_{ads} and S_{ads}) than the streams containing only O_2. In the case of severely poisoned streams, the inclusion of O_3 doubled the achievable current density.

3.2.2. High-Temperature EHS

"Passive" Gas Mixtures

- **H_2/CH_4 Mixtures**

Vermaak et al. [188] investigated the EHS from H_2/CH_4 mixtures. The article reported on performance parameters such as polarization curves, limiting current density, open-circuit voltage, hydrogen permeability, hydrogen selectivity, hydrogen purity and cell efficiencies (current, voltage and power efficiencies). Three compositions were tested: 10%, 50% and 80% CH_4 in hydrogen. When compared to the polarization curves of pure hydrogen, under the same operating conditions, these mixtures showed a diluent effect with no affinity towards the membrane or the catalyst. High-purity grade (>99.9%) hydrogen was generated for all three feed compositions. The authors concluded that high-purity hydrogen can be generated from H_2/CH_4 mixtures in a single stage process regardless of the methane concentration of the feed. Moreover, the authors found that the hydrogen purity was slightly enhanced with an increase in temperature. High hydrogen selectivities of up to ~22,000 were reached. The hydrogen selectivities were inversely related to temperature. They also reported on hydrogen separation from H_2/CO_2 and H_2/NH_3 gas mixtures.

"Active" Gas Mixtures

- **CO and CO_2-Containing mixtures**

Perry et al. [49] investigated and reported on the performance of a HT PBI membrane (processed by the sol-gel process). The electrochemical cell was operated at 120–160 °C with various gas reformate feed streams, with and without humidification. This was done in order to evaluate key parameters, such as the power requirements of the cell, the electrochemical efficiency, the durability, the CO tolerance, and the hydrogen purification efficiency. Almost Faradic flows (correlating to the law of Faraday) were achieved and little power was required to operate the hydrogen pump. In long-term tests, durability was excellent. Polarization curves were constructed to investigate the power requirements. For a typical current density of 0.2 $A \cdot cm^{-2}$, low voltages of ~45 mV were achieved at 160 °C for pure hydrogen. These low voltages are an indication of facile oxidation and reduction of hydrogen and low resistance of the MEA and the cell hardware components. With a stoichiometric hydrogen feed of 1.2 (without humidification), an increase in linearity of the outlet flow rates to current density was recorded.

Thomassen et al. [93] investigated simultaneous hydrogen separation and compression using PBI-based PEM fuel cell technology. The tests were performed using pure hydrogen,

N_2/H_2 mixtures, and reformate feed gas mixtures containing CO, CO_2 and CH_4. The energy efficiency of the compression process was found to be 80–90% (based on the lower heating value of separated hydrogen) at hydrogen fluxes of 5–7 $Nm^3 \cdot m^{-2} \cdot h^{-1}$. Furthermore, a CO tolerance of $\pm 1.5\%$ was reported, with a reduction in the other gas components in the separated hydrogen of up to 99%. Stable compression was reported up to 0.65 bar differential pressure over the PBI membrane, with almost no increase in the energy consumption.

Kim et al. [174] investigated the effects of Pt loading on the cell performance of a PBI-based EHP by means of polarization curves. This was paired with an investigation into electrochemical impedance characteristics, utilizing MEAs with various Pt loadings on the anode and the cathode. In addition, the gas separation of H_2/CO_2 was investigated, with no humidification, at a cell operating temperature of 160 °C. The cell voltage was reported to be a mere 80 mV at 0.8 $A \cdot cm^{-2}$ for a pure hydrogen feed. This was lower than that achieved with perfluorosulphonic acid (PFSA) membranes at a relative humidity of $\leq 43\%$. The cell voltage increased by 72% when the anode Pt loading was decreased from 1.1 $mg \cdot cm^{-2}$ to 0.2 $mg \cdot cm^{-2}$. The effect of various catalyst loadings on the cathode were insignificant.

Chen et al. [175] investigated PBI-based phosphoric acid (PA)-doped fuel cells under simulated reformate gases with various H_2, N_2 and CO concentrations. In the absence of CO, the dilution effect of N_2 had little to no effect on the performance of the cell. However, CO poisoning increased the charge transfer resistance, which substantially decreased the performance. Experimental results indicated that higher operating temperatures supresses the Pt-CO binding reaction, resulting in improved tolerance towards CO.

Huang et al. [182] prepared, characterised, and tested various PA-doped PBI membranes: *para*-PBI, *m/p*-PBI, and *meta*-PBI. Experimental test also included a conventional imbibed *meta*-PBI, for comparison. Various chemistries of PBI were investigated to understand how the chemistry affected the EHS performance, including the voltage requirement, power consumption, efficiency, hydrogen purity, and also long-term durability of the MEAs. Controlling the chemistry and increasing the polymer solids content led to considerable improvement in the creep resistance of the *m/p*-PBI and *meta*-PBI membranes ($<2 \times 10^{-6}$ Pa^{-1}) compared with *para*-PBI (~10×10^{-6} Pa^{-1}). However, the conductivity of *m/p*-PBI and *meta*-PBI exhibited lower proton conductivity (~0.14–0.17 $S \cdot cm^{-1}$) compared with *para*-PBI (0.26 $S \cdot cm^{-1}$). HT PEM cells based on these novel PBI membranes were used to investigate the EHS from various feed gases: Pure hydrogen, and two premixed reformate streams containing H_2, N_2 and CO. The reformate streams were used to validate the increased value of this technique when operated at 160–200 °C due to the increased Pt tolerance to CO. Results indicated that the cell can be operated using a dilute hydrogen feed streams with large amounts of CO (1–3%). Fairly pure hydrogen products (>99.6%) were achieved, with high power efficiencies (up to ~72%).

Vermaak et al. [188] investigated the EHS performance of a TPS membrane (supplied by Advent Technologies Inc., USA) using H_2/CO_2 gas mixtures and comparing it to pure hydrogen, H_2/CH_4 and H_2/NH_3 (refer to the relevant sections). The compositions of the feed mixtures were 10% and 50% CO_2, balance hydrogen. Compared to pure hydrogen and the H_2/CH_4-mixtures, the CO_2-containing mixtures had a significant effect on the performance of the cell. The polarization curves showed a significant decrease in performance, with a maximum current density of 0.16 $A \cdot cm^{-2}$ for a 100 $mL \cdot min^{-1}$ hydrogen flow rate. The decrease in cell performance could attributed to the reduction of CO_2 to CO by the RWGS reaction. However, the authors indicated that no CO was detected in the in-line GC, but rather trace amounts of CH_4, suggesting that CO_2 methanation might have occurred. In terms of hydrogen purity, 98–99.5% was reached with 10% CO_2 in the feed and 96–99.5% with 50% CO_2 in the feed stream. Reasonable separation was achieved with the 1:1 H_2/CO_2 mixture, with selectivities of up to ~200.

- **H_2/NH_3 Mixtures**

Vermaak et al. [188] experimentally investigated the separation performance of a EHS system using H_2/NH_3 gas mixtures at 120–160 °C. The compositions were 1500 and 3000 ppm NH_3, respectively, with the balance hydrogen. The cell performance was poor overall. The highest current density that was reached were 0.12 A·cm^{-2}. The selectivities were <2 for both feed streams (see Figure 5). The authors reported that NH_3 and H_2 competed: In the cases where the selectivity values <1, ammonia transport through the membrane was favoured. Whereas hydrogen transport was favoured at selectivities >1. Ammonia is transferred through the membrane as NH^{4+}. The authors reported severe irreversible damage to the membrane and attributed it to the alkaline nature of NH^{4+}. They further state that NH^{4+} reacts with the phosphoric acid (PO_4^{3-}) that the membrane is doped with, similar to the reaction seen with Nafion membranes in fuel cells [190,191].

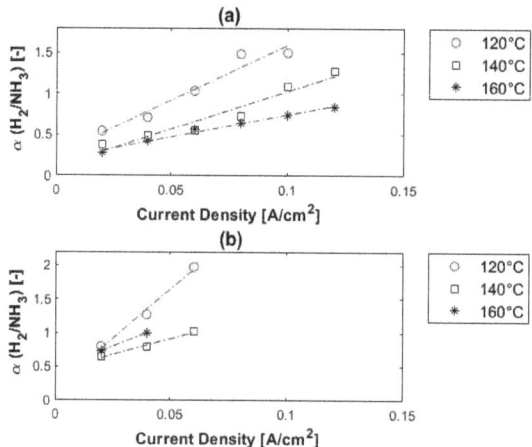

Figure 5. H_2/NH_3 selectivity for: (**a**) 1500 ppm and (**b**) 3000 ppm NH_3 (balance hydrogen) [188].

3.2.3. Review Articles

Granite and O'Brien [167] reviewed some novel methods for CO_2 separation from flue and fuel gas streams, using EHPs, membranes and chemical looping approaches. Whereas, Rhandi et al. [186] reviewed electrochemical hydrogen compression and separation against competing technologies. In their article, the advantages and disadvantages of EHS/compression is highlighted. The review article of Trégaro et al. [187] serves as part 2 of the aforementioned article, where the challenges that EHS face are discussed. Special emphasis is placed on impurities that influence, or degrade, the cell performance, together with electrocatalysts that are commonly used in EHS.

3.2.4. Overall Summary of the State of Electrochemical Hydrogen Separation

In general, very few research articles are available on EHS, as this technology is still relatively new. In summary, the following characteristics can be associated with EHS:

- High selectivity;
- sensitivity to catalyst deactivation (e.g., CO deactivation);
- higher tolerance to "active" impurities (e.g., CO and CO_2) at higher temperatures;
- the hydrogen flux can be controlled by the current and
- simultaneous hydrogen separation and compression is possible.

Very little information is available on failure and degradation mechanisms of the PEMs used in the scope of EHS. However, since fuel cells and electrolyzers are essentially the same technology used for EHS (with different applications), being an electrochemical cell, the degradation mechanisms [192,193] of both these technologies under different operating conditions can be used to partly fill this gap. Some of the phenomena that complicate electrochemical membrane-based processes include the distribution of water in LT application [194], loss of acid in HT PBI membranes [195] and deformation of the membrane and gas diffusion layers by the assembly compression [195–198]. The properties of HT PBI membranes and LT Nafion membranes have also been compared in [199]. Besides, EHS and compression, fuel cells and electrolyzers, other useful applications for electrochemical cells have been reported: The reduction of CO_2 [200], and the electrochemical compression of ammonia [201–203]. However, these applications are in the very early stages of research. Besides the conventional EHS membranes, e.g., Nafion (LT) and PBI (HT), the use of solid acids have also been reported [204] in steam electrolysis. Typical operating temperatures for these materials are ~200 °C. However, this technology is very new and still require immense research effort. Moreover, in the field of EHS, ceramic proton-conducting membranes are also known, but are still at the earlier development stage [108]. This technology has been briefly discussed in Section 2.3.3.

4. Concluding Remarks

There is still much unknown about EHS, largely because the technology is not yet mature. We conclude this overview with mention of the areas that (we consider) require further research in the field of electrochemical hydrogen separation and its further development:

- Little information on component degradation, beside CO catalyst deactivation. However, the degradation studies performed on fuel cells can be used to fill this gap.
- Understanding the life cycle of an electrochemical hydrogen membrane and how an aged membrane's performance compare to that of a membrane at the beginning of its life.
- Contribution/s of various impurities (considered separately) to the performance parameters. Impurities that commonly accompany hydrogen streams generated from various traditional hydrogen generation methods include CH_4, O_2, N_2, CO, CO_2, H_2S, benzene, toluene, xylene and NH_3. However, as is evident from Table 4, research articles are only available on EHS from mixtures containing CO, N_2, CO_2, CH_4, Ar, ethylene and H_2O. Furthermore, HT EHS research mainly included reformate gases. Hence, the contribution of respective impurities, separately, on the performance parameters is largely unknown.
- Further research into HT EHS. From Table 4, it is evident that more information is available on LT separation than HT separation. Further research is required to achieve a broader understanding of the expected extent of separation with respect to the various performance parameters—such as limiting currents, hydrogen recovery, selectivities and fluxes.
- One of the advantages that HT membranes present is the possibility of being able to use catalysts such as iron and cobalt. More information on this topic is required in efforts to determine how beneficial this would be—besides only focusing on the cost reduction.
- Fuel cell application. To date, no studies appear to have been conducted to verify the hydrogen purity of EHS product streams for fuel cell application. Such knowledge could be very beneficial, especially when simulated reformate streams are used from industrial hydrogen production systems.
- EHS from industrial hydrogen streams produced from fuel cells (e.g., product streams from steam methane reforming, partial oxidation and gasification of biomass and coal)
- Simultaneous EHS and electrochemical hydrogen compression, together with the process efficiency in terms of hydrogen purity, hydrogen compression, overall efficiency, etc. Specifically, the simultaneous EHS and compression from H_2/CO_2 streams, where both the hydrogen stream (permeate) and the carbon dioxide (retentate) is purified and compressed. Such study will be beneficial in terms of hydrogen production and CO_2 sequestration (i.e., carbon capture and storage, and even carbon capture and utilization).

- Proton-conducting ceramics could be considered a new and upcoming technology and is also part of EHS. The authors suggest that future reviews be done, similar to the one presented, on this topic.

Author Contributions: L.V.: Writing—Original Draft. H.W.J.P.N.: Writing—Review and Editing. D.G.B.: Writing—Review and Editing. All authors have read and agreed to the published version of the manuscript.

Funding: This work was supported by HySA Infrastructure (via KP5 program) at the North-West University, Potchefstroom, South Africa.

Institutional Review Board Statement: Not applicable.

Informed Consent Statement: Not applicable.

Conflicts of Interest: The authors declare no conflict of interest.

References

1. Alanne, K.; Cao, S. An overview of the concept and technology of ubiquitous energy. *Appl. Energy* **2019**, *238*, 284–302. [CrossRef]
2. Dawood, F.; Anda, M.; Shafiullah, G.M. Hydrogen production for energy: An overview. *Int. J. Hydrogen Energy* **2020**, *45*, 3847–3869. [CrossRef]
3. Zhang, F.; Zhao, P.; Niu, M.; Maddy, J. The survey of key technologies in hydrogen energy storage. *Int. J. Hydrogen Energy* **2016**, *41*, 14535–14552. [CrossRef]
4. Smoliński, A.; Howaniec, N. Hydrogen energy, electrolyzers and fuel cells—The future of modern energy sector. *Int. J. Hydrogen Energy* **2020**, *45*, 5607. [CrossRef]
5. Lund, H. Renewable energy strategies for sustainable development. *Energy* **2007**, *32*, 912–919. [CrossRef]
6. Won, W.; Kwon, H.; Han, J.H.; Kim, J. Design and operation of renewable energy sources based hydrogen supply system: Technology integration and optimization. *Renew. Energy* **2017**, *103*, 226–238. [CrossRef]
7. Scamman, D.; Newborough, M. Using surplus nuclear power for hydrogen mobility and power-to-gas in France. *Int. J. Hydrogen Energy* **2016**, *41*, 10080–10089. [CrossRef]
8. Khodadoost Arani, A.A.; Gharehpetian, G.B.; Abedi, M. Review on Energy Storage Systems Control Methods in Microgrids. *Int. J. Electr. Power Energy Syst.* **2019**, *107*, 745–757. [CrossRef]
9. Gür, T.M. Review of electrical energy storage technologies, materials and systems: Challenges and prospects for large-scale grid storage. *Energy Environ. Sci.* **2018**, *11*, 2696–2767. [CrossRef]
10. Krishan, O.; Suhag, S. An updated review of energy storage systems: Classification and applications in distributed generation power systems incorporating renewable energy resources. *Int. J. Energy Res.* **2019**, *43*, 6171–6210. [CrossRef]
11. Yang, Z.; Zhang, J.; Kintner-Meyer, M.C.W.; Lu, X.; Choi, D.; Lemmon, J.P.; Liu, J. Electrochemical Energy Storage for Green Grid. *Chem. Rev.* **2011**, *111*, 3577–3613. [CrossRef]
12. Smith, W. Role of fuel cells in energy storage. *J. Power Sources* **2000**, *86*, 74–83. [CrossRef]
13. Yan, Z.; Hitt, J.L.; Turner, J.A.; Mallouk, T.E. Renewable electricity storage using electrolysis. *Proc. Natl. Acad. Sci. USA* **2020**, *117*, 12558–12563. [CrossRef] [PubMed]
14. Vanhanen, J.P.; Lund, P.D.; Tolonen, J.S. Electrolyser-metal hydride-fuel cell system for seasonal energy storage. *Int. J. Hydrogen Energy* **1998**, *23*, 267–271. [CrossRef]
15. Hall, P.J.; Mirzaeian, M.; Fletcher, S.I.; Sillars, F.B.; Rennie, A.J.R.; Shitta-Bey, G.O.; Wilson, G.; Cruden, A.; Carter, R. Energy storage in electrochemical capacitors: Designing functional materials to improve performance. *Energy Environ. Sci.* **2010**, *3*, 1238–1251. [CrossRef]
16. Abbas, Q.; Raza, R.; Shabbir, I.; Olabi, A.G. Heteroatom doped high porosity carbon nanomaterials as electrodes for energy storage in electrochemical capacitors: A review. *J. Sci. Adv. Mater. Devices* **2019**, *4*, 341–352. [CrossRef]
17. Gharehpetian, G.B.; Akhavanhejazi, M.; Arani, A.A.K.; Karami, H.; Gharehpetian, G.B.; Hejazi, A. Review of Flywheel Energy Storage Systems structures and applications in power systems and microgrids. *Renew. Sustain. Energy Rev.* **2016**, *69*, 9–18. [CrossRef]
18. Faraji, F.; Majazi, A.; Al-Haddad, K. A comprehensive review of Flywheel Energy Storage System technology. *Renew. Sustain. Energy Rev.* **2017**, *67*, 477–490. [CrossRef]
19. Peña-Alzola, R.; Sebastián, R.; Quesada, J.; Colmenar, A. Review of flywheel based energy storage systems. In Proceedings of the 2011 International Conference on Power Engineering, Energy and Electrical Drives, Malaga, Spain, 11–13 May 2011. [CrossRef]
20. Venkataramani, G.; Parankusam, P.; Ramalingam, V.; Wang, J. A review on compressed air energy storage—A pathway for smart grid and polygeneration. *Renew. Sustain. Energy Rev.* **2016**, *62*, 895–907. [CrossRef]
21. Wang, J.; Lu, K.; Ma, L.; Wang, J.; Dooner, M.; Miao, S.; Li, J.; Wang, D. Overview of compressed air energy storage and technology development. *Energies* **2017**, *10*, 991. [CrossRef]

22. Rehman, S.; Al-Hadhrami, L.M.; Alam, M.M. Pumped hydro energy storage system: A technological review. *Renew. Sustain. Energy Rev.* **2015**, *44*, 586–598. [CrossRef]
23. Ma, T.; Yang, H.; Lu, L.; Peng, J. Technical feasibility study on a standalone hybrid solar-wind system with pumped hydro storage for a remote island in Hong Kong. *Renew. Energy* **2014**, *69*, 7–15. [CrossRef]
24. Agyenim, F.; Hewitt, N.; Eames, P.; Smyth, M. A review of materials, heat transfer and phase change problem formulation for latent heat thermal energy storage systems (LHTESS). *Renew. Sustain. Energy Rev.* **2010**, *14*, 615–628. [CrossRef]
25. Sarbu, I.; Dorca, A. Review on heat transfer analysis in thermal energy storage using latent heat storage systems and phase change materials. *Int. J. Energy Res.* **2019**, *43*, 29–64. [CrossRef]
26. Sarbu, I.; Sebarchievici, C. A comprehensive review of thermal energy storage. *Sustainability* **2018**, *10*, 191. [CrossRef]
27. Languri, E.M.; Cunningham, G. Thermal Energy Storage Systems. *Lect. Notes Energy* **2019**, *70*, 169–176. [CrossRef]
28. Diaz, P. Analysis and Comparison of different types of Thermal Energy Storage Systems: A Review. *J. Adv. Mech. Eng. Sci.* **2016**, *2*, 33–46. [CrossRef]
29. Mukherjee, P.; Rao, V.V. Superconducting magnetic energy storage for stabilizing grid integrated with wind power generation systems. *J. Mod. Power Syst. Clean Energy* **2019**, *7*, 400–411. [CrossRef]
30. Barbir, F.; Veziroglu, T.N. *Hydroger Energy System and Hydrogen Production Methods*; Kluwer Academic Publishers: New York, NY, USA, 1992; pp. 277–278.
31. Veziroglu, T.; Sherif, S.A.; Barbir, F. Chapter 7-Hydrogen Energy Solutions. In *Environmental Solutions*; Agardy, F.J., Nemerow, N.L., Eds.; Academic Press: Cambridge, MA, USA, 2005; pp. 143–180.
32. Barbir, F. *Future of Fuel Cells and Hydrogen*; Academic Press: Waltam, MA, USA, 2013; ISBN 9780123877109.
33. Züttel, A. Hydrogen storage methods. *Naturwissenschaften* **2004**, *91*, 157–172. [CrossRef] [PubMed]
34. Zhou, L. Progress and problems in hydrogen storage methods. *Renew. Sustain. Energy Rev.* **2005**, *9*, 395–408. [CrossRef]
35. Kunowsky, M.; Marco-Lózar, J.P.; Linares-Solano, A. Material Demands for Storage Technologies in a Hydrogen Economy. *J. Renew. Energy* **2013**, *2013*, 878329. [CrossRef]
36. Valenti, G. Hydrogen liquefaction and liquid hydrogen storage. In *Compendium of Hydrogen Energy*; Elsevier: Amsterdam, The Netherlands, 2016; pp. 27–51.
37. Ahluwalia, R.K.; Peng, J.-K.; Hua, T.Q. *Cryo-Compressed Hydrogen Storage*; Elsevier: Amsterdam, The Netherlands, 2016; ISBN 9781782423621.
38. Modisha, P.M.; Ouma, C.N.M.; Garidzirai, R.; Wasserscheid, P.; Bessarabov, D. The Prospect of Hydrogen Storage Using Liquid Organic Hydrogen Carriers. *Energy Fuels* **2019**, *33*, 2778–2796. [CrossRef]
39. Teichmann, D.; Arlt, W.; Wasserscheid, P. Liquid Organic Hydrogen Carriers as an efficient vector for the transport and storage of renewable energy. *Int. J. Hydrogen Energy* **2012**, *37*, 18118–18132. [CrossRef]
40. Preuster, P.; Papp, C.; Wasserscheid, P. Liquid organic hydrogen carriers (LOHCs): Toward a hydrogen-free hydrogen economy. *Acc. Chem. Res.* **2017**, *50*, 74–85. [CrossRef]
41. Makepeace, J.W.; He, T.; Weidenthaler, C.; Jensen, T.R.; Chang, F.; Vegge, T.; Ngene, P.; Kojima, Y.; de Jongh, P.E.; Chen, P.; et al. Reversible ammonia-based and liquid organic hydrogen carriers for high-density hydrogen storage: Recent progress. *Int. J. Hydrogen Energy* **2019**, *44*, 7746–7767. [CrossRef]
42. Corgnale, C.; Hardy, B.J.; Anton, D.L. Structural analysis of metal hydride-based hybrid hydrogen storage systems. *Int. J. Hydrogen Energy* **2012**, *37*, 14223–14233. [CrossRef]
43. Gondal, I.A. *Hydrogen Transportation by Pipelines*; Elsevier: Amsterdam, The Netherlands, 2016; ISBN 9781782423621.
44. Liu, B.; Liu, S.; Guo, S.; Zhang, S. Economic study of a large-scale renewable hydrogen application utilizing surplus renewable energy and natural gas pipeline transportation in China. *Int. J. Hydrogen Energy* **2019**, *45*, 1385–1398. [CrossRef]
45. Kim, J.W.; Boo, K.J.; Cho, J.H.; Moon, I. *Key Challenges in the Development of an Infrastructure for Hydrogen Production, Delivery, Storage and Use*; Woodhead Publishing Limited: Cambridge, UK, 2014; ISBN 9780857097736.
46. Sharma, S.; Ghoshal, S.K. Hydrogen the future transportation fuel: From production to applications. *Renew. Sustain. Energy Rev.* **2015**, *43*, 1151–1158. [CrossRef]
47. Lin, R.H.; Zhao, Y.Y.; Wu, B.D. Toward a hydrogen society: Hydrogen and smart grid integration. *Int. J. Hydrogen Energy* **2020**, *45*, 20164–20175. [CrossRef]
48. Bockris, J.O.; Appleby, A.J. The hydrogen economy—An ultimate economy. *Environ. This Mon.* **1972**, *1*, 29–35. Available online: http://inis.iaea.org/Search/search.aspx?orig_q=RN:3032306 (accessed on 12 August 2020).
49. Perry, K.A.; Eisman, G.A.; Benicewicz, B.C. Electrochemical hydrogen pumping using a high-temperature polybenzimidazole (PBI) membrane. *J. Power Sources* **2008**, *177*, 478–484. [CrossRef]
50. Bessarabov, D.G.; Millet, P. *PEM Water Electrolysis*; Academic Press: Cambridge, MA, USA, 2018; Volume 1, ISBN 9780128111468.
51. Acar, C.; Dincer, I. Review and evaluation of hydrogen production options for better environment. *J. Clean. Prod.* **2019**, *218*, 835–849. [CrossRef]
52. Turner, J.A. Sustainable hydrogen production. *Science (80-)* **2004**, *305*, 972–974. [CrossRef] [PubMed]
53. Kurtz, J.; Sprik, S.; Bradley, T.H. Review of transportation hydrogen infrastructure performance and reliability. *Int. J. Hydrogen Energy* **2019**, *44*, 12010–12023. [CrossRef]
54. Shiva Kumar, S.; Himabindu, V. Hydrogen production by PEM water electrolysis—A review. *Mater. Sci. Energy Technol.* **2019**, *2*, 442–454. [CrossRef]

55. Abdalla, A.M.; Hossain, S.; Nisfindy, O.B.; Azad, A.T.; Dawood, M.; Azad, A.K. Hydrogen production, storage, transportation and key challenges with applications: A review. *Energy Convers. Manag.* **2018**, *165*, 602–627. [CrossRef]
56. da Silva Veras, T.; Mozer, T.S.; da Costa Rubim Messeder dos Santos, D.; da Silva César, A. Hydrogen: Trends, production and characterization of the main process worldwide. *Int. J. Hydrogen Energy* **2017**, *42*, 2018–2033. [CrossRef]
57. López Ortiz, A.; Meléndez Zaragoza, M.J.; Collins-Martínez, V. Hydrogen production research in Mexico: A review. *Int. J. Hydrogen Energy* **2016**, *41*, 23363–23379. [CrossRef]
58. Steinberg, M.; Cheng, H.C. Modern and prospective technologies for hydrogen production from fossil fuels. *Int. J. Hydrogen Energy* **1989**, *14*, 797–820. [CrossRef]
59. Acar, C.; Dincer, I. Impact assessment and efficiency evaluation of hydrogen production methods. *Int. J. Energy Res.* **2015**, *39*, 1757–1768. [CrossRef]
60. Chaubey, R.; Sahu, S.; James, O.O.; Maity, S. A review on development of industrial processes and emerging techniques for production of hydrogen from renewable and sustainable sources. *Renew. Sustain. Energy Rev.* **2013**, *23*, 443–462. [CrossRef]
61. Baykara, S.Z. Hydrogen: A brief overview on its sources, production and environmental impact. *Int. J. Hydrogen Energy* **2018**, *43*, 10605–10614. [CrossRef]
62. Rakib, M.A.; Grace, J.R.; Lim, C.J.; Elnashaie, S.S.E.H.; Ghiasi, B. Steam reforming of propane in a fluidized bed membrane reactor for hydrogen production. *Renew. Energy* **2010**, *35*, 6276–6290. [CrossRef]
63. de Campos Roseno, K.T.; de Brito Alves, R.M.; Giudici, R.; Schmal, M. Syngas Production Using Natural Gas from the Environmental Point of View. In *Biofuels—State of Development*; InTech: London, UK, 2018.
64. Synthesis Gas Chemistry and Synthetic Fuels. Available online: https://www.syncatbeijing.com/syngaschem/ (accessed on 14 August 2020).
65. Dincer, I.; Acar, C. Review and evaluation of hydrogen production methods for better sustainability. *Int. J. Hydrogen Energy* **2014**, *40*, 11094–11111. [CrossRef]
66. Acar, C.; Dincer, I. Comparative assessment of hydrogen production methods from renewable and non-renewable sources. *Int. J. Hydrogen Energy* **2014**, *39*, 1–12. [CrossRef]
67. Nikolaidis, P.; Poullikkas, A. A comparative overview of hydrogen production processes. *Renew. Sustain. Energy Rev.* **2017**, *67*, 597–611. [CrossRef]
68. David, O.C. Membrane Technologies for Hydrogen and Carbon Monoxide Recovery from Residual Gas Streams. PhD Thesis, University of Cantabria, Cantabria, Spain, 2012; pp. 1–183.
69. Holladay, J.D.; Hu, J.; King, D.L.; Wang, Y. An overview of hydrogen production technologies. *Catal. Today* **2009**, *139*, 244–260. [CrossRef]
70. Engelbrecht, N.; Chiuta, S.; Everson, R.C.; Neomagus, H.W.J.P.; Bessarabov, D.G. Experimentation and CFD modelling of a microchannel reactor for carbon dioxide methanation. *Chem. Eng. J.* **2017**, *313*, 847–857. [CrossRef]
71. Wu, H.C.; Chang, Y.C.; Wu, J.H.; Lin, J.H.; Lin, I.K.; Chen, C.S. Methanation of CO_2 and reverse water gas shift reactions on Ni/SiO2 catalysts: The influence of particle size on selectivity and reaction pathway. *Catal. Sci. Technol.* **2015**, *5*, 4154–4163. [CrossRef]
72. Guerra, L.; Rossi, S.; Rodrigues, J.; Gomes, J.; Puna, J.; Santos, M.T. Methane production by a combined Sabatier reaction/water electrolysis process. *J. Environ. Chem. Eng.* **2018**, *6*, 671–676. [CrossRef]
73. Zhu, M.; Ge, Q.; Zhu, X. Catalytic Reduction of CO2 to CO via Reverse Water Gas Shift Reaction: Recent Advances in the Design of Active and Selective Supported Metal Catalysts. *Trans. Tianjin Univ.* **2020**, *26*, 172–187. [CrossRef]
74. Abney, M.B.; Perry, J.L.; Junaedi, C.; Hawley, K.; Walsh, D.; Roychoudhury, S. *Compact Lightweight Sabatier Reaction for Carbon Dioxide Reduction*; American Institute of Aeronautics and Astronautics: Reston, VA, USA, 2011; pp. 1–10.
75. Mayorga, S.G.; Hufton, J.R.; Sircar, S.; Gaffney, T.R. *Sorption Enhanced Reaction Process for Production of Hydrogen. Phase 1 Final Report*; U.S. Department of Energy: Golden, CO, USA, 1997.
76. Andrews, J.W. Hydrogen production and carbon sequestration by steam methane reforming and fracking with carbon dioxide. *Int. J. Hydrogen Energy* **2020**, *45*, 9279–9284. [CrossRef]
77. Ball, M.; Weeda, M. *The Hydrogen Economy—Vision or Reality? The Global Energy Challenge*; Woodhead Publishers: Sawston, UK, 2015. [CrossRef]
78. Shoko, E.; McLellan, B.; Dicks, A.L.; da Costa, J.C.D. Hydrogen from coal: Production and utilisation technologies. *Int. J. Coal Geol.* **2006**, *65*, 213–222. [CrossRef]
79. Ockwig, N.W.; Nenoff, T.M. Membranes for hydrogen separation. *Chem. Rev.* **2007**, *107*, 4078–4110. [CrossRef] [PubMed]
80. Kalamaras, C.M.; Efstathiou, A.M.; Al-Assaf, Y.; Poullikkas, A. Hydrogen Production Technologies: Current State and Future Developments. *Conf. Pap. Energy* **2013**, *2013*, 690627. [CrossRef]
81. Besancon, B.M.; Hasanov, V.; Imbault-Lastapis, R.; Benesch, R.; Barrio, M.; Mølnvik, M.J. Hydrogen quality from decarbonized fossil fuels to fuel cells. *Int. J. Hydrogen Energy* **2009**, *34*, 2350–2360. [CrossRef]
82. Cao, L.; Yu, I.K.; Xiong, X.; Tsang, D.C.; Zhang, S.; Clark, J.H.; Hu, C.; Hau Ng, Y.; Shang, J.; Sik Ok, Y. Biorenewable hydrogen production through biomass gasification: A review and future prospects. *Environ. Res.* **2020**, *186*, 109547. [CrossRef]
83. Shafirovich, E.; Varma, A. Underground Coal Gasification: A Brief Review of Current Status. *Ind. Eng. Chem. Res.* **2009**, *48*, 7865–7875. [CrossRef]

84. Emami-Taba, L.; Faisal Irfan, M.; Ashri, W.M.; Daud, W.; Chakrabarti, M.H. Fuel blending effects on the co-gasification of coal and biomass—A review. *Biomass Bioenergy* **2013**, *57*, 249–263. [CrossRef]
85. Stiegel, G.J.; Ramezan, M. Hydrogen from coal gasification: An economical pathway to a sustainable energy future. *Int. J. Coal Geol.* **2006**, *65*, 173–190. [CrossRef]
86. Yusuf, N.Y.; Masdar, M.S.; Nordin, D.; Husaini, T. Challenges in Biohydrogen Technologies for Fuel Cell Application. *Am. J. Chem.* **2015**, *5*, 40–47. [CrossRef]
87. Pareek, A.; Dom, R.; Gupta, J.; Chandran, J.; Adepu, V.; Borse, P.H. Insights into renewable hydrogen energy: Recent advances and prospects. *Mater. Sci. Energy Technol.* **2020**, *3*, 319–327. [CrossRef]
88. Kalinci, Y.; Hepbasli, A.; Dincer, I. Biomass-based hydrogen production: A review and analysis. *Int. J. Hydrogen Energy* **2009**, *34*, 8799–8817. [CrossRef]
89. Ozbilen, A.; Dincer, I.; Rosen, M.A. Comparative environmental impact and efficiency assessment of selected hydrogen production methods. *Environ. Impact Assess. Rev.* **2013**, *42*, 1–9. [CrossRef]
90. Bhandari, R.; Trudewind, C.A.; Zapp, P. Life cycle assessment of hydrogen production via electrolysis—A review. *J. Clean. Prod.* **2014**, *85*, 151–163. [CrossRef]
91. Shalygin, M.G.; Abramov, S.M.; Netrusov, A.I.; Teplyakov, V.V. Membrane recovery of hydrogen from gaseous mixtures of biogenic and technogenic origin. *Int. J. Hydrogen Energy* **2015**, *40*, 3438–3451. [CrossRef]
92. Adhikari, S.; Fernando, S. Hydrogen membrane separation techniques. *Ind. Eng. Chem. Res.* **2006**, *45*, 875–881. [CrossRef]
93. Thomassen, M.; Sheridan, E.; Kvello, J. Electrochemical hydrogen separation and compression using polybenzimidazole (PBI) fuel cell technology. *J. Nat. Gas Sci. Eng.* **2010**, *2*, 229–234. [CrossRef]
94. Liemberger, W.; Groß, M.; Miltner, M.; Harasek, M. Experimental analysis of membrane and pressure swing adsorption (PSA) for the hydrogen separation from natural gas. *J. Clean. Prod.* **2017**, *167*, 896–907. [CrossRef]
95. Takht Ravanchi, M.; Kaghazchi, T.; Kargari, A. Application of membrane separation processes in petrochemical industry: A review. *Desalination* **2009**, *235*, 199–244. [CrossRef]
96. Mores, P.L.; Arias, A.M.; Scenna, N.J.; Caballero, J.A.; Mussati, S.F.; Mussati, M.C. Membrane-Based Processes: Optimization of Hydrogen Separation by Minimization of Power, Membrane Area, and Cost. *Processes* **2018**, *6*, 221. [CrossRef]
97. Schorer, L.; Schmitz, S.; Weber, A. Membrane based purification of hydrogen system (MEMPHYS). *Int. J. Hydrogen Energy* **2019**, *44*, 12708–12714. [CrossRef]
98. Zhang, Q.; Liu, G.; Feng, X.; Chu, K.H.; Deng, C. Hydrogen networks synthesis considering separation performance of purifiers. *Int. J. Hydrogen Energy* **2014**, *39*, 8357–8373. [CrossRef]
99. Liu, F.; Zhang, N. Strategy of purifier selection and integration in hydrogen networks. *Chem. Eng. Res. Des.* **2004**, *82*, 1315–1330. [CrossRef]
100. Sircar, S.; Golden, T.C. Purification of Hydrogen by Pressure Swing Adsorption. *Sep. Sci. Technol.* **2000**, *35*, 667–687. [CrossRef]
101. Xiao, J.; Peng, Y.; Bénard, P.; Chahine, R. Thermal effects on breakthrough curves of pressure swing adsorption for hydrogen purification. *Int. J. Hydrogen Energy* **2016**, *41*, 8236–8245. [CrossRef]
102. Sircar, S.; Golden, T.C. Pressure Swing Adsorption Technology for Hydrogen Production. In *Hydrogen and Syngas Production and Purification Technologies*; Liu, K., Song, C., Subramani, V., Eds.; John Wiley & Sons: Hoboken, NJ, USA, 2009; pp. 414–450. ISBN 9780471719755.
103. Al-Mufachi, N.A.; Rees, N.V.; Steinberger-Wilkens, R. Hydrogen selective membranes: A review of palladium-based dense metal membranes. *Renew. Sustain. Energy Rev.* **2015**, *47*, 540–551. [CrossRef]
104. Fahim, M.A.; Alsahhaf, T.A.; Elkilani, A. Chapter eleven: Hydrogen production. In *Fundamentals of Petroleum Refining*; Elsevier: Amsterdam, The Netherlands, 2010; pp. 285–302. ISBN 9780444527851.
105. Grande, C.A.; Lopes, F.V.S.; Ribeiro, A.M.; Loureiro, J.M.; Rodrigues, A.E. Adsorption of Off-Gases from Steam Methane Reforming (H_2, CO_2, CH_4, CO and N_2) on Activated Carbon. *Sep. Sci. Technol.* **2008**, *43*, 1338–1364. [CrossRef]
106. Marković, N.M.; Schmidt, T.J.; Grgur, B.N.; Gasteiger, H.A.; Behm, R.J.; Ross, P.N. Effect of Temperature on Surface Processes at the Pt(111)—Liquid Interface: Hydrogen Adsorption, Oxide Formation, and CO Oxidation. *J. Phys. Chem. B* **1999**, *103*, 8568–8577. [CrossRef]
107. Mulder, M. *Basic Principles of Membrane Technology*, 2nd ed.; Kluwer Academic Publishers: Dordrecht, The Netherlands, 2003.
108. Cardoso, S.P.; Azenha, I.S.; Lin, Z.; Rodrigues, A.E.; Silva, C.M. Inorganic Membranes for Hydrogen Separation. *Sep. Purif. Rev.* **2018**, *47*, 229–266. [CrossRef]
109. Li, P.; Wang, Z.; Qiao, Z.; Liu, Y.; Cao, X.; Li, W.; Wang, J.; Wang, S. Recent developments in membranes for efficient hydrogen purification. *J. Membr. Sci.* **2015**, *495*, 130–168. [CrossRef]
110. Brinkmann, T.; Shishatskiy, S. Hydrogen Separation with Polymeric Membranes. *Hydrogen Sci. Eng. Mater. Process. Syst. Technol.* **2016**, *1*, 509–541. [CrossRef]
111. Lu, G.Q.; Diniz Da Costa, J.C.; Duke, M.; Giessler, S.; Socolow, R.; Williams, R.H.; Kreutz, T. Inorganic membranes for hydrogen production and purification: A critical review and perspective. *J. Colloid Interface Sci.* **2007**, *314*, 589–603. [CrossRef]
112. Escorihuela, S.; Tena, A.; Shishatskiy, S.; Escolástico, S.; Brinkmann, T.; Serra, J.M.; Abetz, V. Gas separation properties of polyimide thin films on ceramic supports for high temperature applications. *Membranes* **2018**, *8*, 16. [CrossRef]
113. Hu, X.; Lee, W.H.; Bae, J.Y.; Kim, J.S.; Jung, J.T.; Wang, H.H.; Park, H.J.; Lee, Y.M. Thermally rearranged polybenzoxazole copolymers incorporating Tröger's base for high flux gas separation membranes. *J. Membr. Sci.* **2020**, *612*, 118437. [CrossRef]

114. Zornoza, B.; Casado, C.; Navajas, A. *Advances in Hydrogen Separation and Purification with Membrane Technology*; Elsevier: Amsterdam, The Netherlands, 2016; pp. 245–268. [CrossRef]
115. Sanchez Marcano, J.G.; Tsotsis, T.T. *Catalytic Membranes and Membrane Reactors*; Wiley-VCH: Weiheim, Germany, 2002; ISBN 3527302778.
116. Phair, J.W.; Badwal, S.P.S. Materials for separation membranes in hydrogen and oxygen production and future power generation Materials for separation membranes in hydrogen and oxygen production and future power generation. *Sci. Technol. Adv. Mater.* **2006**, *7*, 792. [CrossRef]
117. Oyama, S.T.; Yamada, M.; Sugawara, T.; Takagaki, A.; Kikuchi, R. Review on mechanisms of gas permeation through inorganic membranes. *J. Japan Pet. Inst.* **2011**, *54*, 298–309. [CrossRef]
118. Gilron, J.; Soffer, A. Knudsen diffusion in microporous carbon membranes with molecular sieving character. *J. Membr. Sci.* **2002**, *209*, 339–352. [CrossRef]
119. Sazali, N.; Mohamed, M.A.; Norharyati, W.; Salleh, W. Membranes for hydrogen separation: A significant review. *Int. J. Adv. Manuf. Technol.* **2020**, *107*, 1859–1881. [CrossRef]
120. Iulianelli, A.; Basile, A.; Li, H.; Van den Brink, R.W. Inorganic membranes for pre-combustion carbon dioxide. In *Advanced Membrane Science and Technology for Sustainable Energy and Environmental Applications*; Woodhead Publishing: Cambridge, UK, 2011; pp. 184–213.
121. Singh, R.P.; Berchtold, K.A. H$_2$ Selective Membranes for Precombustion Carbon Capture. In *Novel Materials for Carbon Dioxide Mitigation Technology*; Shi, F., Morreale, B., Eds.; Elsevier: Amsterdam, The Netherlands, 2015; pp. 117–206.
122. Phair, J.W.; Badwal, S.P.S. Review of proton conductors for hydrogen separation. *Ionics (Kiel)* **2006**, *12*, 103–115. [CrossRef]
123. Yun, S.; Oyama, S.T. Correlations in palladium membranes for hydrogen separation: A review. *J. Membr. Sci.* **2011**, *375*, 28–45. [CrossRef]
124. Uemiya, S. Brief review of steam reforming using a metal membrane reactor. *Top. Catal.* **2004**, *29*, 79–84. [CrossRef]
125. Ryi, S.K.; Park, J.S.; Kim, S.H.; Cho, S.H.; Park, J.S.; Kim, D.W. Development of a new porous metal support of metallic dense membrane for hydrogen separation. *J. Membr. Sci.* **2006**, *279*, 439–445. [CrossRef]
126. Teplyakov, V.; Meares, P. Correlation aspects of the selective gas permeabilities of polymeric materials and membranes. *Gas Sep. Purif.* **1990**, *4*, 66–74. [CrossRef]
127. Nenoff, T.M.; Spontak, R.J.; Aberg, C.M. Membranes for hydrogen purification: An important step toward a hydrogen-based economy. *MRS Bull.* **2006**, *31*, 735–741. [CrossRef]
128. Yampolskii, Y.; Ryzhikh, V. Polymeric membrane materials for hydrogen separation. In *Hydrogen Production, Separation and Purification for Energy*; Basile, A., Dalena, F., Tong, J., Nejat Veziroglu, T., Eds.; Institution of Engineering and Technology: London, UK, 2017; pp. 319–341. ISBN 9781785611001.
129. Yampolskii, Y. Polymeric Gas Separation Membranes. *Macromolecules* **2012**, *45*, 3298–3311. [CrossRef]
130. Perry, J.D.; Nagai, K.; Koros, W.J. Polymer Membranes for Hydrogen Separations. *MRS Bull.* **2020**, *31*, 745–749. [CrossRef]
131. Tanaka, Y.; Yoshinari, B. Hydrogen-Metal Systems: Basic Properties (2). In *Encyclopedia of Materials: Science and Technology*, 2nd ed.; Elsevier Ltd.: Amsterdam, The Netherlands, 2001; pp. 3919–3923. [CrossRef]
132. Nagy, E. Mass Transport Through a Membrane Layer. In *Basic Equations of Mass Transport Through a Membrane Layer*; Elsevier Inc.: Amsterdam, The Netherlands, 2019; pp. 21–68.
133. Burggraaf, A.J. Single gas permeation of thin zeolite (MFI) membranes: Theory and analysis of experimental observations. *J. Membr. Sci.* **1999**, *155*, 45–65. [CrossRef]
134. Bhandarkar, M.; Shelekhin, A.B.; Dixon, A.G.; Ma, Y.H. Adsorption, permeation, and diffusion of gases in microporous membranes. I. Adsorption of gases on microporous glass membranes. *J. Membr. Sci.* **1992**, *75*, 221–231. [CrossRef]
135. Hägg, M.-B.; He, X. Chapter 15. Carbon Molecular Sieve Membranes for Gas Separation. In *Membrane Engineering for the Treatment of Gases*; The Royal Society of Chemistry: London, UK, 2011; pp. 162–191.
136. Yin, H.; Yip, A.C.K. A review on the production and purification of biomass-derived hydrogen using emerging membrane technologies. *Catalysts* **2017**, *7*, 297. [CrossRef]
137. Doong, S.J. Advanced hydrogen (H$_2$) gas separation membrane development for power plants. In *Advanced Power Plant Materials, Design and Technology*; Woodhead Publishing Limited: Cambridge, UK, 2010; pp. 111–142.
138. Steward, S.A. *Review of Hydrogen Isotope Permeability Through Materials*; Lawrence Livermore National Lab. (LLNL): Livermore, CA, USA, 1983. [CrossRef]
139. Gallucci, F. Richardson Law. In *Encyclopedia of Membrane Science and Technology*; John Wiley & Sons, Inc.: Hoboken, NJ, USA, 2012. [CrossRef]
140. Vadrucci, M.; Borgognoni, F.; Moriani, A.; Santucci, A.; Tosti, S. Hydrogen permeation through Pd-Ag membranes: Surface effects and Sieverts' law. *Int. J. Hydrogen Energy* **2013**, *38*, 4144–4152. [CrossRef]
141. Hara, S.; Ishitsuka, M.; Suda, H.; Mukaida, M.; Haraya, K. Pressure-Dependent Hydrogen Permeability Extended for Metal Membranes Not Obeying the Square-Root Law. *J. Phys. Chem. B* **2009**, *113*, 9795–9801. [CrossRef]
142. Gugliuzza, A.; Basile, A. Membrane processes for biofuel separation: An introductionNo Title. In *Membranes for Clean and Renewable Power Applications*; Woodhead Publishing Limited: Cambridge, UK, 2014; pp. 65–103.
143. Li, W.; Cao, Z.; Cai, L.; Zhang, L.; Zhu, X.; Yang, W. H2S-tolerant oxygen-permeable ceramic membranes for hydrogen separation with a performance comparable to those of palladium-based membranes. *Energy Environ. Sci.* **2017**, *10*, 101–106. [CrossRef]

144. Escolastico, S.; Solis, C.; Serra, J.M. Hydrogen separation and stability study of ceramic membranes based on the system Nd 5 LnWO 12. *Int. J. Hydrogen Energy* **2011**, *36*, 11946–11954. [CrossRef]
145. Ivanova, M.E.; Serra, J.M.; Roitsch, S. *Proton-Conducting Ceramic Membranes for Solid Oxide Fuel Cells and Hydrogen (H_2) Processing*; Woodhead Publishing: Cambridge, UK, 2011; pp. 541–567. [CrossRef]
146. Hashim, S.S.; Somalu, M.R.; Loh, K.S.; Liu, S.; Zhao, W.; Sunarso, J. Perovskite-based proton conducting membranes for hydrogen separation: A review. *Int. J. Hydrogen Energy* **2018**, *43*, 15281–15305. [CrossRef]
147. Iwahara, H. Hydrogen pumps using proton-conducting ceramics and their applications. *Solid State Ion.* **1999**, *125*, 271–278. [CrossRef]
148. Cheng, S.; Gupta, V.K.; Lin, J.Y.S. Synthesis and hydrogen permeation properties of asymmetric proton-conducting ceramic membranes. *Solid State Ion.* **2005**, *176*, 2653–2662. [CrossRef]
149. Tong, Y.; Meng, X.; Luo, T.; Cui, C.; Wang, Y.; Wang, S.; Peng, R.; Xie, B.; Chen, C.; Zhan, Z. Protonic Ceramic Electrochemical Cell for Efficient Separation of Hydrogen. *Appl. Mater. Interfaces* **2020**, *12*, 25809–25817. [CrossRef]
150. Leonard, K.; Deibert, W.; Ivanova, M.E.; Meulenberg, W.A.; Ishihara, T.; Matsumoto, H. Processing Ceramic Proton Conductor Membranes for Use in Steam Electrolysis. *Membranes* **2020**, *12*, 339. [CrossRef] [PubMed]
151. Tao, Z.; Yan, L.; Qiao, J.; Wang, B.; Zhang, L.; Zhang, J. A review of advanced proton-conducting materials for hydrogen separation. *Prog. Mater. Sci.* **2015**, *74*, 1–50. [CrossRef]
152. Norby, T.; Haugsrud, R. Dense Ceramic Membranes for Hydrogen Separation Oxide thermoelectric materials View project Metal supported proton conducting electrolyser cell for renewable hydrogen production (METALLICA) View project. In *Nonporous Inorganic Membranes: For Chemical Processing*; Sammells, A.F., Mundschau, M.V., Eds.; Wiley-VCH: Weinheim, Germany, 2014; pp. 1–48. ISBN 3527313427.
153. Kreuer, K.D. On the complexity of proton conduction phenomena. *Solid State Ion.* **2000**, *136–137*, 149–160. [CrossRef]
154. Fontaine, M.L.; Norby, T.; Larring, Y.; Grande, T.; Bredesen, R. Oxygen and Hydrogen Separation Membranes Based on Dense Ceramic Conductors. *Membr. Sci. Technol.* **2008**, *13*, 401–458. [CrossRef]
155. Gallucci, F.; Fernandez, E.; Corengia, P.; van Sint Annaland, M. Recent advances on membranes and membrane reactors for hydrogen production. *Chem. Eng. Sci.* **2013**, *92*, 40–66. [CrossRef]
156. Gardner, C.L.; Ternan, M. Electrochemical separation of hydrogen from reformate using PEM fuel cell technology. *J. Power Sources* **2007**, *171*, 835–841. [CrossRef]
157. Bessarrabov, D. Electrochemically-aided membrane separation and catalytic processes. *Membr. Technol.* **1998**, *93*, 8–11. [CrossRef]
158. Sakai, T.; Matsumoto, H.; Kudo, T.; Yamamoto, R.; Niwa, E.; Okada, S.; Hashimoto, S.; Sasaki, K.; Ishihara, T. High performance of electroless-plated platinum electrode for electrochemical hydrogen pumps using strontium-zirconate-based proton conductors. *Electrochim. Acta* **2008**, *53*, 8172–8177. [CrossRef]
159. Nguyen, M.T.; Grigoriev, S.A.; Kalinnikov, A.A.; Filippov, A.A.; Millet, P.; Fateev, V.N. Characterisation of a electrochemical hydrogen pump using electrochemical impedance spectroscopy. *J. Appl. Electrochem.* **2011**, *41*, 1033–1042. [CrossRef]
160. Wu, X.; Benziger, J.; He, G. Comparison of Pt and Pd catalysts for hydrogen pump separation from reformate. *J. Power Sources* **2012**, *218*, 424–434. [CrossRef]
161. Grigoriev, S.A.; Shtatniy, I.G.; Millet, P.; Porembsky, V.I.; Fateev, V.N. Description and characterization of an electrochemical hydrogen compressor/concentrator based on solid polymer electrolyte technology. *Int. J. Hydrogen Energy* **2011**, *36*, 4148–4155. [CrossRef]
162. Hydrogen Mobility Europe. Available online: https://h2me.eu/ (accessed on 21 April 2020).
163. Speers, P. Hydrogen Mobility Europe (H2ME): Vehicle and hydrogen refuelling station deployment results. *World Electr. Veh. J.* **2018**, *9*, 2. [CrossRef]
164. Bard, A.J.; Faulkner, L.R. *Electrochemical Methods: Fundamentals and Applications*; Wiley: Hoboken, NJ, USA, 2001; pp. 261–304. ISBN 978-0-471-04372-0. Available online: https://books.google.com.mx/books?id=kv56QgAACAAJ (accessed on 13 January 2020).
165. Barbir, F.; Görgün, H. Electrochemical hydrogen pump for recirculation of hydrogen in a fuel cell stack. *J. Appl. Electrochem.* **2007**, *37*, 359–365. [CrossRef]
166. Lee, H.K.; Choi, H.Y.; Choi, K.H.; Park, J.H.; Lee, T.H. Hydrogen separation using electrochemical method. *J. Power Sources* **2004**, *132*, 92–98. [CrossRef]
167. Granite, E.J.; O'Brien, T. Review of novel methods for carbon dioxide separation from flue and fuel gases. *Fuel Process. Technol.* **2005**, *86*, 1423–1434. [CrossRef]
168. Ibeh, B.; Gardner, C.; Ternan, M. Separation of hydrogen from a hydrogen/methane mixture using a PEM fuel cell. *Int. J. Hydrogen Energy* **2007**, *32*, 908–914. [CrossRef]
169. Onda, K.; Ichihara, K.; Nagahama, M.; Minamoto, Y.; Araki, T. Separation and compression characteristics of hydrogen by use of proton exchange membrane. *J. Power Sources* **2007**, *164*, 1–8. [CrossRef]
170. Casati, C.; Longhi, P.; Zanderighi, L.; Bianchi, F. Some fundamental aspects in electrochemical hydrogen purification/compression. *J. Power Sources* **2008**, *180*, 103–113. [CrossRef]
171. Doucet, R.; Gardner, C.L.; Ternan, M. Separation of hydrogen from hydrogen/ethylene mixtures using PEM fuel cell technology. *Int. J. Hydrogen Energy* **2009**, *34*, 998–1007. [CrossRef]

172. Onda, K.; Araki, T.; Ichihara, K.; Nagahama, M. Treatment of low concentration hydrogen by electrochemical pump or proton exchange membrane fuel cell. *J. Power Sources* **2009**, *188*, 1–7. [CrossRef]
173. Abdulla, A.; Laney, K.; Padilla, M.; Sundaresan, S.; Benziger, J. Efficiency of hydrogen recovery from reformate with a polymer electrolyte hydrogen pump. *Am. Inst. Chem. Eng. J.* **2011**, *57*, 1767–1779. [CrossRef]
174. Kim, S.J.; Lee, B.S.; Ahn, S.H.; Han, J.Y.; Park, H.Y.; Kim, S.H.; Yoo, S.J.; Kim, H.J.; Cho, E.; Henkensmeier, D.; et al. Characterizations of polybenzimidazole based electrochemical hydrogen pumps with various Pt loadings for H_2/CO_2 gas separation. *Int. J. Hydrogen Energy* **2013**, *38*, 14816–14823. [CrossRef]
175. Chen, C.Y.; Lai, W.H.; Chen, Y.K.; Su, S.S. Characteristic studies of a PBI/H3PO4 high temperature membrane PEMFC under simulated reformate gases. *Int. J. Hydrogen Energy* **2014**, *39*, 13757–13762. [CrossRef]
176. Kim, S.J.; Park, H.Y.; Ahn, S.H.; Lee, B.S.; Kim, H.J.; Cho, E.A.; Henkensmeier, D.; Nam, S.W.; Kim, S.H.; Yoo, S.J.; et al. Highly active and CO_2 tolerant Ir nanocatalysts for H_2/CO_2 separation in electrochemical hydrogen pumps. *Appl. Catal. B Environ.* **2014**, *158–159*, 348–354. [CrossRef]
177. Bouwman, P.J. Advances in Electrochemical Hydrogen Compression and Purification. *ECS Trans.* **2016**, *75*, 503–510. [CrossRef]
178. Huang, S.; Wang, T.; Wu, X.; Xiao, W.; Yu, M.; Chen, W.; Zhang, F.; He, G. Coupling hydrogen separation with butanone hydrogenation in an electrochemical hydrogen pump with sulfonated poly (phthalazinone ether sulfone ketone) membrane. *J. Power Sources* **2016**, *327*, 178–186. [CrossRef]
179. Ru, F.Y.; Zulkefli, N.N.; Yusra, N.; Yusuf, M.; Masdar, M.S. Effect of Operating Parameter on H_2/CO_2 Gas Separation using Electrochemical Cell. *Int. J. Appl. Eng. Res.* **2018**, *13*, 505–510.
180. Nordio, M.; Rizzi, F.; Manzolini, G.; Mulder, M.; Raymakers, L.; Van Sint Annaland, M.; Gallucci, F. Experimental and modelling study of an electrochemical hydrogen compressor. *Chem. Eng. J.* **2019**, *369*, 432–442. [CrossRef]
181. Nordio, M.; Eguaras Barain, M.; Raymakers, L.; Van Sint Annaland, M.; Mulder, M.; Gallucci, F. Effect of CO_2 on the performance of an electrochemical hydrogen compressor. *Chem. Eng. J.* **2019**, *392*, 123647. [CrossRef]
182. Huang, F.; Pingitore, A.T.; Benicewicz, B.C. Electrochemical Hydrogen Separation from Reformate Using High-Temperature Polybenzimidazole (PBI) Membranes: The Role of Chemistry. *ACS Sustain. Chem. Eng.* **2020**, *8*, 6234–6242. [CrossRef]
183. Jackson, C.; Raymakers, L.F.J.M.; Mulder, M.J.J.; Kucernak, A.R.J. Assessing electrocatalyst hydrogen activity and CO tolerance: Comparison of performance obtained using the high mass transport 'floating electrode' technique and in electrochemical hydrogen pumps. *Appl. Catal. B Environ.* **2020**, *268*, 118734. [CrossRef]
184. Jackson, C.; Raymakers, L.F.J.M.; Mulder, M.J.J.; Kucernak, A.R.J. Poison mitigation strategies for the use of impure hydrogen in electrochemical hydrogen pumps and fuel cells. *J. Power Sources* **2020**, *472*, 228476. [CrossRef]
185. Ohs, B.; Abduly, L.; Krödel, M.; Wessling, M. Combining electrochemical hydrogen separation and temperature vacuum swing adsorption for the separation of N_2, H_2 and CO_2. *Int. J. Hydrogen Energy* **2020**, *45*, 9811–9820. [CrossRef]
186. Rhandi, M.; Trégaro, M.; Druart, F.; Deseure, J.; Chatenet, M. Electrochemical hydrogen compression and purification versus competing technologies: Part I. Pros and cons. *Chinese J. Catal.* **2020**, *41*, 756–769. [CrossRef]
187. Trégaro, M.; Rhandi, M.; Druart, F.; Deseure, J.; Chatenet, M. Electrochemical hydrogen compression and purification versus competing technologies: Part II. Challenges in electrocatalysis. *Chinese J. Catal.* **2020**, *41*, 770–782. [CrossRef]
188. Vermaak, L.; Neomagus, H.W.J.P.; Bessarabov, D.G. *Hydrogen Separation and Purification from Various Gas Mixtures by Means of Electrochemical Membrane Technology in the Temperature Range 100–160 °C*, 2021; Unpublished.
189. Zhang, J. Investigation of CO Tolerance in Proton Exchange Membrane Fuel Cells. Ph.D. Thesis, Worcester Polytechnic Institute, Worcester, MA, USA, 2004; pp. 1–219.
190. Uribe, F.A.; Gottesfeld, S.; Zawodzinski, T.A. Effect of Ammonia as Potential Fuel Impurity on Proton Exchange Membrane Fuel Cell Performance. *J. Electrochem. Soc.* **2002**, *149*, A293. [CrossRef]
191. Halseid, R.; Vie, P.J.S.; Tunold, R. Influence of Ammonium on Conductivity and Water Content of Nafion 117 Membranes. *J. Electrochem. Soc.* **2004**, *151*, A381. [CrossRef]
192. Kirsten, W.; Krüger, A.; Neomagus, H.; Bessarabov, D. Effect of Relative Humidity and Temperature on the Mechanical Properties of PFSA Nafion^{TM}-cation-exchanged membranes for Electrochemical Applications. *Int. J. Electrochem. Sci.* **2017**, *12*, 2573–2582. [CrossRef]
193. Friend, P.J. Modelling and experimental characterization of an ionic polymer metal composite actuator PJ Friend Supervisor. Ph.D. Thesis, North-West University, Potchefstroom, South Africa, 2018.
194. Bessarabov, D. Chapter 8: Other Polymer Membrane Electrolysis Processes. In *RSC Energy and Environment Series*; Royal Society of Chemistry: London, UK, 2020; pp. 286–305.
195. Araya, S.S.; Zhou, F.; Liso, V.; Sahlin, S.L.; Vang, J.R.; Thomas, S.; Gao, X.; Jeppesen, C.; Kær, S.K. A comprehensive review of PBI-based high temperature PEM fuel cells. *Int. J. Hydrogen Energy* **2016**, *41*, 21310–21344. [CrossRef]
196. Dafalla, A.M.; Jiang, F. Stresses and their impacts on proton exchange membrane fuel cells: A review. *Int. J. Hydrogen Energy* **2018**, *43*, 2327–2348. [CrossRef]
197. Taymaz, I.; Benli, M. Numerical study of assembly pressure effect on the performance of proton exchange membrane fuel cell. *Energy* **2010**, *35*, 2134–2140. [CrossRef]
198. Bouwman, P. Fundamental of Electrochemical Hydrogen Compression. In *PEM Electrolysis for Hydrog. Production: Principles and Applications*; Bessarabov, D., Wang, H., Li, H., Zhao, N., Eds.; CRC Press: Boca Raton, FL, USA, 2015; pp. 269–299.

199. Haque, M.A.; Sulong, A.B.; Loh, K.S.; Majlan, E.H.; Husaini, T.; Rosli, R.E. Acid doped polybenzimidazoles based membrane electrode assembly for high temperature proton exchange membrane fuel cell: A review. *Int. J. Hydrogen Energy* **2017**, *42*, 9156–9179. [CrossRef]
200. Gao, D.; Cai, F.; Xu, Q.; Wang, G.; Pan, X.; Bao, X. Gas-phase electrocatalytic reduction of carbon dioxide using electrolytic cell based on phosphoric acid-doped polybenzimidazole membrane. *J. Energy Chem.* **2014**, *23*, 674–700. [CrossRef]
201. Tao, Y.; Hwang, Y.; Wang, C.; Radermacher, R. The Integration of Ammonia Electrochemical Compressor in Vapor Compression System; In Proceedings of the 12th IEA Heat Pump Conference, Rotterdam, The Netherlands, 15–18 May 2017; Volume 4.
202. Tao, Y.; Gibbons, W.; Hwang, Y.; Radermacher, R.; Wang, C. Electrochemical ammonia compression. *Chem. Commun.* **2017**, *53*, 5637. [CrossRef]
203. Tao, Y.; Hwang, Y.; Radermacher, R.; Wang, C. Experimental study on electrochemical compression of ammonia and carbon dioxide for vapor compression refrigeration system. *Int. J. Refrig.* **2019**, *104*, 180–188. [CrossRef]
204. Fujiwara, N.; Nagase, H.; Tada, S.; Kikuchi, R. Hydrogen Production by Steam Electrolysis in Solid Acid Electrolysis Cells Naoya. *Chem. Sustain. Energy Mater.* **2020**, *14*, 417–427. [CrossRef]

MDPI
St. Alban-Anlage 66
4052 Basel
Switzerland
Tel. +41 61 683 77 34
Fax +41 61 302 89 18
www.mdpi.com

Membranes Editorial Office
E-mail: membranes@mdpi.com
www.mdpi.com/journal/membranes